Multiple Choice Questions in Electronics and Electrical Engineering

To my grandchildren Shelley Louise and Michael John Davies

By the same author

Newnes Circuit Calculations Pocket Book

Multiple Choice Questions in Electronics and Electrical Engineering

Thomas J Davies

BUTTERWORTH
HEINEMANN

Butterworth-Heinemann Ltd
Linacre House, Jordan Hill, Oxford OX2 8DP

A member of the Reed Elsevier group

OXFORD LONDON BOSTON
MUNICH NEW DELHI SINGAPORE SYDNEY
TOKYO TORONTO WELLINGTON

First published 1994

British Library Cataloguing in Publication Data
Davies, Thomas J.
Multiple Choice Questions in Electronics and Electrical Engineering
I. Title
537.5076

ISBN 0 7506 1677 6

Library of Congress Cataloguing in Publication Data
A catalogue record for this book is available from the Library of Congress

ISBN 0 7506 1677 6

Composition by Genesis Typesetting, Rochester, Kent
Printed and bound in Great Britain

Contents

Preface

Multiple choice questions offer an excellent guide to progress during courses at school and technical colleges and they also provide a very good revision tool prior to examinations. Some examination boards set multiple choice papers as part of the end of year examinations.

This book has been written for students embarking on courses in electronics and electrical engineering and it should be useful during the first three years. The 20 chapters are devoted to individual topics and the appendices provide some sample papers suitable for the first year.

The study of electronics requires some understanding of both mathematics and physics. It is a requirement to be able to use indices and transpose formulas, for example, and Chapters 18, 19 and 20 are included for this purpose.

Acknowledgements

I would like to thank the staff of the Bournemouth and Poole College of Further Education for their support and also the publishers and in particular Miss Bridget Buckley for all her helpful advice. I would also like to thank my family Sue, Glyn, Jude and Carol for all their help and especially my wife Enid for typing the complete work.

1 Resistors

1.1 Resistance is measured in

- ❑ *a* Volts
- ❑ *b* Amperes
- ❑ *c* Ohms
- ❑ *d* Watts

1.2 The resistance of a conductor is directly proportional to

- ❑ *a* The length
- ❑ *b* The cross-sectional area
- ❑ *c* The velocity
- ❑ *d* The pressure

1.3 The resistance of a conductor is inversely proportional to

- ❑ *a* The length
- ❑ *b* The resistivity
- ❑ *c* The cross-sectional area
- ❑ *d* The pressure

1.4 Which of the following is a good conductor?

- ❑ *a* Porcelain
- ❑ *b* Mica
- ❑ *c* Copper
- ❑ *d* Rubber

1.5 Which of the following is a good insulator?

- ❑ *a* Aluminium
- ❑ *b* Ebonite
- ❑ *c* Iron
- ❑ *d* Steel

1.6 The resistance of a pure metal

- ❑ *a* Is unaffected by temperature
- ❑ *b* Increases with temperature rise
- ❑ *c* Decreases with temperature rise
- ❑ *d* Always remains constant

1.7 Carbon has a negative temperature coefficient. This means that the resistance of carbon

- ❑ *a* Increases with temperature rise
- ❑ *b* Decreases with temperature rise
- ❑ *c* Is unaffected by temperature changes
- ❑ *d* Is the same as the resistance of copper

1.8 Doped silicon is classed as

- ❑ *a* A conductor
- ❑ *b* A semiconductor
- ❑ *c* An insulator
- ❑ *d* An impurity

1.9 The diagram shows a 3300 ± 10% resistor. The colour bands 1 to 4 are

- ❑ *a* Brown red blue silver
- ❑ *b* Orange orange red silver
- ❑ *c* Orange red orange gold
- ❑ *d* Yellow red green gold

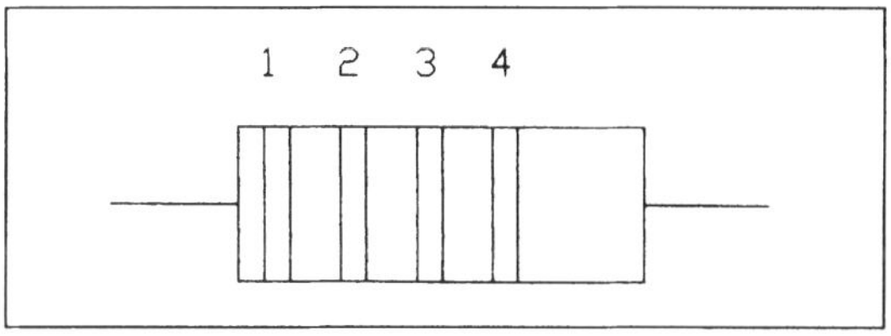

1.10 If the resistor in Question 1.9 has a value of 15 ± 2%, the colour bands 1 to 4 are

- ❑ *a* Brown green brown brown
- ❑ *b* Brown green brown red
- ❑ *c* Brown green brown gold
- ❑ *d* Brown green black red

1.11 If the resistor in Question 1.9 has a value of 0.82 ± 1%, the colour bands 1 to 4 are
- ❑ *a* Grey red silver brown
- ❑ *b* Red grey brown silver
- ❑ *c* Grey red brown red
- ❑ *d* Red grey brown red

1.12 If the resistor in Question 1.9 has a value of 1000 ± 5%, the colour bands 1 to 4 are
- ❑ *a* Black brown red brown
- ❑ *b* Brown black red gold
- ❑ *c* Red black brown red
- ❑ *d* Brown red black silver

1.13 A resistor is shown on a circuit diagram as 1k5F. The ohmic value is
- ❑ *a* 150 ± 10%
- ❑ *b* 15 ± 5%
- ❑ *c* 15 ± 2%
- ❑ *d* 1500 ± 1%

1.14 A 1M8M resistor has a value of
- ❑ *a* 180 ± 5%
- ❑ *b* 180 000 ± 10%
- ❑ *c* 1 800 000 ± 20%
- ❑ *d* 18 000 ± 2%

1.15 If a resistor is shown as R56F, the ohmic value is
- ❑ *a* 56 ± 10%
- ❑ *b* 0.56 ± 1%
- ❑ *c* 5.6 ± 2%
- ❑ *d* 560 ± 1%

1.16 A 120RJ resistor has a value of
- ❑ *a* 1.2 ± 1%
- ❑ *b* 12 ± 10%
- ❑ *c* 120 ± 5%
- ❑ *d* 1200 ± 2%

1.17 Which one of the following is not a preferred value in the ±10% resistor range?
- ❑ *a* 820
- ❑ *b* 2700
- ❑ *c* 4800
- ❑ *d* 5600

1.18 The lower limit value of a 150R ± 10% resistor is
- ❑ *a* 115
- ❑ *b* 125
- ❑ *c* 135
- ❑ *d* 145

1.19 The upper limit value of a 56kK resistor is
- ❑ *a* 58 200
- ❑ *b* 61 600
- ❑ *c* 65 000
- ❑ *d* 68 400

1.20 An 820R resistor with a tolerance of ±10% has a resistance spread of
- ❑ *a* 690–854
- ❑ *b* 700–864
- ❑ *c* 738–902
- ❑ *d* 780–944

1.21 The physical size of a resistor determines
- ❑ *a* The voltage rating
- ❑ *b* The current rating
- ❑ *c* The power rating
- ❑ *d* The temperature rating

1.22 The diagram shows two resistors connected in series. The combined resistance *R* is
- ❑ *a* 540R
- ❑ *b* 620R
- ❑ *c* 640R
- ❑ *d* 700R

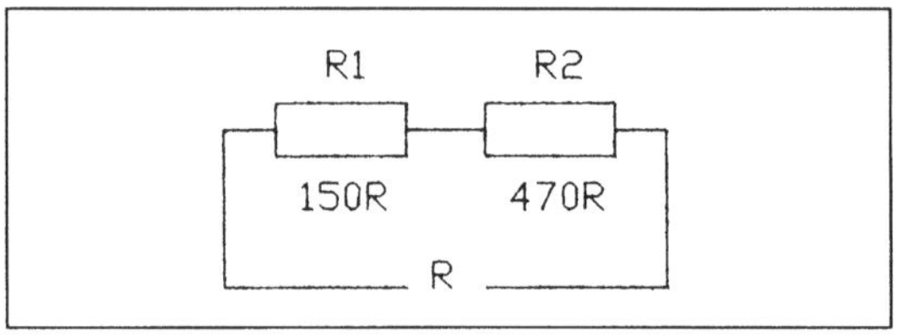

1.23 If R_1 in question 1.22 is changed to 3k3, the combined resistance *R* is
- ❑ *a* 347R
- ❑ *b* 3470R
- ❑ *c* 3747R
- ❑ *d* 3770R

1.24 The diagram shows two resistors connected in parallel. The combined resistance R is

❑ *a* 24R
❑ *b* 26R
❑ *c* 28R
❑ *d* 30R

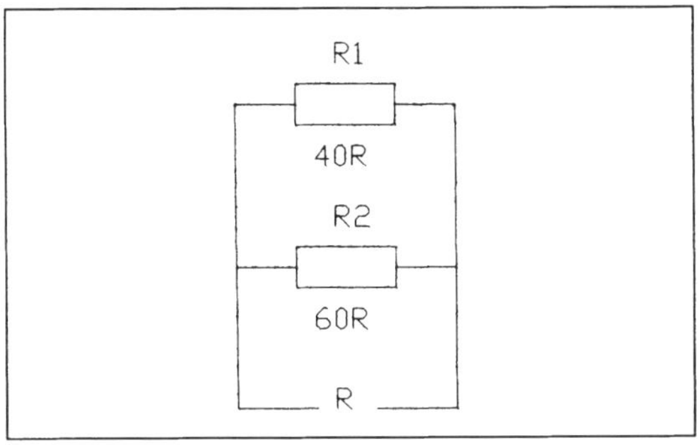

1.25 If R_2 in Question 1.24 is changed to 10R, the combined resistance R is

❑ *a* 2R
❑ *b* 4.5R
❑ *c* 8R
❑ *d* 9.5R

1.26 The resistance of a cable 9133 m long having a radius of 5 mm and a specific resistance of 1.72×10^{-8} Ω m is approximately

❑ *a* 16 Ω
❑ *b* 10 Ω
❑ *c* 2 Ω
❑ *d* 0.6 Ω

1.27 If a cable has a resistance of 0.5 Ω, a radius of 2 mm and a specific resistance of 1.72×10^{-8} Ω m, the length is approximately

❑ *a* 240 m
❑ *b* 365 m
❑ *c* 406 m
❑ *d* 485 m

1.28 If the resistance of the cable in Question 1.27 is 2 Ω, the length is approximately

❑ *a* 1461 m
❑ *b* 1524 m
❑ *c* 1595 m
❑ *d* 1611 m

1.29 If the radius of the cable in Question 1.27 is increased to 3 mm, the length is approximately

❑ *a* 1020 m
❑ *b* 628 m
❑ *c* 822 m
❑ *d* 426 m

1.30 A cable is 400 m long and has a resistance of 0.2 Ω. If the specific resistance of the conductor is 1.72×10^{-8} Ω m, the radius is approximately

❑ *a* 2.6 mm
❑ *b* 3.3 mm
❑ *c* 4.0 mm
❑ *d* 4.7 mm

1.31 If the cable in Question 1.30 was doubled in length, and it was a requirement to maintain the same parameters, the radius of the cable would need to be approximately

❑ *a* 3.48 mm
❑ *b* 4.68 mm
❑ *c* 10.68 mm
❑ *d* 12.68 mm

1.32 The upper limit resistance value of a 10% resistor is 5170 Ω. The nominal value is

❑ *a* 4500 Ω
❑ *b* 4600 Ω
❑ *c* 4700 Ω
❑ *d* 4800 Ω

1.33 The lower limit of a 5% resistor is 209 Ω. The nominal value is

❑ *a* 215 Ω
❑ *b* 220 Ω
❑ *c* 225 Ω
❑ *d* 230 Ω

1.34 The upper limit of a 2% resistor is 10 200 Ω. The nominal value is

❑ *a* 9200 Ω
❑ *b* 9600 Ω
❑ *c* 9800 Ω
❑ *d* 10 000 Ω

1.35 The lower limit of a 20% resistor is 544 Ω. The nominal value is
- ❑ *a* 680 Ω
- ❑ *b* 710 Ω
- ❑ *c* 580 Ω
- ❑ *d* 744 Ω

1.36 The upper limit of a 1% resistor is 101 Ω. The nominal value is
- ❑ *a* 85 Ω
- ❑ *b* 90 Ω
- ❑ *c* 95 Ω
- ❑ *d* 100 Ω

1.37 The lower limit of a 2% resistor is 147 Ω. The nominal value is
- ❑ *a* 150 Ω
- ❑ *b* 160 Ω
- ❑ *c* 148 Ω
- ❑ *d* 152 Ω

1.38 A copper conductor has a resistance of 4 Ω at 24 °C. If the temperature coefficient at 24 °C is 0.0039/°C, the resistance at 48 °C is
- ❑ *a* 4.25 Ω
- ❑ *b* 4.29 Ω
- ❑ *c* 4.33 Ω
- ❑ *d* 4.37 Ω

1.39 If the resistor in Question 1.38 had a resistance of 12 Ω, the resistance at 48 °C would be
- ❑ *a* 12.09 Ω
- ❑ *b* 13.12 Ω
- ❑ *c* 14.16 Ω
- ❑ *d* 14.29 Ω

1.40 A copper conductor has a resistance of 3 Ω at 16 °C and a temperature coefficient of 0.0039/°C. If the temperature is raised until the resistance is 3.16 Ω the approximate value of the temperature then is
- ❑ *a* 18 °C
- ❑ *b* 22 °C
- ❑ *c* 26 °C
- ❑ *d* 30 °C

1.41 If the resistance of the conductor in Question 1.40 had reached 3.5 Ω, the approximate value of the temperature would be
- ❑ *a* 59.2 °C
- ❑ *b* 58.7 °C
- ❑ *c* 54.2 °C
- ❑ *d* 53.6 °C

1.42 A conductor has a resistance of 10 Ω at 15 °C and a resistance of 10.6 Ω at 30 °C. The temperature coefficient at 15 °C is
- ❑ *a* 0.0039/°C
- ❑ *b* 0.004/°C
- ❑ *c* 0.0042/°C
- ❑ *d* 0.0043/°C

1.43 Four resistors in series with values of 1k5, 3k3, 3k9 and 4k7 will have a combined value of
- ❑ *a* 9k2
- ❑ *b* 10k4
- ❑ *c* 13k4
- ❑ *d* 15k6

1.44 Four resistors connected in series have a combined resistance value of 885R. If three of the resistors have values of 150R, 330R and 390R, the value of the fourth is
- ❑ *a* 10R
- ❑ *b* 15R
- ❑ *c* 68R
- ❑ *d* 120R

1.45 Two resistors connected in parallel have a combined resistance value of 21R. One resistor has a value of 70R. The value of the second is
- ❑ *a* 30R
- ❑ *b* 35R
- ❑ *c* 60R
- ❑ *d* 70R

1.46 Three resistors having values of 120R, 150R and 330R are connected in parallel. The combined resistance is approximately
- ❑ *a* 40.46R
- ❑ *b* 46.46R
- ❑ *c* 50.46R
- ❑ *d* 55.46R

1.47 Three resistors are connected in parallel. The combined resistance is 120R. Two of the resistors have values of 240R and 600R. The value of the third is

- ❑ *a* 120R
- ❑ *b* 240R
- ❑ *c* 360R
- ❑ *d* 400R

1.48 If the combined resistance in Question 1.47 is 82R76, the value of the third is

- ❑ *a* 100R
- ❑ *b* 160R
- ❑ *c* 180R
- ❑ *d* 200R

1.49 Four 680R resistors are connected in parallel. The combined resistance is

- ❑ *a* 170R
- ❑ *b* 280R
- ❑ *c* 340R
- ❑ *d* 400R

1.50 Two 150R ± 10% resistors are connected in series. The overall resistance is

- ❑ *a* 300R ± 5%
- ❑ *b* 300R ± 10%
- ❑ *c* 300R ± 15%
- ❑ *d* 300R ± 20%

1.51 A 100R ± 5% resistor is connected in series with a 150R ± 10% resistor. The overall resistance value is

- ❑ *a* 250R ± 5%
- ❑ *b* 250R ± 10%
- ❑ *c* 250R ± 8%
- ❑ *d* 250R ± 7.5%

1.52 Three resistors having values of 100R ± 1%, 100R ± 2% and 100R ± 10% are connected in series. The overall resistance value is

- ❑ *a* 300R ± 1%
- ❑ *b* 300R ± 3.33%
- ❑ *c* 300R ± 2%
- ❑ *d* 300R ± 4.33%

1.53 Three resistors having values of 1k0J, 1k0K and 1k0M are connected in series. The overall value of the resistance is

- ❑ *a* 3000 Ω ± 10.66%
- ❑ *b* 3000 Ω ± 11.33%
- ❑ *c* 3000 Ω ± 11.66%
- ❑ *d* 3000 Ω ± 12.66%

1.54 Two 100R ± 10% resistors are connected in parallel. The overall resistance is

- ❑ *a* 45R ± 5%
- ❑ *b* 50R ± 5%
- ❑ *c* 45R ± 10%
- ❑ *d* 50R ± 10%

1.55 A 100R ± 10% resistor is connected in parallel with a 100 R ± 20%. The overall tolerance of the combined resistance is approximately

- ❑ *a* ±10%
- ❑ *b* ±15%
- ❑ *c* ±20%
- ❑ *d* ±25%

1.56 Three 150R ± 10% resistors are connected in parallel. The combined resistance is

- ❑ *a* 50R ± 3.33%
- ❑ *b* 50R ± 5%
- ❑ *c* 50R ± 6.67%
- ❑ *d* 50R ± 10%

1.57 Four resistors connected in parallel have an equivalent resistance value of 75R. Three of the resistors have values of 400R, 300R and 150R. The value of the fourth is

- ❑ *a* 800R
- ❑ *b* 1k0
- ❑ *c* 1k2
- ❑ *d* 1k4

1.58 In the diagram the value of the overall resistance R is

- ❑ *a* 30R
- ❑ *b* 36R
- ❑ *c* 40R
- ❑ *d* 66R

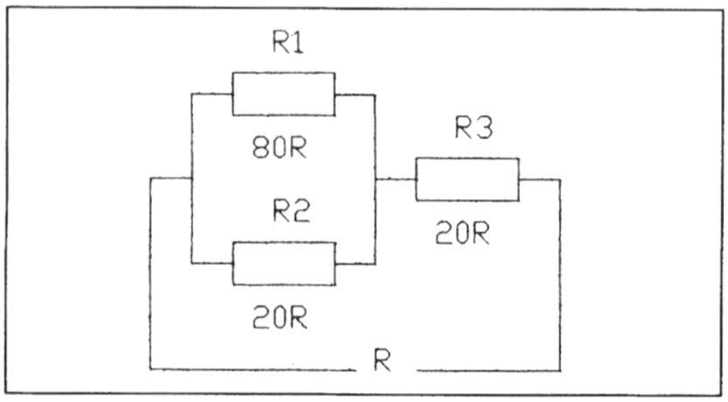

1.59 Resistor R_2 in Question 1.58 is changed in order to make the value of R 84R. The value of R_2 is

- ❑ *a* 100R
- ❑ *b* 200R
- ❑ *c* 240R
- ❑ *d* 320R

1.60 In the diagram the value of the overall resistance R is

- ❑ *a* 200R
- ❑ *b* 150R
- ❑ *c* 450R
- ❑ *d* 300R

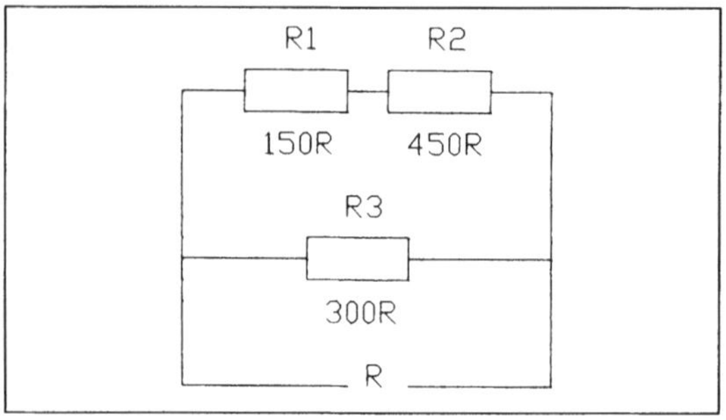

1.61 Resistor R_2 in Question 1.60 is changed in order to make the value of R 210R. The value of R_2 is

- ❑ *a* 330R
- ❑ *b* 450R
- ❑ *c* 750R
- ❑ *d* 550R

1.62 Resistor R_1 in Question 1.60 is changed in order to make the value of R 187R5. The value of R_1 is

- ❑ *a* 50R
- ❑ *b* 75R
- ❑ *c* 100R
- ❑ *d* 150R

1.63 In the diagram the value of the overall resistance R is

- ❑ *a* 75R
- ❑ *b* 90R
- ❑ *c* 98R
- ❑ *d* 100R

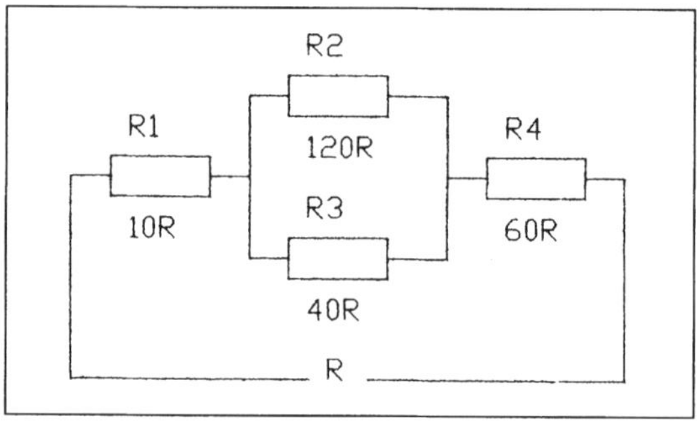

1.64 Resistor R_1 in Question 1.63 is changed in order to make the value of R 680R. The value of R_1 is

- ❑ *a* 660R
- ❑ *b* 590R
- ❑ *c* 470R
- ❑ *d* 330R

1.65 Resistor R_2 in Question 1.63 is changed in order to make the value of R 94R. The value of R_2 is

- ❑ *a* 60R
- ❑ *b* 70R
- ❑ *c* 80R
- ❑ *d* 82R

1.66 Resistor R_3 in Question 1.63 is changed in order to make the value of R 166R. The value of R_3 is

- ❑ *a* 440R
- ❑ *b* 480R
- ❑ *c* 520R
- ❑ *d* 600R

1.67 Resistor R_4 in Question 1.63 is changed in order to make the value of R 470R. The value of R_4 is

❑ *a* 390R
❑ *b* 410R
❑ *c* 430R
❑ *d* 450R

1.68 In the diagram the value of the overall resistance R is

❑ *a* 300R
❑ *b* 400R
❑ *c* 500R
❑ *d* 600R

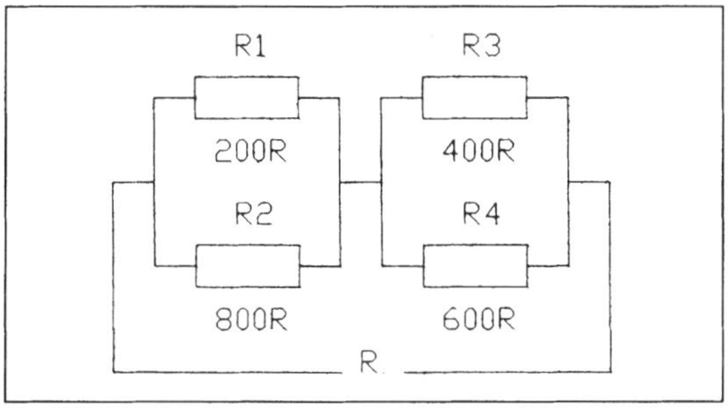

1.69 Resistor R_1 in Question 1.68 is changed in order to make the value of R 640R. The value of R_1 is

❑ *a* 500R
❑ *b* 600R
❑ *c* 700R
❑ *d* 800R

1.70 Resistor R_2 in Question 1.68 is changed in order to make the value of R 280R. The value of R_2 is

❑ *a* 45R
❑ *b* 50R
❑ *c* 60R
❑ *d* 65R

1.71 Resistor R_3 in Question 1.68 is changed in order to make the value of R 560R. The value of R_3 is

❑ *a* 1k0
❑ *b* 1k1
❑ *c* 1k2
❑ *d* 1k4

1.72 In the diagram the value of the overall resistance is

❑ *a* 80R25
❑ *b* 87R35
❑ *c* 91R15
❑ *d* 93R75

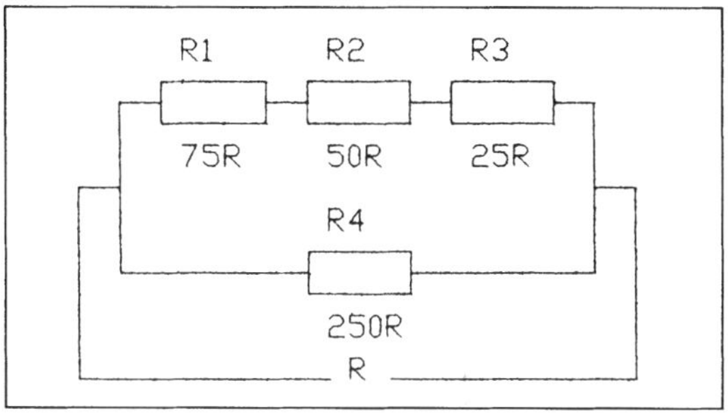

1.73 Resistor R_1 in Question 1.72 is changed in order to make the value of R 187R5. The value of R_1 is

❑ *a* 675R
❑ *b* 700R
❑ *c* 750R
❑ *d* 800R

1.74 Resistor R_2 in Question 1.73 is changed in order to make the value of R 218R75. The value of R_2 is

❑ *a* 1k5
❑ *b* 1k6
❑ *c* 1k65
❑ *d* 1k75

1.75 Resistor R_3 in Question 1.73 is changed in order to make the value of R 225R. The value of R_3 is

❑ *a* 2k15
❑ *b* 2k125
❑ *c* 4k125
❑ *d* 4k25

2 DC voltages and circuits

2.1 A voltage of 0.0025 V expressed in microvolts is
- ❑ *a* 25 μV
- ❑ *b* 250 μV
- ❑ *c* 2.5 μV
- ❑ *d* 2500 μV

2.2 A voltage of 800 mV expressed in volts is
- ❑ *a* 0.008 V
- ❑ *b* 0.08 V
- ❑ *c* 0.8 V
- ❑ *d* 8.0 V

2.3 A voltage of 0.36 mV expressed in microvolts is
- ❑ *a* 360 μV
- ❑ *b* 36 μV
- ❑ *c* 3.6 μV
- ❑ *d* 3600 μV

2.4 A voltage of 800×10^3 μV expressed in millivolts is
- ❑ *a* 80 mV
- ❑ *b* 800 mV
- ❑ *c* 8.0 mV
- ❑ *d* 8000 mV

2.5 A voltage of 0.25 kV expressed in volts is
- ❑ *a* 2.5 V
- ❑ *b* 25 V
- ❑ *c* 250 V
- ❑ *d* 2500 V

2.6 If 2.4 J are required to move 15 C from point A to point B in a circuit, the potential difference between the two points is
- ❑ *a* 0.12 V
- ❑ *b* 0.16 V
- ❑ *c* 6.25 V
- ❑ *d* 6.75 V

2.7 A current of 0.5 A is measured at a circuit point over a period of 2 min. The charge that has passed that point is
- ❑ *a* 10 C
- ❑ *b* 20 C
- ❑ *c* 40 C
- ❑ *d* 60 C

2.8 If the current at a circuit point is 0.25 A, the time for 7.5 C to flow is
- ❑ *a* 30 s
- ❑ *b* 35 s
- ❑ *c* 3.5 s
- ❑ *d* 2.5 s

2.9 If the parameters in Question 2.8 are 100 mA and 0.5 C, the time is
- ❑ *a* 0.5 s
- ❑ *b* 5.0 s
- ❑ *c* 50 s
- ❑ *d* 500 s

2.10 If there is a flow of 360 C in a resistor over a period of 30 min, the current is
- ❑ *a* 200 mA
- ❑ *b* 360 mA
- ❑ *c* 2 A
- ❑ *d* 20 A

2.11 If the parameters for Question 2.10 are 60 mC and 2 hours, the current is
- ❑ *a* 4.33 μA
- ❑ *b* 6.67 μA
- ❑ *c* 7.33 μA
- ❑ *d* 8.33 μA

2.12 A current of 20 mA expressed in amps is
- ❑ *a* 0.002 A
- ❑ *b* 0.02 A
- ❑ *c* 0.2 A
- ❑ *d* 2.0 A

2.13 A current of 60 μA expressed in milliamps is
- ❑ *a* 600 mA
- ❑ *b* 6 mA
- ❑ *c* 0.6 mA
- ❑ *d* 0.06 mA

2.14 A power of 0.05 MW expressed in kilowatts is
- ❑ *a* 0.5 kW
- ❑ *b* 5 kW
- ❑ *c* 50 kW
- ❑ *d* 500 kW

2.15 A power of 12 mW is equivalent to
- ❑ *a* 12×10^{-3} W
- ❑ *b* 12×10^{-6} W
- ❑ *c* 12×10^{3} W
- ❑ *d* 12×10^{6} W

2.16 The diagram shows a single resistor connected to a DC supply. The current I is
- ❑ *a* 2 A
- ❑ *b* 0.2 A
- ❑ *c* 0.5 A
- ❑ *d* 2 mA

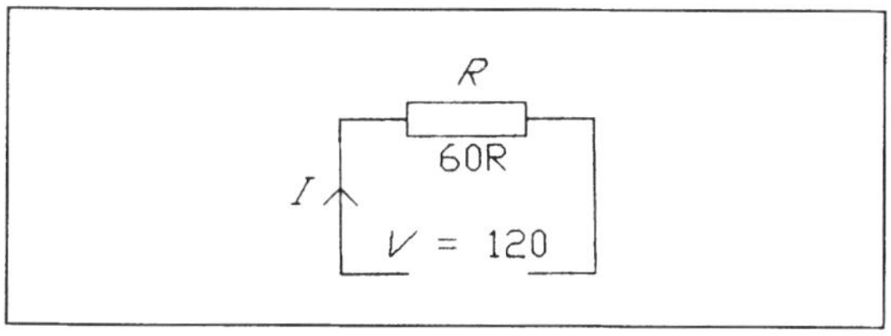

2.17 The input power in Question 2.16 is
- ❑ *a* 240 W
- ❑ *b* 30 W
- ❑ *c* 120 W
- ❑ *d* 2 W

2.18 If the resistor in Question 2.16 is changed to 24R, the input current is then
- ❑ *a* 20 μA
- ❑ *b* 200 μA
- ❑ *c* 20 mA
- ❑ *d* 5 A

2.19 The input power in Question 2.18 is
- ❑ *a* 12 W
- ❑ *b* 20 W
- ❑ *c* 22 W
- ❑ *d* 600 W

2.20 The diagram shows two resistors connected in parallel and driven by a DC supply. The current I is
- ❑ *a* 0.2 A
- ❑ *b* 2.4 A
- ❑ *c* 3.6 A
- ❑ *d* 4.0 A

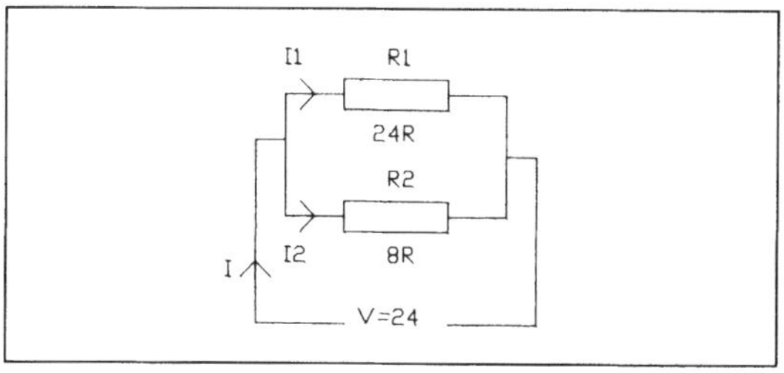

2.21 In Question 2.20 the value of current I_1 is
- ❑ *a* 10 mA
- ❑ *b* 100 mA
- ❑ *c* 1.0 A
- ❑ *d* 2.0 A

2.22 In Question 2.20 the value of current I_2 is
- ❑ *a* 2.0 A
- ❑ *b* 3.0 A
- ❑ *c* 200 mA
- ❑ *d* 300 mA

2.23 In Question 2.20 the input power is
- ❑ *a* 20 W
- ❑ *b* 46 W
- ❑ *c* 80 W
- ❑ *d* 96 W

2.24 In Question 2.20 the power dissipated in R_1 is
- ❑ *a* 24 W
- ❑ *b* 48 W
- ❑ *c* 60 W
- ❑ *d* 80 W

2.25 In Question 2.20 the power dissipated in R_2 is
- ❑ *a* 60 W
- ❑ *b* 72 W
- ❑ *c* 78 W
- ❑ *d* 82 W

2.26 The diagram shows a series circuit. The value of I is
- ❑ *a* 6 A
- ❑ *b* 4 A
- ❑ *c* 2 A
- ❑ *d* 1 A

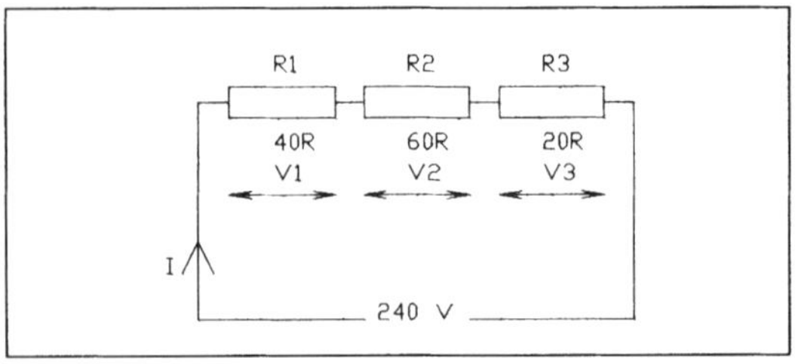

2.27 The value of V_3 in Question 2.26 is
- ❑ *a* 40 V
- ❑ *b* 60 V
- ❑ *c* 20 V
- ❑ *d* 10 V

2.28 The value of the input power in Question 2.26 is
- ❑ *a* 70 W
- ❑ *b* 140 W
- ❑ *c* 200 W
- ❑ *d* 480 W

2.29 The value of the power consumed in R_1 in Question 2.26 is
- ❑ *a* 100 W
- ❑ *b* 160 W
- ❑ *c* 180 W
- ❑ *d* 200 W

2.30 The value of the power consumed by R_2 in Question 2.26 is
- ❑ *a* 240 W
- ❑ *b* 260 W
- ❑ *c* 280 W
- ❑ *d* 300 W

2.31 The diagram shows a parallel circuit. The value of current I is
- ❑ *a* 1.5 A
- ❑ *b* 2.0 A
- ❑ *c* 2.5 A
- ❑ *d* 3.0 A

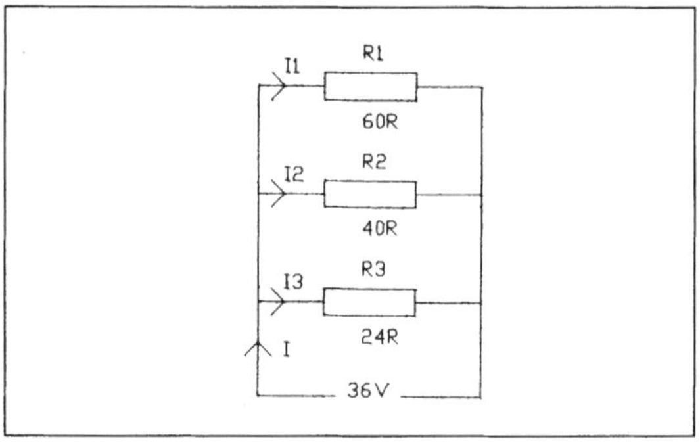

2.32 The value of the current I_2 in Question 2.31 is
- ❑ *a* 600 mA
- ❑ *b* 750 mA
- ❑ *c* 900 mA
- ❑ *d* 920 mA

2.33 The value of the current I_3 in Question 2.31 is
- ❑ *a* 1.5 A
- ❑ *b* 1.75 A
- ❑ *c* 2.25 A
- ❑ *d* 2.5 A

2.34 The value of the power consumed by R_1 in Question 2.31 is

- ❑ *a* 20.5 W
- ❑ *b* 21.6 W
- ❑ *c* 22.4 W
- ❑ *d* 24.2 W

2.35 The value of the power consumed by R_2 in Question 2.31 is

- ❑ *a* 20.5 W
- ❑ *b* 30.2 W
- ❑ *c* 32.4 W
- ❑ *d* 33.6 W

2.36 Four resistors having values of 150R, 250R, 1k0 and 1k5 are connected in parallel. The combined resistance is

- ❑ *a* 33.33R
- ❑ *b* 47.67R
- ❑ *c* 81.08R
- ❑ *d* 120.33R

2.37 The combined value of four resistors connected in parallel is 40R. If three of the resistors have values of 80R, 160R and 200R, the value of the fourth is

- ❑ *a* 360R
- ❑ *b* 480R
- ❑ *c* 640R
- ❑ *d* 800R

2.38 A meter gives a full scale deflection with a current of 0.5 mA, and it has a resistance of 0.5 Ω. The series resistance needed for it to read up to 2 V is

- ❑ *a* 3999R5
- ❑ *b* 4000R
- ❑ *c* 5000R5
- ❑ *d* 5500R

2.39 If the meter in Question 2.38 is required to read up to 10 V, the series resistance would need to be

- ❑ *a* 1999R
- ❑ *b* 1999R5
- ❑ *c* 19999R
- ❑ *d* 19999R5

2.40 An ammeter gives full scale deflection with a current of 20 mA, and it has a coil resistance of 4 Ω. The value of the shunt resistance required to enable it to read up to 2 A is

- ❑ *a* 4.8×10^{2} Ω
- ❑ *b* 4.04×10^{-2} Ω
- ❑ *c* 6.2×10^{-2} Ω
- ❑ *d* 5.4×10^{-3} Ω

2.41 If the meter in Question 2.40 is required to read up to 10 A, the shunt resistance required is

- ❑ *a* 8.0×10^{-3} Ω
- ❑ *b* 6.0×10^{-2} Ω
- ❑ *c* 9.0×10^{-4} Ω
- ❑ *d* 3.0×10^{-3} Ω

2.42 A voltmeter having a resistance of 20k0 is used to measure the voltage between points A and B in the diagram shown. The meter will indicate a voltage of

- ❑ *a* 20 V
- ❑ *b* 60 V
- ❑ *c* 80 V
- ❑ *d* 100 V

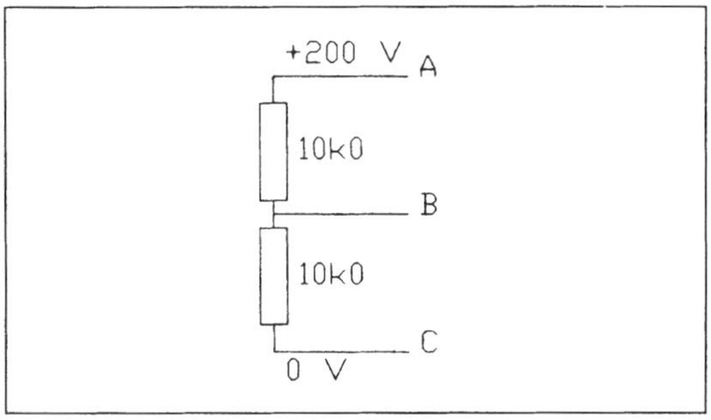

2.43 If the voltmeter in Question 2.42 has a resistance of 100k0, the voltage indicated is

- ❑ *a* 75.3 V
- ❑ *b* 89.2 V
- ❑ *c* 90.43 V
- ❑ *d* 95.23 V

2.44 The diagram shows a series-parallel circuit. The value of I is

❑ *a* 1.5 A
❑ *b* 2.4 A
❑ *c* 2.8 A
❑ *d* 3.0 A

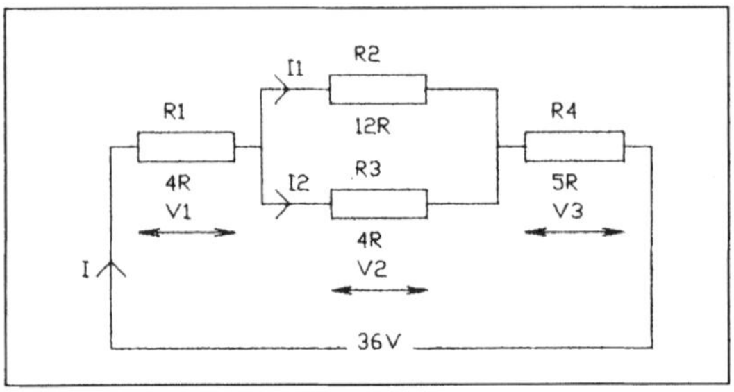

2.45 In Question 2.44 the value of V_2 is

❑ *a* 9 V
❑ *b* 10 V
❑ *c* 12 V
❑ *d* 22 V

2.46 In Question 2.44 the value of V_3 is

❑ *a* 6 V
❑ *b* 10 V
❑ *c* 12 V
❑ *d* 15 V

2.47 In Question 2.44 the value of I_1 is

❑ *a* 0.5 A
❑ *b* 0.75 A
❑ *c* 1.0 A
❑ *d* 1.25 A

2.48 In Question 2.44 the value of I_2 is

❑ *a* 1.75 A
❑ *b* 2.0 A
❑ *c* 2.25 A
❑ *d* 2.5 A

2.49 In Question 2.44 the power consumed by R_2 is

❑ *a* 2.4 W
❑ *b* 6.75 W
❑ *c* 8.2 W
❑ *d* 9.5 W

2.50 In Question 2.44 the power consumed by R_3 is

❑ *a* 20.25 W
❑ *b* 24.25 W
❑ *c* 24.75 W
❑ *d* 30.25 W

2.51 In Question 2.44 the power consumed by R_4 is

❑ *a* 25 W
❑ *b* 35 W
❑ *c* 45 W
❑ *d* 55 W

2.52 The diagram shows a Wheatstone bridge which is in balance. The value of the unknown resistor R_4 is

❑ *a* 15k0
❑ *b* 22k0
❑ *c* 33k0
❑ *d* 47k0

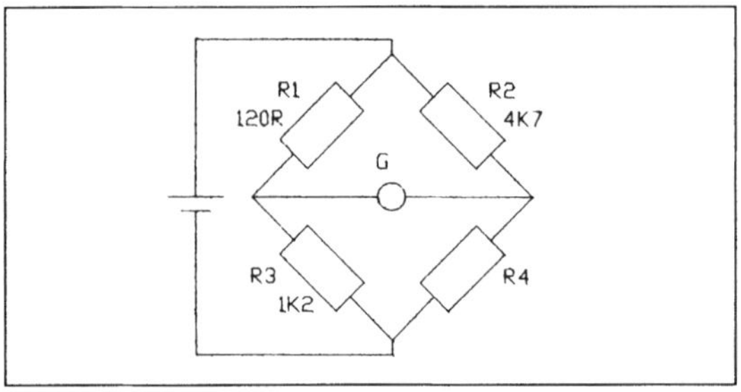

2.53 If the bridge in Question 2.52 is balanced, with R_2 changed to 680R, the value of R_4 is

❑ *a* 4k7
❑ *b* 6k8
❑ *c* 5k6
❑ *d* 3k3

2.54 Six resistors are connected in parallel. Three have values of 1k44 and three have values of 480R. The combined resistance is

❑ *a* 100R
❑ *b* 120R
❑ *c* 200R
❑ *d* 240R

2.55 Five 1k5 resistors are connected in parallel. It is a requirement to add a sixth resistor in parallel with the network to give a combined resistance of 210R. The value of the additional resistor is

❑ *a* 550R
❑ *b* 650R
❑ *c* 700R
❑ *d* 800R

2.56 The diagram shows a series-parallel circuit. The value of current I is

❑ *a* 6.67 A
❑ *b* 8.33 A
❑ *c* 10.67 A
❑ *d* 13.33 A

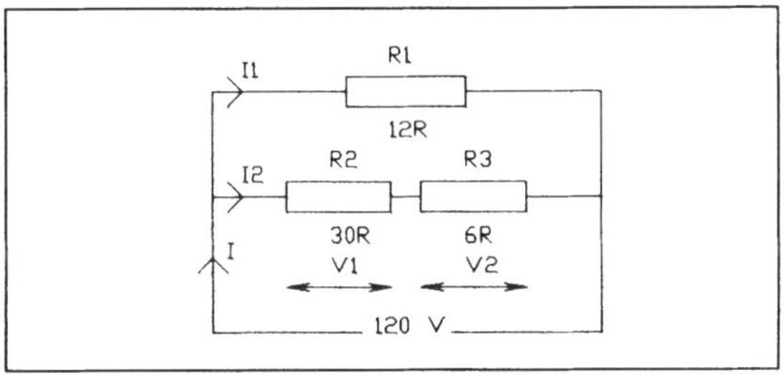

2.57 The value of current I_2 in Question 2.56 is

❑ *a* 3.33 A
❑ *b* 3.93 A
❑ *c* 4.33 A
❑ *d* 4.67 A

2.58 The value of the voltage V_1 in Question 2.56 is

❑ *a* 50 V
❑ *b* 70 V
❑ *c* 90 V
❑ *d* 100 V

2.59 The value of the power dissipated in R_1 in Question 2.56 is

❑ *a* 900 W
❑ *b* 1000 W
❑ *c* 1100 W
❑ *d* 1200 W

2.60 The value of the power dissipated in R_3 in Question 2.56 is

❑ *a* 50.33 W
❑ *b* 66.67 W
❑ *c* 69.33 W
❑ *d* 72.67 W

2.61 The diagram shows a series-parallel circuit. The value of R_4 is

❑ *a* 16R
❑ *b* 24R
❑ *c* 36R
❑ *d* 48R

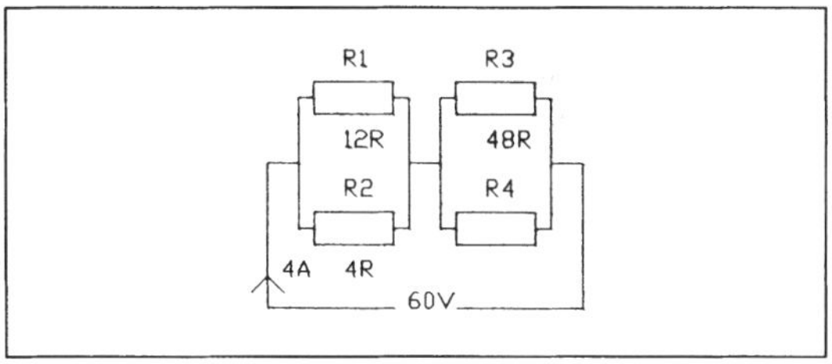

2.62 The voltage across resistor R_1 in Question 2.61 is

❑ *a* 18 V
❑ *b* 16 V
❑ *c* 20 V
❑ *d* 12 V

2.63 The power consumed by R_2 in Question 2.61 is

❑ *a* 12 W
❑ *b* 15 W
❑ *c* 28 W
❑ *d* 36 W

2.64 The power consumed by R_4 in Question 2.61 is

❑ *a* 20 W
❑ *b* 144 W
❑ *c* 240 W
❑ *d* 280 W

2.65 If R_4 in Question 2.61 is changed so that the input current is 1.714 A, the value of R_4 is

❑ *a* 24R
❑ *b* 48R
❑ *c* 64R
❑ *d* 96R

2.66 The diagram shows a four-resistor network. The value of the current flowing in the 6R resistor is

❑ *a* 2 A
❑ *b* 4 A
❑ *c* 6 A
❑ *d* 8 A

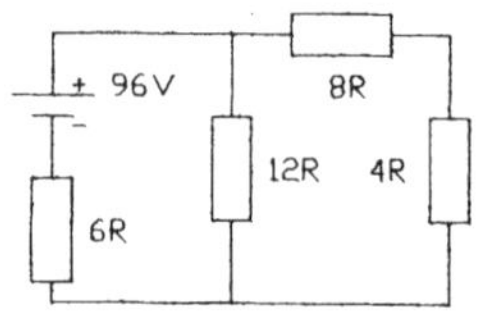

2.67 In Question 2.66 the current flowing in the 12R resistor is

❑ *a* 2 A
❑ *b* 3 A
❑ *c* 4 A
❑ *d* 5 A

2.68 In Question 2.66 the current flowing in the 8R resistor is

❑ *a* 4 A
❑ *b* 6 A
❑ *c* 8 A
❑ *d* 10 A

2.69 If the battery voltage in Question 2.66 is increased to 108 V, the current flowing in the 6R resistor becomes

❑ *a* 8.5 A
❑ *b* 9.0 A
❑ *c* 10.0 A
❑ *d* 10.5 A

2.70 If the battery voltage in Question 2.66 is increased to 108 V, the current flowing in the 4R resistor becomes

❑ *a* 6.5 A
❑ *b* 3.5 A
❑ *c* 7.5 A
❑ *d* 4.5 A

2.71 The diagram shows a network containing two batteries. The current flowing through the 5R resistor is

❑ *a* 0.623 A
❑ *b* 0.658 A
❑ *c* 0.706 A
❑ *d* 0.714 A

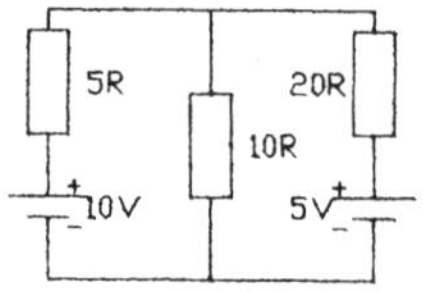

2.72 The current flowing through the 10R resistor in Question 2.71 is

❑ *a* 0.643 A
❑ *b* 0.683 A
❑ *c* 0.723 A
❑ *d* 0.753 A

2.73 The current flowing through the 20R resistor in Question 2.71 is

❑ *a* 0.06 A
❑ *b* 0.064 A
❑ *c* 0.069 A
❑ *d* 0.071 A

2.74 The voltage between the top and bottom rails of the network in Question 2.71 is

❑ *a* 2.62 V
❑ *b* 4.53 V
❑ *c* 6.43 V
❑ *d* 7.28 V

2.75 A 240-V 60-W electric light bulb has a filament resistance of

❑ *a* 120 Ω
❑ *b* 240 Ω
❑ *c* 480 Ω
❑ *d* 960 Ω

3 AC voltages

3.1 Alternating waveforms can be plotted against degrees or radians; 45° expressed in radians is

- ❑ *a* 0.4954 rad
- ❑ *b* 0.7855 rad
- ❑ *c* 0.8661 rad
- ❑ *d* 0.9232 rad

3.2 360° expressed in radians is

- ❑ *a* π/4 rad
- ❑ *b* π rad
- ❑ *c* 2π rad
- ❑ *d* 4π rad

3.3 175° expressed in radians is

- ❑ *a* 3.055 rad
- ❑ *b* 4.024 rad
- ❑ *c* 5.033 rad
- ❑ *d* 6.128 rad

3.4 30° expressed in radians is

- ❑ *a* π/3 rad
- ❑ *b* π/6 rad
- ❑ *c* π/2 rad
- ❑ *d* 4π/3 rad

3.5 450° expressed in radians is

- ❑ *a* 7.855
- ❑ *b* 8.824
- ❑ *c* 8.932
- ❑ *d* 9.461

3.6 The sin of 5π/6 rad is

- ❑ *a* 0.1
- ❑ *b* 0.5
- ❑ *c* 0.866
- ❑ *d* 1.0

3.7 The sin of 3π/2 rad is

- ❑ *a* 1.0
- ❑ *b* −0.5
- ❑ *c* 0.5
- ❑ *d* −1.0

3.8 π rad expressed in degrees is

- ❑ *a* 60°
- ❑ *b* 90°
- ❑ *c* 150°
- ❑ *d* 180°

3.9 1.8 rad expressed in degrees is

- ❑ *a* 100.24°
- ❑ *b* 103.119°
- ❑ *c* 106.27°
- ❑ *d* 204.43°

3.10 6.284 rad expressed in degrees is

- ❑ *a* 270°
- ❑ *b* 105°
- ❑ *c* 360°
- ❑ *d* 120°

3.11 7.0 rad expressed in degrees is approximately

- ❑ *a* 302°
- ❑ *b* 401°
- ❑ *c* 506°
- ❑ *d* 550°

3.12 5π/6 rad expressed in degrees is

- ❑ *a* 65°
- ❑ *b* 85°
- ❑ *c* 125°
- ❑ *d* 150°

3.13 The sin of 2π rad is
- ❑ *a* 0
- ❑ *b* 1.0
- ❑ *c* 2.0
- ❑ *d* 3.0

3.14 The sin of $11\pi/6$ rad is
- ❑ *a* 0
- ❑ *b* 0.5
- ❑ *c* −0.5
- ❑ *d* −1.0

3.15 A voltage sine wave having a maximum value of 24 V is plotted from 0 ° to 360 °. The instantaneous value at 30 ° is
- ❑ *a* 1.62 V
- ❑ *b* 1.86 V
- ❑ *c* 2.02 V
- ❑ *d* 12 V

3.16 If the waveform in Question 3.15 had a maximum value of 240 V, the instantaneous value at 90 ° would be
- ❑ *a* 0 V
- ❑ *b* 240 V
- ❑ *c* −120 V
- ❑ *d* −240 V

3.17 If the waveform in Question 3.15 had a maximum value of 50 V, the instantaneous value at 200 ° would be
- ❑ *a* −17.1 V
- ❑ *b* 34.2 V
- ❑ *c* −64.2 V
- ❑ *d* 120 V

3.18 If the waveform in Question 3.15 had an instantaneous value of −31.18 V at 240 °, the maximum value would be
- ❑ *a* 20 V
- ❑ *b* 30 V
- ❑ *c* 36 V
- ❑ *d* 40 V

3.19 A sinusoidal voltage has an instantaneous value of −240 V at 270 °. The maximum value is
- ❑ *a* 200 V
- ❑ *b* 60 V
- ❑ *c* 240 V
- ❑ *d* 120 V

3.20 A voltage sine wave having a maximum value of 100 V is plotted from 0 ° to 360 °. The waveform will have an instantaneous value of 86.6 V at
- ❑ *a* 20 °
- ❑ *b* 40 °
- ❑ *c* 45 °
- ❑ *d* 60 °

3.21 If the waveform in Question 3.20 has a maximum value of 300 V, an instantaneous value of −259.8 V will occur at
- ❑ *a* 190 °
- ❑ *b* 270 °
- ❑ *c* 300 °
- ❑ *d* 360 °

3.22 A sine wave has an r.m.s. value of 45 V. The peak value is
- ❑ *a* 52.26 V
- ❑ *b* 59.18 V
- ❑ *c* 60.23 V
- ❑ *d* 63.64 V

3.23 A sine wave has an r.m.s. value of 100 V. The peak to peak value is
- ❑ *a* 282.84 V
- ❑ *b* 300 V
- ❑ *c* 322.42 V
- ❑ *d* 340 V

3.24 The domestic mains supply has a value of 240 V r.m.s. The peak value is approximately
- ❑ *a* 280 V
- ❑ *b* 300 V
- ❑ *c* 320 V
- ❑ *d* 340 V

3.25 The peak to peak value of the domestic mains supply is approximately
- ❑ *a* 560 V
- ❑ *b* 600 V
- ❑ *c* 640 V
- ❑ *d* 680 V

3.26 An oscillator produces a sine wave with an r.m.s. value of 1.5 mV. The peak value is
- ❑ *a* 2.0 mV
- ❑ *b* 2.12 mV
- ❑ *c* 4.24 mV
- ❑ *d* 5.2 mV

3.27 If a sine wave has an r.m.s. value of 7.07 V, the peak to peak value is
- ❑ *a* 10 V
- ❑ *b* 15 V
- ❑ *c* 20 V
- ❑ *d* 25 V

3.28 A sinusoidal voltage having a value of 2000 V r.m.s. would have a peak to peak value of approximately
- ❑ *a* 4000 V
- ❑ *b* 5656 V
- ❑ *c* 6000 V
- ❑ *d* 6666 V

3.29 A sine wave has an r.m.s. value of 150 μV. The peak value in millivolts is
- ❑ *a* 0.21
- ❑ *b* 0.26
- ❑ *c* 0.8
- ❑ *d* 1.2

3.30 A sinusoidal voltage has a peak to peak value of 4.0 V. The r.m.s. value is
- ❑ *a* 1.0 V
- ❑ *b* 1.02 V
- ❑ *c* 1.414 V
- ❑ *d* 2.0 V

3.31 A sine wave has a peak value of 240 V. The r.m.s. value is
- ❑ *a* 169.7 V
- ❑ *b* 180 V
- ❑ *c* 192.7 V
- ❑ *d* 200 V

3.32 A sine wave has a peak value of 0.85 mV. The r.m.s. value in microvolts is approximately
- ❑ *a* 200
- ❑ *b* 300
- ❑ *c* 600
- ❑ *d* 800

3.33 A sine wave has a peak to peak value of 0.5 mV. The r.m.s. value in microvolts is
- ❑ *a* 177
- ❑ *b* 200
- ❑ *c* 217
- ❑ *d* 225

3.34 A sine wave having a peak voltage of 7.5 mV will have an r.m.s. value in microvolts of approximately
- ❑ *a* 2062
- ❑ *b* 4612
- ❑ *c* 4823
- ❑ *d* 5303

3.35 A sine wave having a peak voltage of 0.004 mV will have an r.m.s. value in nanovolts of
- ❑ *a* 480
- ❑ *b* 2400
- ❑ *c* 2828
- ❑ *d* 3000

3.36 A sine wave has a peak to peak value of 0.3 mV. The r.m.s. value in microvolts is approximately
- ❑ *a* 106
- ❑ *b* 126
- ❑ *c* 146
- ❑ *d* 206

3.37 A waveform has a frequency of 250 Hz. The periodic time in seconds is
- ❑ *a* 0.02
- ❑ *b* 0.04
- ❑ *c* 0.004
- ❑ *d* 0.008

3.38 The periodic time of the waveform in Question 3.37 in milliseconds is

❑ *a* 20
❑ *b* 40
❑ *c* 8
❑ *d* 4

3.39 A sine wave has a frequency of 1 kHz. The periodic time in microseconds is

❑ *a* 250
❑ *b* 500
❑ *c* 750
❑ *d* 1000

3.40 The domestic mains has a frequency of 50 Hz. The periodic time is

❑ *a* 0.5 s
❑ *b* 2.5 s
❑ *c* 20 ms
❑ *d* 20 μs

3.41 A voltage waveform has a frequency of 150 kHz. The periodic time in milliseconds is

❑ *a* 6.67×10^3
❑ *b* 6.67×10^{-3}
❑ *c* 6.67×10^4
❑ *d* 6.67×10^{-4}

3.42 The output of an oscillator has a frequency of 2.5 MHz. The periodic time in micro-seconds is

❑ *a* 0.4
❑ *b* 40
❑ *c* 140
❑ *d* 400

3.43 The periodic time for Question 3.42 in nanoseconds is

❑ *a* 1400
❑ *b* 4000
❑ *c* 200
❑ *d* 400

3.44 A waveform has a periodic time of 12.5 ms. The frequency is

❑ *a* 25 Hz
❑ *b* 75 Hz
❑ *c* 80 Hz
❑ *d* 125 Hz

3.45 A sine wave has a periodic time of 400 μs. The frequency is

❑ *a* 800 Hz
❑ *b* 2.5 kHz
❑ *c* 3.4 kHz
❑ *d* 10 kHz

3.46 A waveform has a periodic time of 0.25 ms. The frequency is

❑ *a* 1000 Hz
❑ *b* 2000 Hz
❑ *c* 3000 Hz
❑ *d* 4000 Hz

3.47 An oscillator produccs a sinc wave with a periodic time of 100 ns. The frequency is

❑ *a* 10^7 Hz
❑ *b* 10^6 Hz
❑ *c* 10^5 Hz
❑ *d* 10^4 Hz

3.48 If the periodic time in Question 3.47 is halved, the frequency will then be

❑ *a* 0.5×10^6 Hz
❑ *b* 2×10^7 Hz
❑ *c* 0.5×10^7 Hz
❑ *d* 2×10^5 Hz

3.49 Radio 4 has a wavelength of 1500 m. The frequency is

❑ *a* 200 Hz
❑ *b* 800 Hz
❑ *c* 100 kHz
❑ *d* 200 kHz

3.50 A wavelength of 25 m corresponds to a frequency of

❑ *a* 10^6 Hz
❑ *b* 1.5×10^5 Hz
❑ *c* 1.2×10^7 Hz
❑ *d* 10^8 Hz

3.51 A wavelength of 375 m is equivalent to a frequency of

❑ *a* 200 kHz
❑ *b* 400 kHz
❑ *c* 800 kHz
❑ *d* 900 kHz

3.52 A sine wave has a wavelength of 20 cm. The frequency is
- ❑ *a* 1.5×10^9 Hz
- ❑ *b* 4.5×10^8 Hz
- ❑ *c* 10^7 Hz
- ❑ *d* 10^4 Hz

3.53 An infra-red signal has a wavelength of 3×10^{-5} m. The frequency is
- ❑ *a* 10^5 Hz
- ❑ *b* 10^8 Hz
- ❑ *c* 10^{10} Hz
- ❑ *d* 10^{13} Hz

3.54 A transmitter is radiating a signal with a wavelength of 0.25 m. The frequency is
- ❑ *a* 100 MHz
- ❑ *b* 200 kHz
- ❑ *c* 1200 MHz
- ❑ *d* 1500 kHz

3.55 The frequency of the transmitter in Question 3.54 in gigahertz is
- ❑ *a* 0.1
- ❑ *b* 1.2
- ❑ *c* 12
- ❑ *d* 100

3.56 An ultraviolet ray has a wavelength of 10^{-8} m. The frequency is
- ❑ *a* 10^{10} Hz
- ❑ *b* 2×10^{12} Hz
- ❑ *c* 3×10^{16} Hz
- ❑ *d* 4×10^2 Hz

3.57 A cosmic ray has a wavelength of 3×10^{-12} m. The frequency is
- ❑ *a* 10^{20} Hz
- ❑ *b* 10^{15} Hz
- ❑ *c* 10^8 Hz
- ❑ *d* 10^6 Hz

3.58 A frequency of 100 Hz has a wavelength of
- ❑ *a* 10^6 m
- ❑ *b* 3×10^6 m
- ❑ *c* 10^7 m
- ❑ *d* 4×10^8 m

3.59 A frequency of 2.5 kHz has a wavelength of
- ❑ *a* 120 m
- ❑ *b* 250 m
- ❑ *c* 250 km
- ❑ *d* 120 km

3.60 A frequency of 10^6 Hz corresponds to a wavelength of
- ❑ *a* 10 m
- ❑ *b* 150 m
- ❑ *c* 300 m
- ❑ *d* 600 m

3.61 A frequency of 95 MHz has a wavelength of
- ❑ *a* 1.64 m
- ❑ *b* 2.53 m
- ❑ *c* 2.92 m
- ❑ *d* 3.16 m

3.62 A waveform of 1 GHz has a wavelength of
- ❑ *a* 0.3 m
- ❑ *b* 1.3 m
- ❑ *c* 2.3 m
- ❑ *d* 2.6 m

3.63 A sine wave has a peak voltage of 25 V. The average value is approximately
- ❑ *a* 16 V
- ❑ *b* 17 V
- ❑ *c* 18 V
- ❑ *d* 19 V

3.64 The form factor for the waveform in Question 3.63 is
- ❑ *a* 0.8
- ❑ *b* 1.1
- ❑ *c* 2.1
- ❑ *d* 2.3

3.65 A sine wave has an r.m.s. value of 100 V. The average value is approximately
- ❑ *a* 50 V
- ❑ *b* 70 V
- ❑ *c* 80 V
- ❑ *d* 90 V

3.66 The average value of a sine wave is 152.88 V. The peak value is
- ❑ *a* 160 V
- ❑ *b* 220 V
- ❑ *c* 240 V
- ❑ *d* 300 V

3.67 The average value of a sine wave is 135.13 V. The r.m.s. value is
- ❑ *a* 150 V
- ❑ *b* 200 V
- ❑ *c* 180 V
- ❑ *d* 160 V

3.68 A sine wave with a peak voltage of 30 V is applied to a half-wave rectifier. Assuming no losses, the mean DC output voltage is
- ❑ *a* 6.32 V
- ❑ *b* 7.63 V
- ❑ *c* 8.26 V
- ❑ *d* 9.55 V

3.69 The output r.m.s. voltage in Question 3.68 is
- ❑ *a* 25 V
- ❑ *b* 15 V
- ❑ *c* 10 V
- ❑ *d* 7.5 V

3.70 If the voltage in Question 3.68 is applied to a full-wave rectifier, the mean DC output voltage is
- ❑ *a* 19.1 V
- ❑ *b* 20.2 V
- ❑ *c* 23.4 V
- ❑ *d* 24.5 V

3.71 The output r.m.s. voltage in Question 3.70 is
- ❑ *a* 16.41 V
- ❑ *b* 17.67 V
- ❑ *c* 21.22 V
- ❑ *d* 23.41 V

3.72 A square wave comprises
- ❑ *a* The fundamental plus all odd harmonics
- ❑ *b* The fundamental plus all even harmonics
- ❑ *c* All odd harmonics
- ❑ *d* All even harmonics

3.73 A waveform which includes the fundamental and all even harmonics is
- ❑ *a* Sinusoidal
- ❑ *b* Square
- ❑ *c* Sawtooth
- ❑ *d* Triangular

3.74 A square wave has a fundamental frequency of 450 Hz. The fifth harmonic is
- ❑ *a* 2000 Hz
- ❑ *b* 2250 Hz
- ❑ *c* 4500 Hz
- ❑ *d* 5500 Hz

3.75 Ten cycles of a voltage waveform occur in 2 ms. The frequency is
- ❑ *a* 100 Hz
- ❑ *b* 200 Hz
- ❑ *c* 1000 Hz
- ❑ *d* 5000 Hz

4 Capacitors

4.1 A 2-μF capacitor expressed in nanofarads is

- ❑ *a* 0.02 nF
- ❑ *b* 20 nF
- ❑ *c* 200 nF
- ❑ *d* 2000 nF

4.2 4.5×10^7 pF expressed in microfarads is

- ❑ *a* 45 μF
- ❑ *b* 450 μF
- ❑ *c* 4500 μF
- ❑ *d* 45×10^3 μF

4.3 10^5 pF expressed in nanofarads is

- ❑ *a* 1.0 nF
- ❑ *b* 10 nF
- ❑ *c* 100 nF
- ❑ *d* 1000 nF

4.4 1000 pF expressed in microfarads is

- ❑ *a* 10 μF
- ❑ *b* 1.0 μF
- ❑ *c* 0.1 μF
- ❑ *d* 0.001 μF

4.5 A 150-pF capacitor has 12 V across the plates. The charge stored is

- ❑ *a* 1.2 nC
- ❑ *b* 1.8 nC
- ❑ *c* 12 μC
- ❑ *d* 18 μC

4.6 A voltage of 25 V is applied to a 1.8-nF capacitor. The charge stored is

- ❑ *a* 25 μC
- ❑ *b* 45 μC
- ❑ *c* 25 nC
- ❑ *d* 45 nC

4.7 200 V is applied to a 0.86-nF capacitor. The charge stored in nanocoulombs is

- ❑ *a* 160
- ❑ *b* 172
- ❑ *c* 200
- ❑ *d* 286

4.8 A 1.5-μF capacitor stores a charge of 2.4×10^{-4} C. The terminal voltage is

- ❑ *a* 100 V
- ❑ *b* 160 V
- ❑ *c* 1000 V
- ❑ *d* 1600 V

4.9 A 1000-μF capacitor stores a charge of 0.45 C. The terminal voltage is

- ❑ *a* 4.5 V
- ❑ *b* 45 V
- ❑ *c* 450 V
- ❑ *d* 4500 V

4.10 A charge of 1.2 C is stored in a 250-μF capacitor. The terminal voltage is

- ❑ *a* 0.48 kV
- ❑ *b* 1.2 kV
- ❑ *c* 4.8 kV
- ❑ *d* 6.4 kV

4.11 A capacitor stores 0.5 mC with 10 V across the plates. The capacitor value is

- ❑ *a* 50 nF
- ❑ *b* 50 μF
- ❑ *c* 100 nF
- ❑ *d* 100 μF

4.12 A capacitor has a charge of 120 μC with a terminal voltage of 1.5 V. The capacitor value is

❑ *a* 800 pf
❑ *b* 80 pf
❑ *c* 80 nF
❑ *d* 80 μF

4.13 A capacitor has a charge of 0.26 C. The terminal voltage is 200 V. The capacitor value is

❑ *a* 800 μF
❑ *b* 1020 μF
❑ *c* 1300 μF
❑ *d* 1500 μF

4.14 The energy stored in a 3.6-μF capacitor with an applied voltage of 15 V is

❑ *a* 0.405 mJ
❑ *b* 0.625 mJ
❑ *c* 0.705 mJ
❑ *d* 0.835 mJ

4.15 A capacitor has a terminal voltage of 12 V and has a value of 400 μF. The energy stored is

❑ *a* 20.6 mJ
❑ *b* 26.8 mJ
❑ *c* 28.8 mJ
❑ *d* 30.2 mJ

4.16 The energy stored in a 500-μF capacitor with an applied voltage of 100 V is

❑ *a* 1.5 J
❑ *b* 2.5 J
❑ *c* 3.5 J
❑ *d* 4.5 J

4.17 A 15-μF capacitor stores 200 J. The terminal voltage is approximately

❑ *a* 2468 V
❑ *b* 3246 V
❑ *c* 4000 V
❑ *d* 5164 V

4.18 28 J is stored in a 100-μF capacitor. The applied voltage is approximately

❑ *a* 700 V
❑ *b* 748 V
❑ *c* 766 V
❑ *d* 7480 V

4.19 2000 mJ is stored in a 1.8-μF capacitor. The voltage between the plates is approximately

❑ *a* 259 V
❑ *b* 1491 V
❑ *c* 1582 V
❑ *d* 1671 V

4.20 250 J is stored in a capacitor when the applied voltage is 1500 V. The capacitor value is

❑ *a* 160 μF
❑ *b* 200 μF
❑ *c* 222 μF
❑ *d* 262 μF

4.21 2.5 J is stored in a capacitor when the applied voltage is 140 V. The capacitor value is

❑ *a* 255 μF
❑ *b* 125 μF
❑ *c* 600 μF
❑ *d* 650 μF

4.22 A capacitor stores 50 mJ when the applied voltage is 12 V. The capacitor value is approximately

❑ *a* 20 μF
❑ *b* 294 μF
❑ *c* 600 μF
❑ *d* 694 μF

4.23 The diagram shows two capacitors connected in series. The value of the overall capacitance *C* is

❑ *a* 6.5 μF
❑ *b* 10 μF
❑ *c* 4 μF
❑ *d* 12.5 μF

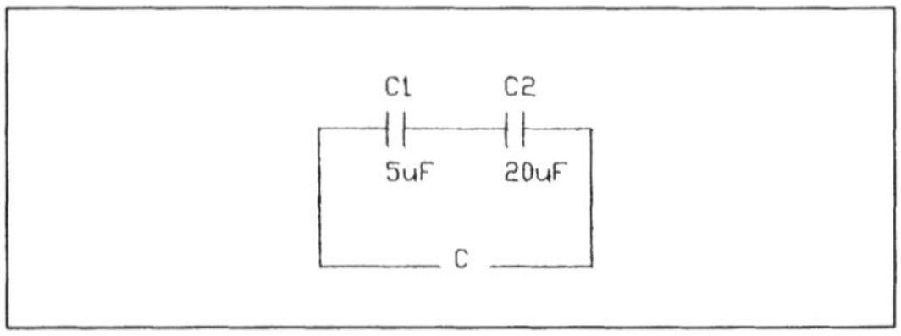

4.24 If the two capacitors in Question 4.23 are connected in parallel, the overall capacitance value is
- ❑ *a* 5 μF
- ❑ *b* 20 μF
- ❑ *c* 25 μF
- ❑ *d* 100 μF

4.25 If an additional 1.0-μF capacitor is added in series with the circuit in Question 4.23, the overall capacitance then becomes
- ❑ *a* 600 nF
- ❑ *b* 800 nF
- ❑ *c* 0.9 μF
- ❑ *d* 1.25 μF

4.26 Two capacitor plates are separated by a dielectric 0.5 mm thick. If the terminal voltage is 6 V, the field strength in volts per metre is
- ❑ *a* 600
- ❑ *b* 6×10^3
- ❑ *c* 12×10^3
- ❑ *d* 15×10^4

4.27 If the dielectric thickness in Question 4.26 is 2.4 mm and the applied voltage is 2.5 V, the field strength is approximately
- ❑ *a* 1041 $V\,m^{-1}$
- ❑ *b* 750 $V\,m^{-1}$
- ❑ *c* 1250 $V\,m^{-1}$
- ❑ *d* 600 $V\,m^{-1}$

4.28 If the dielectric thickness in Question 4.26 is 0.1 mm and the terminal voltage is 1.2 V, the field strength is
- ❑ *a* 0.6 $kV\,m^{-1}$
- ❑ *b* 6 $kV\,m^{-1}$
- ❑ *c* 12 $kV\,m^{-1}$
- ❑ *d* 18 $kV\,m^{-1}$

4.29 If 100 V applied to a capacitor produces a field strength of 80 $kV\,m^{-1}$, the dielectric thickness is
- ❑ *a* 1.2 mm
- ❑ *b* 1.25 mm
- ❑ *c* 2.0 mm
- ❑ *d* 2.1 mm

4.30 If 50 V applied to a capacitor produces a field strength of 50 $kV\,m^{-1}$, the separation between the plates is
- ❑ *a* 1.0 mm
- ❑ *b* 0.1 mm
- ❑ *c* 0.1 cm
- ❑ *d* 0.5 cm

4.31 If the voltage is 60 V and the field strength is 100 $kV\,m^{-1}$ in Question 4.30, the separation between the plates is
- ❑ *a* 0.06 cm
- ❑ *b* 0.6 cm
- ❑ *c* 0.62 mm
- ❑ *d* 0.06 mm

4.32 A capacitor has a dielectric which is 0.2 mm thick. If the field strength is 250 $kV\,m^{-1}$, the terminal voltage is
- ❑ *a* 10 V
- ❑ *b* 25 V
- ❑ *c* 40 V
- ❑ *d* 50 V

4.33 The plates of a capacitor are separated by 0.15 mm. If the field strength is 500 $kV\,m^{-1}$, the terminal voltage is
- ❑ *a* 75 V
- ❑ *b* 750 V
- ❑ *c* 800 V
- ❑ *d* 850 V

4.34 A capacitor has a dielectric which is 0.05 cm thick. If the field strength is 400 $kV\,m^{-1}$, the terminal voltage is
- ❑ *a* 100 V
- ❑ *b* 200 V
- ❑ *c* 300 V
- ❑ *d* 400 V

4.35 A capacitor consists of two metal plates 50 mm in length and 100 mm wide. The capacitor holds a charge of 0.2 μC. The electric flux density in microcoulombs per square metre is
- ❑ *a* 10
- ❑ *b* 20
- ❑ *c* 30
- ❑ *d* 40

4.36 If the plates in Question 4.35 are 40 mm long and 40 mm wide and the charge is 1.5 μC, the flux density is approximately

❑ *a* 468 $\mu C\,m^{-2}$
❑ *b* 792 $\mu C\,m^{-2}$
❑ *c* 938 $\mu C\,m^{-2}$
❑ *d* 998 $\mu C\,m^{-2}$

4.37 If the plates in Question 4.35 are 2.0 cm long and 4.0 cm wide and the charge is 500 nC, the flux density is

❑ *a* 625 $\mu C\,m^{-2}$
❑ *b* 700 $\mu C\,m^{-2}$
❑ *c* 725 $\mu C\,m^{-2}$
❑ *d* 800 $\mu C\,m^{-2}$

4.38 The diagram shows two capacitors connected in series. The voltage V_2 is

❑ *a* 10 V
❑ *b* 15 V
❑ *c* 25 V
❑ *d* 30 V

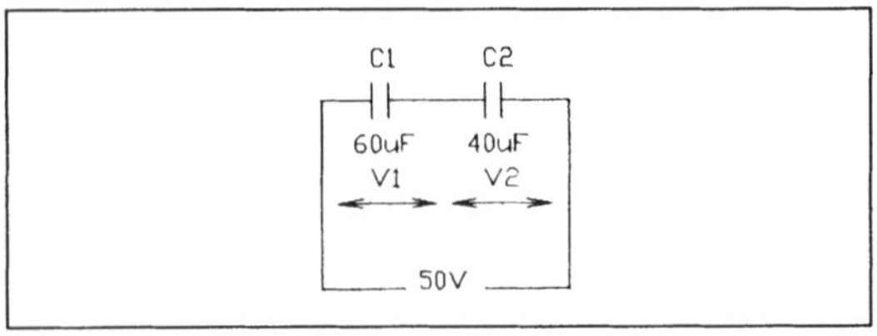

4.39 The charge in capacitor C_1 in Question 4.38 is

❑ *a* 400 μC
❑ *b* 1200 μC
❑ *c* 600 μC
❑ *d* 800 μC

4.40 The total charge in the circuit of Question 4.38 is

❑ *a* 400 μC
❑ *b* 800 μC
❑ *c* 1200 μC
❑ *d* 2400 μC

4.41 The energy stored in capacitor C_1 in Question 4.38 is

❑ *a* 12 mJ
❑ *b* 14 mJ
❑ *c* 16 mJ
❑ *d* 18 mJ

4.42 The energy stored in capacitor C_2 in Question 4.38 is

❑ *a* 14 mJ
❑ *b* 16 mJ
❑ *c* 18 mJ
❑ *d* 20 mJ

4.43 The total energy stored in the circuit of Question 4.38 is

❑ *a* 16 mJ
❑ *b* 28 mJ
❑ *c* 30 mJ
❑ *d* 38 mJ

4.44 The diagram shows two capacitors connected in parallel. The combined capacitance is

❑ *a* 2.4 μF
❑ *b* 4.2 μF
❑ *c* 2.0 μF
❑ *d* 10 μF

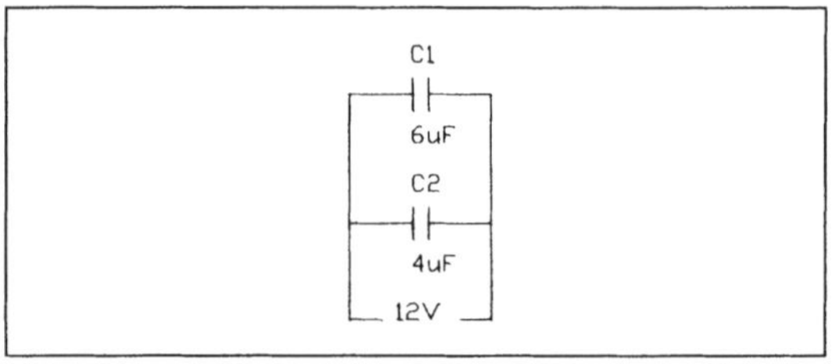

4.45 In Question 4.44 the total charge in the circuit is

❑ *a* 102 μC
❑ *b* 120 μC
❑ *c* 140 μC
❑ *d* 160 μC

4.46 In Question 4.44 the charge on capacitor C_1 is

❑ *a* 60 μC
❑ *b* 64 μC
❑ *c* 70 μC
❑ *d* 72 μC

4.47 In Question 4.44 the charge on capacitor C_2 is

❑ *a* 48 μC
❑ *b* 50 μC
❑ *c* 60 μC
❑ *d* 64 μC

4.48 The total energy stored in Question 4.44 is
- ❑ *a* 680 μJ
- ❑ *b* 720 μJ
- ❑ *c* 800 μJ
- ❑ *d* 870 μJ

4.49 The energy stored in capacitor C_1 in Question 4.44 is
- ❑ *a* 210 μJ
- ❑ *b* 242 μJ
- ❑ *c* 420 μJ
- ❑ *d* 432 μJ

4.50 The energy stored in capacitor C_2 in Question 4.44 is
- ❑ *a* 200 μJ
- ❑ *b* 244 μJ
- ❑ *c* 288 μJ
- ❑ *d* 290 μJ

4.51 A capacitor has a flux density of 800 $\mu C\,m^{-2}$ and a charge of 0.1 μC. The area of the plate in square millimetres is
- ❑ *a* 100
- ❑ *b* 125
- ❑ *c* 200
- ❑ *d* 250

4.52 A capacitor has two circular plates with diameters of 5 mm. If the flux density is 20 $\mu C\,m^{-2}$, the charge in nanocoulombs is approximately
- ❑ *a* 0.87
- ❑ *b* 1.23
- ❑ *c* 1.41
- ❑ *d* 1.57

4.53 The reactance of a 0.2-μF capacitor at 1 kHz is approximately
- ❑ *a* 800 Ω
- ❑ *b* 796 Ω
- ❑ *c* 656 Ω
- ❑ *d* 420 Ω

4.54 A capacitor has a reactance of 2 kΩ at a frequency of 2 kHz. The capacitor value is approximately
- ❑ *a* 39.78 nF
- ❑ *b* 50.68 nF
- ❑ *c* 62.63 nF
- ❑ *d* 82.07 nF

4.55 A 150-pF capacitor has a reactance of 2500 Ω at a frequency of
- ❑ *a* 4244 Hz
- ❑ *b* 42 480 Hz
- ❑ *c* 42 441 Hz
- ❑ *d* 424 413 Hz

4.56 The diagram shows three capacitors connected in parallel and driven by 20 V. The total capacitance is
- ❑ *a* 80 μF
- ❑ *b* 100 μF
- ❑ *c* 110 μF
- ❑ *d* 112 μF

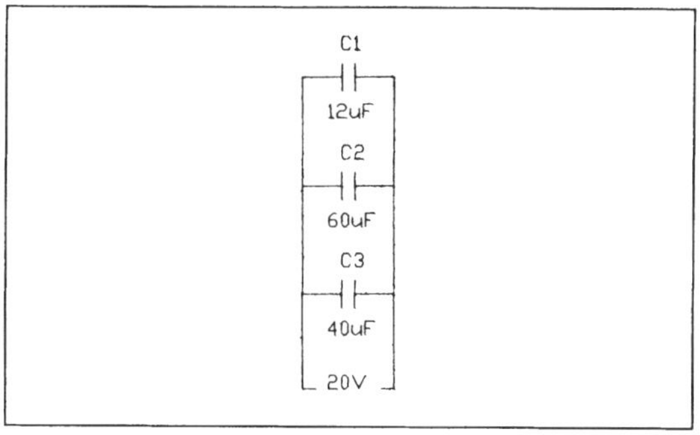

4.57 The charge stored in capacitor C_2 in Question 4.56 is
- ❑ *a* 1000 μC
- ❑ *b* 1200 μC
- ❑ *c* 2000 μC
- ❑ *d* 2400 μC

4.58 The total charge stored in the network in Question 4.56 is
- ❑ *a* 1680 μC
- ❑ *b* 1800 μC
- ❑ *c* 2240 μC
- ❑ *d* 2800 μC

4.59 The energy stored in capacitor C_3 in Question 4.56 is

- ❑ *a* 8 mJ
- ❑ *b* 8.6 mJ
- ❑ *c* 9 mJ
- ❑ *d* 10 mJ

4.60 The total energy stored in the network in Question 4.56 is

- ❑ *a* 20.2 mJ
- ❑ *b* 22.4 mJ
- ❑ *c* 32 mJ
- ❑ *d* 40 mJ

4.61 The diagram shows a series-parallel capacitor network. The combined capacitance is

- ❑ *a* 2 μF
- ❑ *b* 4 μF
- ❑ *c* 6 μF
- ❑ *d* 8 μF

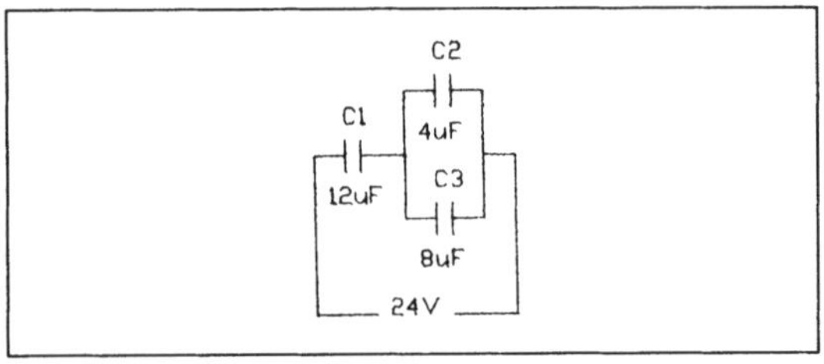

4.62 The voltage drop across capacitor C_1 in Question 4.61 is

- ❑ *a* 6 V
- ❑ *b* 12 V
- ❑ *c* 16 V
- ❑ *d* 20 V

4.63 The charge in capacitor C_2 in Question 4.61 is

- ❑ *a* 40 μC
- ❑ *b* 48 μC
- ❑ *c* 400 μC
- ❑ *d* 480 μC

4.64 The energy stored in capacitor C_1 in Question 4.61 is

- ❑ *a* 864 μJ
- ❑ *b* 900 μJ
- ❑ *c* 940 μJ
- ❑ *d* 964 μJ

4.65 The total energy stored in the network in Question 4.61 is

- ❑ *a* 0.662 mJ
- ❑ *b* 0.826 mJ
- ❑ *c* 1.624 mJ
- ❑ *d* 1.728 mJ

4.66 The diagram shows a four-capacitor series-parallel network. The combined capacitance is

- ❑ *a* 10 μF
- ❑ *b* 15 μF
- ❑ *c* 20 μF
- ❑ *d* 40 μF

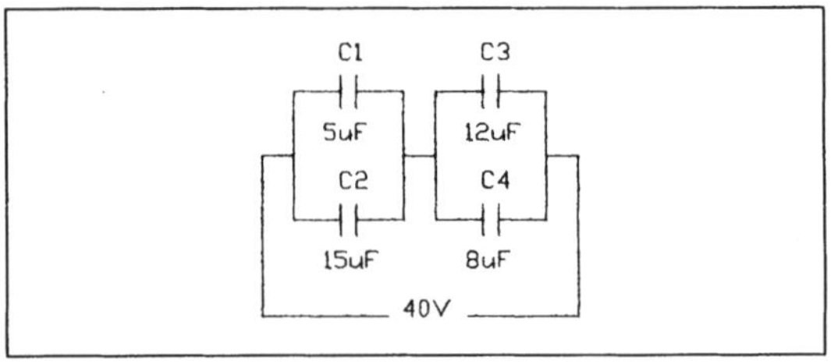

4.67 The total charge stored in the network in Question 4.66 is

- ❑ *a* 200 μC
- ❑ *b* 400 μC
- ❑ *c* 600 μC
- ❑ *d* 800 μC

4.68 The charge stored in capacitor C_2 in Question 4.66 is

- ❑ *a* 300 μC
- ❑ *b* 400 μC
- ❑ *c* 500 μC
- ❑ *d* 600 μC

4.69 The total energy stored in the network in Question 4.66 is

- ❑ *a* 2 mJ
- ❑ *b* 4 mJ
- ❑ *c* 6 mJ
- ❑ *d* 8 mJ

4.70 The energy stored in capacitor C_4 in Question 4.66 is

- ❑ *a* 1.2 mJ
- ❑ *b* 1.6 mJ
- ❑ *c* 2.2 mJ
- ❑ *d* 4.6 mJ

4.71 The diagram shows a three-capacitor series-parallel network. The combined capacitance is

- ❑ *a* 10 μF
- ❑ *b* 20 μF
- ❑ *c* 30 μF
- ❑ *d* 40 μF

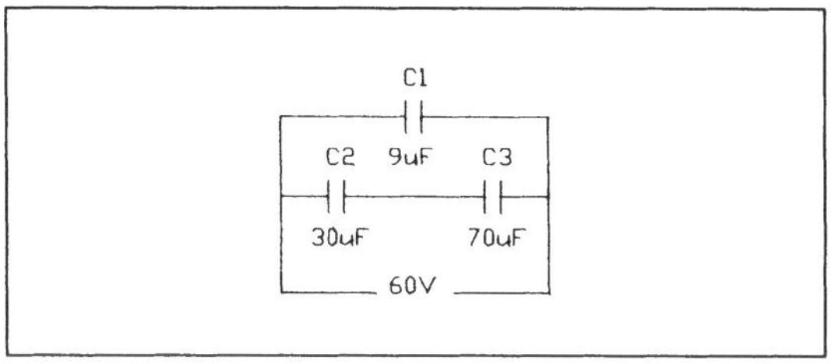

4.72 The voltage drop across C_2 in Question 4.71 is

- ❑ *a* 30 V
- ❑ *b* 42 V
- ❑ *c* 48 V
- ❑ *d* 50 V

4.73 The total charge in the network in Question 4.71 is

- ❑ *a* 1.8 mC
- ❑ *b* 2.8 mC
- ❑ *c* 4.6 mC
- ❑ *d* 5.4 mC

4.74 The energy stored in C_1 in Question 4.71 is

- ❑ *a* 10.2 mC
- ❑ *b* 12.4 mC
- ❑ *c* 16.2 mC
- ❑ *d* 18.6 mC

4.75 The total energy stored in the network in Question 4.71 is

- ❑ *a* 54 mJ
- ❑ *b* 100 mJ
- ❑ *c* 220 mJ
- ❑ *d* 284 mJ

5 Inductors

5.1 A 0.04-H inductance expressed in millihenrys is
- ❑ *a* 0.4 mH
- ❑ *b* 4 mH
- ❑ *c* 40 mH
- ❑ *d* 400 mH

5.2 An inductor has a value of 0.002 H. The value converted to microhenrys is
- ❑ *a* 2 μH
- ❑ *b* 20 μH
- ❑ *c* 200 μH
- ❑ *d* 2000 μH

5.3 If an inductor has an inductance of 22 H, this value expressed in microhenrys is
- ❑ *a* 2.2×10^7 μH
- ❑ *b* 22×10^3 μH
- ❑ *c* 2.2×10^3 μH
- ❑ *d* 22×10^2 μH

5.4 A 230-μH inductance expressed in millihenrys is
- ❑ *a* 0.0023
- ❑ *b* 0.023
- ❑ *c* 0.23
- ❑ *d* 2.3

5.5 A 0.84-mH inductance expressed in microhenrys is
- ❑ *a* 8.4 μH
- ❑ *b* 84 μH
- ❑ *c* 840 μH
- ❑ *d* 8400 μH

5.6 An inductance of 1200 μH expressed in henrys is
- ❑ *a* 0.012
- ❑ *b* 0.0012
- ❑ *c* 0.12
- ❑ *d* 1.2

5.7 A coil has 400 turns. If a flux of 100 mWb acting through the coil is reversed in 0.04 s, the induced voltage is
- ❑ *a* 200 V
- ❑ *b* 500 V
- ❑ *c* 1000 V
- ❑ *d* 2000 V

5.8 If a flux of 240 mWb is reversed in a 60-turn coil in 0.2 s, the induced voltage is
- ❑ *a* 144 V
- ❑ *b* 240 V
- ❑ *c* 288 V
- ❑ *d* 480 V

5.9 A voltage of 12 V is induced in a coil when a flux of 150 mWb is reversed in 80 s. The number of turns on the coil is
- ❑ *a* 1800
- ❑ *b* 2200
- ❑ *c* 3200
- ❑ *d* 400

5.10 A voltage of 24 V is induced in a coil when a flux of 78 mWb is reversed in 3.4 s. The coil has
- ❑ *a* 417 turns
- ❑ *b* 523 turns
- ❑ *c* 600 turns
- ❑ *d* 629 turns

5.11 A voltage of 200 V is induced in a coil having 1400 turns when the flux is reversed in 0.6 s. The value of the flux in milliwebers is

- ❑ *a* 42.86
- ❑ *b* 63.92
- ❑ *c* 74.13
- ❑ *d* 82.47

5.12 If 12 V are induced in a coil which has 100 turns, when the flux is reversed in 2 s the value of the flux in webers is

- ❑ *a* 0.06
- ❑ *b* 0.08
- ❑ *c* 0.12
- ❑ *d* 1.2

5.13 A coil has an inductance of 12 H. If the current through it is increased from 2 A to 8 A in 130 ms, the induced voltage is approximately

- ❑ *a* 282 V
- ❑ *b* 324 V
- ❑ *c* 462 V
- ❑ *d* 554 V

5.14 A current is increased from 1.5 A to 6.5 A in 100 ms through a coil having an inductance of 5 H. The induced voltage is

- ❑ *a* 240 V
- ❑ *b* 250 V
- ❑ *c* 260 V
- ❑ *d* 300 V

5.15 A current of 12 A is reversed in a 3-H coil in 24 ms. The induced voltage is

- ❑ *a* 600 V
- ❑ *b* 1500 V
- ❑ *c* 3000 V
- ❑ *d* 3500 V

5.16 If a current of 5 A is reversed in a coil whose inductance is 2.5 H in 0.05 s, the induced voltage is

- ❑ *a* 500 V
- ❑ *b* 700 V
- ❑ *c* 800 V
- ❑ *d* 900 V

5.17 A voltage of 300 V is induced in a coil with an inductance of 6 H when a current is reversed in 60 ms. The value of the current in amps is

- ❑ *a* 1.2
- ❑ *b* 1.5
- ❑ *c* 2.2
- ❑ *d* 2.5

5.18 If a voltage of 480 V is induced in a coil whose inductance is 8 H, when a current is reversed in 96 ms the value of the current is approximately

- ❑ *a* 0.88 A
- ❑ *b* 1.92 A
- ❑ *c* 2.88 A
- ❑ *d* 3.24 A

5.19 When a current of 11.5 A is reversed in a coil in 230 ms, a voltage of 1500 V is induced. The value of the inductance in henrys is

- ❑ *a* 15
- ❑ *b* 20
- ❑ *c* 25
- ❑ *d* 30

5.20 If a current of 15 A is reversed in a coil in 0.15 s and the induced voltage is 6 kV, the value of the coil inductance is

- ❑ *a* 20 H
- ❑ *b* 30 H
- ❑ *c* 40 H
- ❑ *d* 45 H

5.21 A coil having 1800 turns is wound on a non-magnetic former. If a current of 6 A produces a flux of 400 μWb, the coil inductance in henrys is

- ❑ *a* 0.06
- ❑ *b* 0.08
- ❑ *c* 0.11
- ❑ *d* 0.12

5.22 If the current in Question 5.21 is reversed in 40 ms, the induced voltage is

- ❑ *a* 12 V
- ❑ *b* 24 V
- ❑ *c* 36 V
- ❑ *d* 48 V

5.23 A coil having 4500 turns is wound on a non-magnetic former. If a current of 18 A produces a flux of 0.9 mWb, the coil inductance is

- ❑ *a* 0.225 H
- ❑ *b* 0.3 H
- ❑ *c* 0.325 H
- ❑ *d* 0.4 H

5.24 If the current in Question 5.23 is reversed in 180 ms, the induced voltage is

- ❑ *a* 36 V
- ❑ *b* 42 V
- ❑ *c* 45 V
- ❑ *d* 50 V

5.25 If the current in Question 5.23 is reversed in 360 ms, the induced voltage is

- ❑ *a* 22.5 V
- ❑ *b* 25 V
- ❑ *c* 30 V
- ❑ *d* 32.5 V

5.26 If a current of 5.4 A flows in a 14.6-H inductor, the energy stored in joules is approximately

- ❑ *a* 165
- ❑ *b* 197
- ❑ *c* 200
- ❑ *d* 213

5.27 If a current of 4 A flows in a 2-H inductor, the energy stored in joules is

- ❑ *a* 12
- ❑ *b* 14
- ❑ *c* 16
- ❑ *d* 20

5.28 If a current of 12 A flows in a 20-H inductor, the energy stored in joules is

- ❑ *a* 650
- ❑ *b* 800
- ❑ *c* 1000
- ❑ *d* 1440

5.29 If 3.8 J of energy is stored in a coil when the current is 4 A, the coil inductance is

- ❑ *a* 0.42 H
- ❑ *b* 0.475 H
- ❑ *c* 0.525 H
- ❑ *d* 0.6 H

5.30 28 J of energy is stored in a coil when the current flowing is 7 A. The coil inductance in henrys is

- ❑ *a* 1.04
- ❑ *b* 1.14
- ❑ *c* 2.4
- ❑ *d* 2.48

5.31 500 mJ of energy is stored in a coil when the current flowing is 2 A. The coil inductance in millihenrys is

- ❑ *a* 250
- ❑ *b* 400
- ❑ *c* 420
- ❑ *d* 450

5.32 A coil with an inductance of 0.075 H stores 15 J. The coil current is

- ❑ *a* 18 A
- ❑ *b* 20 A
- ❑ *c* 30 A
- ❑ *d* 32 A

5.33 A coil with an inductance of 400 mH stores 200 J. The coil current is

- ❑ *a* 5.6 A
- ❑ *b* 10 A
- ❑ *c* 31.6 A
- ❑ *d* 32 A

5.34 16 J is stored in a coil with an inductance of 12 H, when the coil current is

- ❑ *a* 1.63 A
- ❑ *b* 1.82 A
- ❑ *c* 2.1 A
- ❑ *d* 2.42 A

5.35 The mutual inductance between two coils is 0.4 H. The average voltage induced in one coil when the current in the other coil increases from 2 A to 8 A in 140 ms is

- ❑ *a* 14.23 V
- ❑ *b* 15.06 V
- ❑ *c* 17.14 V
- ❑ *d* 19.21 V

5.36 Two coils have a mutual inductance of 1.2 H. If the current in one of the coils increases from 0 A to 10 A in 30 ms, the average voltage induced in the other coil is

- ❑ *a* 60 V
- ❑ *b* 180 V
- ❑ *c* 200 V
- ❑ *d* 400 V

5.37 Two coils have a mutual inductance of 1.5 H. If the current in one of the coils increases from 0 A to 5 A in 45 ms, the average voltage induced in the other coil is

- ❑ *a* 133.33 V
- ❑ *b* 166.67 V
- ❑ *c* 203.32 V
- ❑ *d* 233.33 V

5.38 400 V is induced in a coil which has an inductance of 2 H, when the current increases from 0 A to 10 A. The time for the increase in milliseconds is

- ❑ *a* 5
- ❑ *b* 10
- ❑ *c* 40
- ❑ *d* 50

5.39 A coil has an inductance of 1.05 H. If 240 V is induced in the coil when the current increases from 1 A to 7 A, the time for the increase in milliseconds is approximately

- ❑ *a* 16.28
- ❑ *b* 18.34
- ❑ *c* 20.22
- ❑ *d* 26.25

5.40 A coil has an inductance of 50 mH. If 150 V is induced in the coil when the current increases from 1.34 A to 3.68 A, the time for the increase in milliseconds is approximately

- ❑ *a* 0.78
- ❑ *b* 1.22
- ❑ *c* 2.56
- ❑ *d* 3.29

5.41 The reactance of a 0.46-H inductor at a frequency of 2 kHz is approximately

- ❑ *a* 4828 Ω
- ❑ *b* 5780 Ω
- ❑ *c* 6263 Ω
- ❑ *d* 7426 Ω

5.42 The reactance of a 2.8-mH inductor at a frequency of 500 Hz is approximately

- ❑ *a* 8.8R
- ❑ *b* 10.2R
- ❑ *c* 14.5R
- ❑ *d* 20.9R

5.43 A 0.98-H inductor has a reactance at a frequency of 150 Hz of approximately

- ❑ *a* 257 Ω
- ❑ *b* 332 Ω
- ❑ *c* 680 Ω
- ❑ *d* 924 Ω

5.44 A 0.98-mH coil has a reactance of 1.4R at an approximate frequency of

- ❑ *a* 142 Hz
- ❑ *b* 227 Hz
- ❑ *c* 361 Hz
- ❑ *d* 473 Hz

5.45 A 400-μH inductor has a reactance of 10R at a frequency of approximately

- ❑ *a* 2624 Hz
- ❑ *b* 3979 Hz
- ❑ *c* 4783 Hz
- ❑ *d* 5500 Hz

5.46 A 1.2-mH inductor has a reactance of 60R at a frequency of approximately
- ❑ *a* 4644 Hz
- ❑ *b* 5472 Hz
- ❑ *c* 6200 Hz
- ❑ *d* 7958 Hz

5.47 A coil with a reactance of 1200 Ω at a frequency of 1 kHz has an inductance in henrys of approximately
- ❑ *a* 0.191
- ❑ *b* 0.522
- ❑ *c* 0.604
- ❑ *d* 0.822

5.48 A coil with a reactance of 56 kΩ at a frequency of 1.5 kHz has an inductance of approximately
- ❑ *a* 2.62 H
- ❑ *b* 4.81 H
- ❑ *c* 5.94 H
- ❑ *d* 6.04 H

5.49 A coil has a reactance of 10 kΩ at a frequency of 600 Hz. The inductance is approximately
- ❑ *a* 0.86 H
- ❑ *b* 1.62 H
- ❑ *c* 2.65 H
- ❑ *d* 2.98 H

5.50 Three coils having inductances of 2.4 H, 6 H and 7.2 H are connected in series. The overall inductance is
- ❑ *a* 1.2 H
- ❑ *b* 3 H
- ❑ *c* 9 H
- ❑ *d* 15.6 H

5.51 A coil has 240 turns and is wound on an air-cored bobbin 4.5 cm long and 0.2 cm in diameter. Given that the permeability of free space is $4\pi \times 10^{-7}$ H m^{-1}, the inductance in microhenrys is
- ❑ *a* 4.05
- ❑ *b* 4.15
- ❑ *c* 5.05
- ❑ *d* 6.25

5.52 If the coil in Question 5.51 had 100 turns, the same diameter and was 14 cm long, the inductance in microhenrys would have been
- ❑ *a* 0.16
- ❑ *b* 0.282
- ❑ *c* 0.364
- ❑ *d* 0.42

5.53 If the coil in Question 5.51 had 6000 turns, a length of 10 cm and diameter 0.5 cm, the inductance would be
- ❑ *a* 8.88 mH
- ❑ *b* 9.2 mH
- ❑ *c* 10.8 mH
- ❑ *d* 11.92 mH

5.54 A coil has 120 turns and is wound on a ferromagnetic bobbin. The bobbin is 4 cm long and 10 mm in diameter. If the core has a relative permeability of 8400, and the permeability of free space is taken as $4\pi \times 10^{-7}$ H m^{-1}, the coil inductance is approximately
- ❑ *a* 98 μH
- ❑ *b* 198 μH
- ❑ *c* 240 μH
- ❑ *d* 298 μH

5.55 If the coil in Question 5.54 had a relative permeability of 10 000, 400 turns, a length of 5 cm, and a diameter of 20 mm, the inductance would be approximately
- ❑ *a* 8.43 mH
- ❑ *b* 12.63 mH
- ❑ *c* 15.72 mH
- ❑ *d* 18.04 mH

5.56 If the coil in Question 5.54 had a relative permeability of 450, 200 turns, a length of 6 cm, and a diameter of 24 mm, the inductance in microhenrys would be approximately
- ❑ *a* 170
- ❑ *b* 190
- ❑ *c* 210
- ❑ *d* 230

5.57 The diagram shows three inductances connected in series. The overall inductance is
- ❑ *a* 18 H
- ❑ *b* 24 H
- ❑ *c* 48 H
- ❑ *d* 42 H

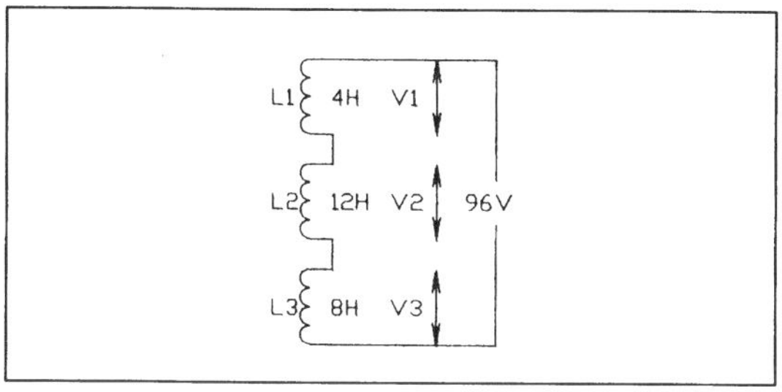

5.58 The current growth in Question 5.57 is
- ❑ *a* 2 A s^{-1}
- ❑ *b* 4 A s^{-1}
- ❑ *c* 9 A s^{-1}
- ❑ *d* 12 A s^{-1}

5.59 The value of V_3 in Question 5.57 is
- ❑ *a* 12 V
- ❑ *b* 16 V
- ❑ *c* 24 V
- ❑ *d* 32 V

5.60 The diagram shows three inductances connected in series. The current growth is
- ❑ *a* 8 A s^{-1}
- ❑ *b* 10 A s^{-1}
- ❑ *c* 15 A s^{-1}
- ❑ *d* 20 A s^{-1}

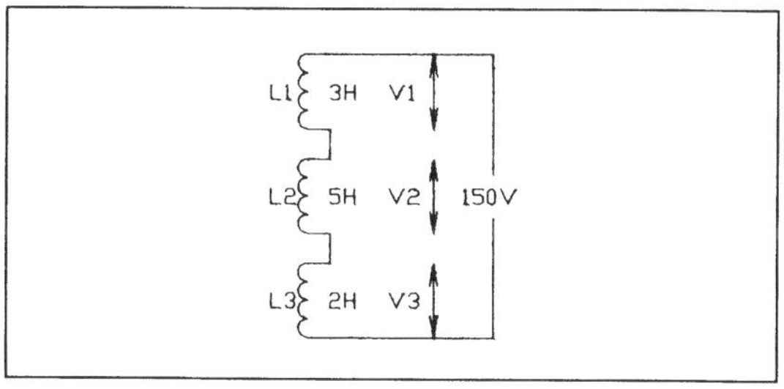

5.61 The voltage V_2 in Question 5.60 is
- ❑ *a* 30 V
- ❑ *b* 50 V
- ❑ *c* 60 V
- ❑ *d* 75 V

5.62 The voltage V_3 in Question 5.60 is
- ❑ *a* 30 V
- ❑ *b* 50 V
- ❑ *c* 60 V
- ❑ *d* 75 V

5.63 The current after 12 s in Question 5.60 is
- ❑ *a* 170 A
- ❑ *b* 180 A
- ❑ *c* 190 A
- ❑ *d* 200 A

5.64 The energy stored in L1 after 12 s in Question 5.60 is
- ❑ *a* 34 000 J
- ❑ *b* 36 000 J
- ❑ *c* 40 000 J
- ❑ *d* 48 600 J

5.65 The energy stored in L2 after 12 s in Question 5.60 is
- ❑ *a* 60 000 J
- ❑ *b* 72 000 J
- ❑ *c* 81 000 J
- ❑ *d* 90 000 J

5.66 The energy stored in L3 after 12 s in Question 5.60 is
- ❑ *a* 26 400 J
- ❑ *b* 28 400 J
- ❑ *c* 30 400 J
- ❑ *d* 32 400 J

5.67 The diagram shows three inductors connected in parallel. The combined inductance is
- ❑ *a* 1.2 H
- ❑ *b* 2.4 H
- ❑ *c* 9.6 H
- ❑ *d* 12.4 H

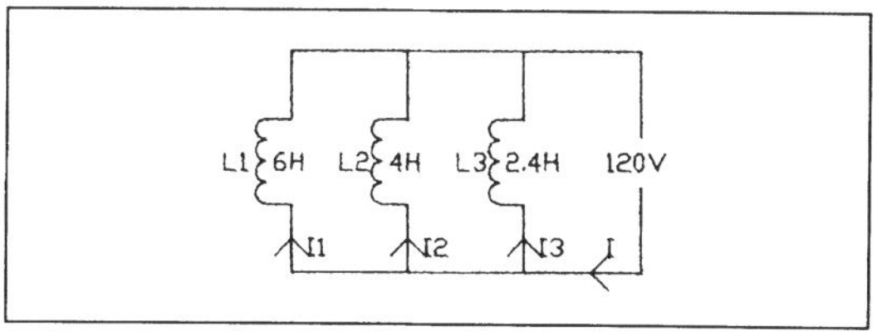

5.68 The current growth in L1 in Question 5.67 is
- ❑ *a* 5 $A\,s^{-1}$
- ❑ *b* 10 $A\,s^{-1}$
- ❑ *c* 20 $A\,s^{-1}$
- ❑ *d* 22 $A\,s^{-1}$

5.69 The current growth in L2 in Question 5.67 is
- ❑ *a* 30 $A\,s^{-1}$
- ❑ *b* 34 $A\,s^{-1}$
- ❑ *c* 40 $A\,s^{-1}$
- ❑ *d* 50 $A\,s^{-1}$

5.70 The current growth in L3 in Question 5.67 is
- ❑ *a* 30 $A\,s^{-1}$
- ❑ *b* 40 $A\,s^{-1}$
- ❑ *c* 50 $A\,s^{-1}$
- ❑ *d* 60 $A\,s^{-1}$

5.71 The total current growth in the circuit in Question 5.67 is
- ❑ *a* 40 $A\,s^{-1}$
- ❑ *b* 60 $A\,s^{-1}$
- ❑ *c* 80 $A\,s^{-1}$
- ❑ *d* 100 $A\,s^{-1}$

5.72 The current in L2 in Question 5.67 after 5 s is
- ❑ *a* 150 A
- ❑ *b* 160 A
- ❑ *c* 190 A
- ❑ *d* 200 A

5.73 The current in L3 in Question 5.67 after 5 s is
- ❑ *a* 100 A
- ❑ *b* 150 A
- ❑ *c* 200 A
- ❑ *d* 250 A

5.74 The energy stored after 1 s in L1 in Question 5.67 is
- ❑ *a* 900 J
- ❑ *b* 1100 J
- ❑ *c* 1200 J
- ❑ *d* 1400 J

5.75 The total energy stored after 1 s in the circuit in Question 5.67 is
- ❑ *a* 4000 J
- ❑ *b* 6000 J
- ❑ *c* 7000 J
- ❑ *d* 8000 J

6 Capacitors and inductors in DC circuits

6.1 In the diagram the time constant is

❑ *a* 47 ms
❑ *b* 47 μs
❑ *c* 0.47 μs
❑ *d* 470 ms

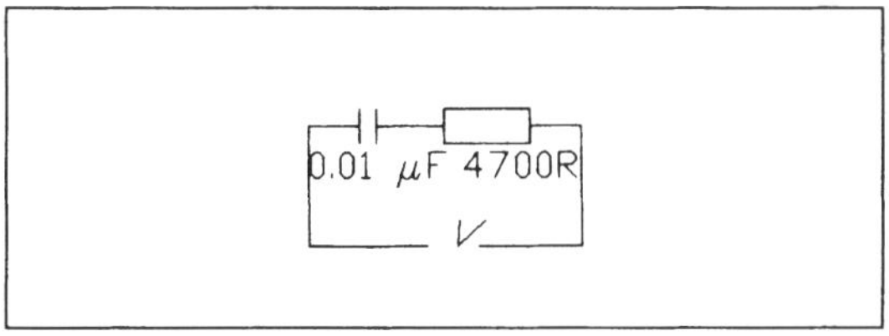

6.2 If the component values in Question 6.1 are 5 μF and 100 kΩ, the time constant is

❑ *a* 0.5 μs
❑ *b* 5 ms
❑ *c* 0.5 ms
❑ *d* 0.5 s

6.3 If the component values in Question 6.1 are 47 nF and 1 MΩ, the time constant is

❑ *a* 4.7 ms
❑ *b* 47 ms
❑ *c* 470 μs
❑ *d* 470 s

6.4 If the component values in Question 6.1 are 10 pF and 2 MΩ, the time constant is

❑ *a* 20 ms
❑ *b* 200 ms
❑ *c* 20 μs
❑ *d* 200 μs

6.5 If the component values in Question 6.1 are 800 pF and 5 MΩ, the time constant is

❑ *a* 4 ms
❑ *b* 40 ms
❑ *c* 4 μs
❑ *d* 400 μs

6.6 In the diagram the time constant is

❑ *a* 0.6 ms
❑ *b* 64 ms
❑ *c* 1.6 s
❑ *d* 16 μs

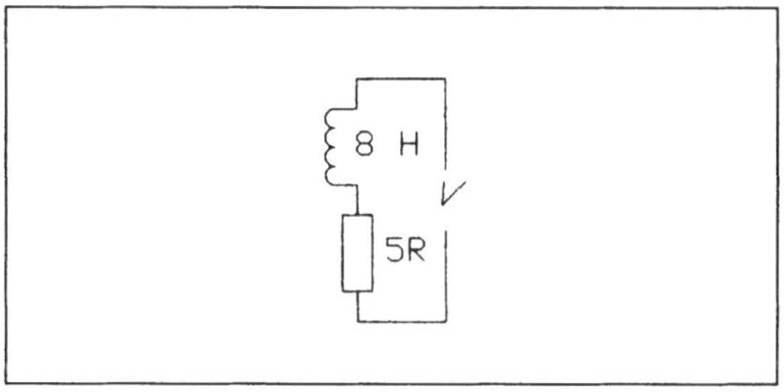

6.7 If the component values in Question 6.6 are 0.075 H and 250 Ω, the time constant is

❑ *a* 2.5 s
❑ *b* 25 ms
❑ *c* 250 μs
❑ *d* 300 μs

6.8 If the component values in Question 6.6 are 60 mH and 300 Ω, the time constant is

❑ *a* 0.2 ms
❑ *b* 2.2 ms
❑ *c* 3.6 μs
❑ *d* 300 μs

6.9 If the component values in Question 6.6 are 9 mH and 3 MΩ, the time constant is

❑ *a* 3 s
❑ *b* 3 ms
❑ *c* 3 μs
❑ *d* 3 ns

6.10 If the component values in Question 6.6 are 84 mH and 210 kΩ, the time constant is
- ❑ *a* 40 ms
- ❑ *b* 4 μs
- ❑ *c* 400 ns
- ❑ *d* 4 s

6.11 In the diagram the time constant is
- ❑ *a* 1 ms
- ❑ *b* 1 μs
- ❑ *c* 1 s
- ❑ *d* 10 s

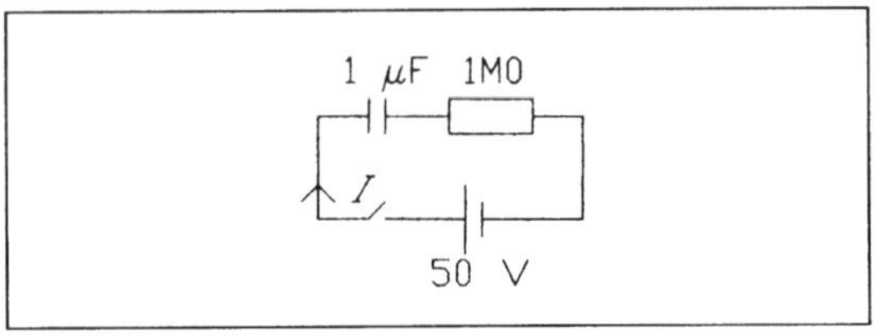

6.12 In Question 6.11 the current at $t = 0$, i.e. the instant of switching on, is
- ❑ *a* 50 mA
- ❑ *b* 500 μA
- ❑ *c* 100 mA
- ❑ *d* 50 μA

6.13 The current in Question 6.11 after 4 s is approximately
- ❑ *a* 0.92 μA
- ❑ *b* 1.62 mA
- ❑ *c* 7.3 mA
- ❑ *d* 8.16 μA

6.14 In Question 6.11 the capacitor voltage after 2 s is approximately
- ❑ *a* 15 V
- ❑ *b* 36 V
- ❑ *c* 43 V
- ❑ *d* 49 V

6.15 In Question 6.11 the voltage across the resistor after 3 s is approximately
- ❑ *a* 36.24 V
- ❑ *b* 18.81 V
- ❑ *c* 6.92 V
- ❑ *d* 2.49 V

6.16 The diagram shows a charged capacitor disconnected from a 24-V supply and short-circuited through the 500-kΩ resistor. The current at $t = 0$ s is
- ❑ *a* −24 mA
- ❑ *b* −48 μA
- ❑ *c* −60 mA
- ❑ *d* −100 μA

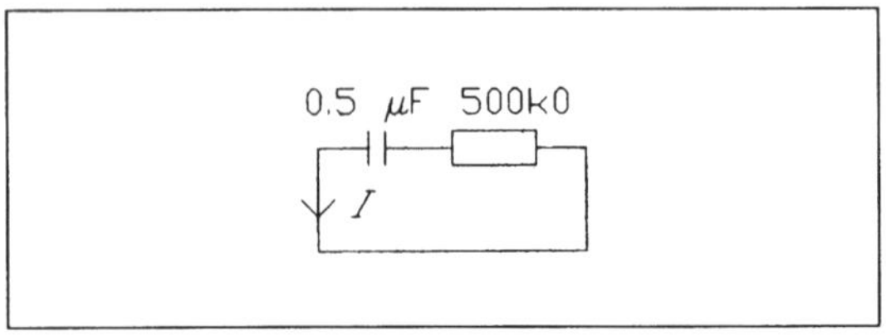

6.17 In Question 6.16 the current flowing at $t =$ 1 s is approximately
- ❑ *a* −0.06 mA
- ❑ *b* −0.72 mA
- ❑ *c* −0.88 μA
- ❑ *d* −1.22 μA

6.18 In Question 6.16 the capacitor voltage at $t =$ 1 s is approximately
- ❑ *a* 0.44 V
- ❑ *b* 0.82 V
- ❑ *c* 1.64 V
- ❑ *d* 10.22 V

6.19 In Question 6.16 the voltage across the resistor at $t = 1$ s is approximately
- ❑ *a* −10.44 V
- ❑ *b* −6.26 V
- ❑ *c* −0.92 V
- ❑ *d* −0.44 V

6.20 In Question 6.16 the capacitor voltage at $t =$ 4 s is
- ❑ *a* 2.7×10^{-6} V
- ❑ *b* 4.3×10^{-4} V
- ❑ *c* 3.6×10^{-2} V
- ❑ *d* 4.2×10^{-3} V

6.21 In the diagram the time constant is
- ❑ *a* 25 ms
- ❑ *b* 250 ms
- ❑ *c* 25 μs
- ❑ *d* 250 μs

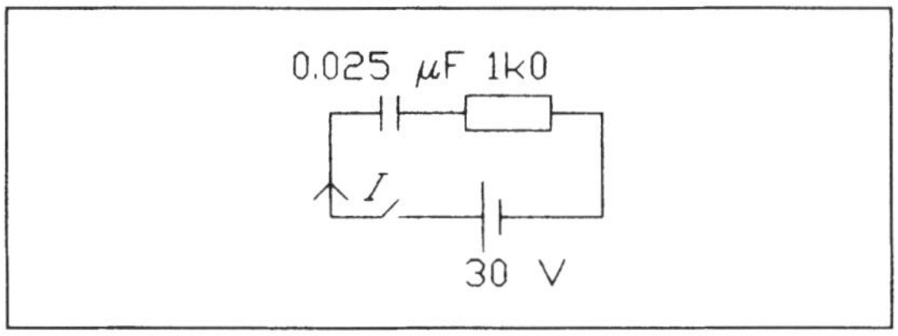

6.22 In Question 6.21 the current at t = 0 s, i.e. the instant of switching on, is
- ❑ *a* 30 mA
- ❑ *b* 60 mA
- ❑ *c* 120 μA
- ❑ *d* 180 μA

6.23 In Question 6.21 the capacitor voltage at t = 50 μs is
- ❑ *a* 16.28 V
- ❑ *b* 21.32 V
- ❑ *c* 25.94 V
- ❑ *d* 28.86 V

6.24 In Question 6.21 the voltage across the resistor at t = 75 μs is approximately
- ❑ *a* 1.12 V
- ❑ *b* 1.49 V
- ❑ *c* 6.33 V
- ❑ *d* 7.94 V

6.25 In Question 6.21 the capacitor voltage at t = 100 μs is approximately
- ❑ *a* 16.34 V
- ❑ *b* 29.45 V
- ❑ *c* 15.66 V
- ❑ *d* 12.88 V

6.26 The diagram shows a charged capacitor disconnected from a 125-V supply and short-circuited through the 1-kΩ resistor. The current at t = 0 s is
- ❑ *a* −50 mA
- ❑ *b* −75 mA
- ❑ *c* −100 mA
- ❑ *d* −125 mA

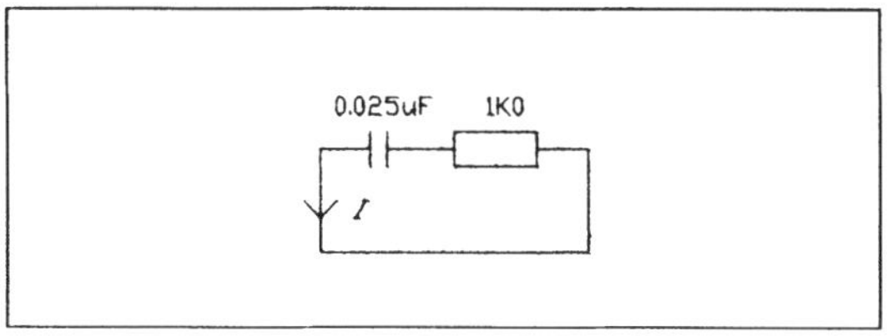

6.27 In Question 6.26 the capacitor voltage at t = 25 μs is approximately
- ❑ *a* 45.98 V
- ❑ *b* 66.74 V
- ❑ *c* 100.6 V
- ❑ *d* 110.76 V

6.28 In Question 6.26 the capacitor voltage at t = 50 μs is approximately
- ❑ *a* 120.61 V
- ❑ *b* 75.54 V
- ❑ *c* 16.92 V
- ❑ *d* 9.34 V

6.29 In Question 6.26 the voltage across the resistor at t = 75 μs is approximately
- ❑ *a* −4.31 V
- ❑ *b* −6.22 V
- ❑ *c* −12.98 V
- ❑ *d* −20.67 V

6.30 In Question 6.26 the current at t = 100 μs is approximately
- ❑ *a* −100.42 mA
- ❑ *b* −66.33 mA
- ❑ *c* −40.26 mA
- ❑ *d* −2.29 mA

6.31 In the diagram the time constant is

- ❑ *a* 4 ms
- ❑ *b* 40 ms
- ❑ *c* 400 ms
- ❑ *d* 40 μs

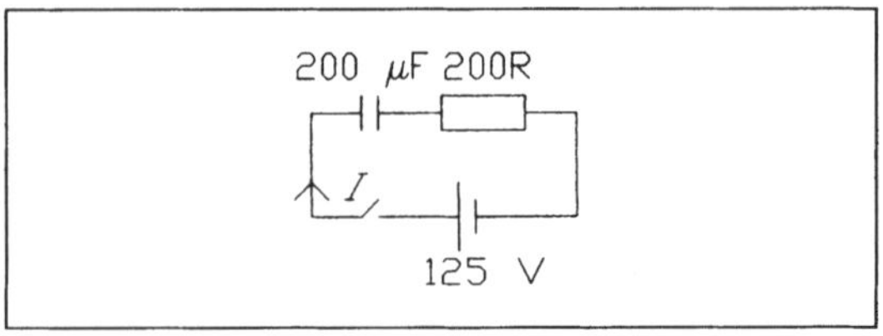

6.32 In Question 6.31 the current at the instant of switching on, $t = 0$ s, is

- ❑ *a* 200 mA
- ❑ *b* 225 mA
- ❑ *c* 625 mA
- ❑ *d* 800 mA

6.33 In Question 6.31 the capacitor voltage at $t =$ 40 ms is approximately

- ❑ *a* 14 V
- ❑ *b* 28 V
- ❑ *c* 66 V
- ❑ *d* 79 V

6.34 In Question 6.31 the voltage across the resistor at $t = 80$ ms is approximately

- ❑ *a* 16.9 V
- ❑ *b* 28.8 V
- ❑ *c* 46.2 V
- ❑ *d* 50.6 V

6.35 In Question 6.31 the current at $t = 200$ ms is approximately

- ❑ *a* 2.4 mA
- ❑ *b* 4.2 mA
- ❑ *c* 6.6 mA
- ❑ *d* 10.3 mA

6.36 The diagram shows a charged capacitor disconnected from a 125-V supply and short-circuited through the 200-Ω resistor. The current at $t = 40$ ms is approximately

- ❑ *a* –230 mA
- ❑ *b* –400 mA
- ❑ *c* –470 mA
- ❑ *d* –560 mA

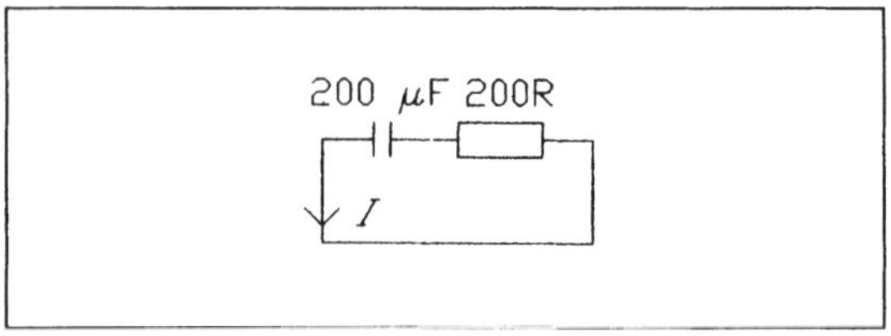

6.37 In Question 6.36 the capacitor voltage at $t =$ 80 ms is approximately

- ❑ *a* 12.5 V
- ❑ *b* 13.6 V
- ❑ *c* 14.2 V
- ❑ *d* 16.9 V

6.38 In Question 6.36 the capacitor voltage at $t =$ 120 ms is approximately

- ❑ *a* 4.82 V
- ❑ *b* 6.22 V
- ❑ *c* 10.54 V
- ❑ *d* 20.62 V

6.39 In Question 6.36 the voltage across the resistor at $t = 160$ ms is approximately

- ❑ *a* –2.29 V
- ❑ *b* –4.39 V
- ❑ *c* –7.48 V
- ❑ *d* –8.36 V

6.40 In Question 6.36 the current at $t = 200$ ms is approximately

- ❑ *a* –1.5 mA
- ❑ *b* –3.4 mA
- ❑ *c* –4.2 mA
- ❑ *d* –6.8 mA

6.41 In the diagram the time constant is

- ❑ *a* 2.2 s
- ❑ *b* 4 s
- ❑ *c* 0.2 s
- ❑ *d* 0.4 s

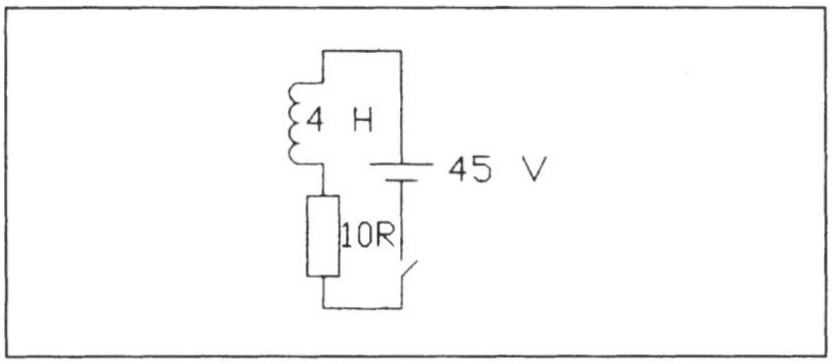

6.42 In Question 6.41 the current at the instant of switching on, $t = 0$ s, is

- ❑ *a* 0 A
- ❑ *b* 4 A
- ❑ *c* 10 A
- ❑ *d* 14 A

6.43 In Question 6.41 the voltage across the inductor at $t = 0$ is

- ❑ *a* 0 V
- ❑ *b* 10 V
- ❑ *c* 22.5 V
- ❑ *d* 45 V

6.44 In Question 6.41 the voltage across the resistor at $t = 0$ is

- ❑ *a* 45 V
- ❑ *b* 20 V
- ❑ *c* 2 V
- ❑ *d* 0 V

6.45 The voltage across the inductor in Question 6.41 at $t = 1.2$ s is approximately

- ❑ *a* 1.61 V
- ❑ *b* 2.24 V
- ❑ *c* 3.82 V
- ❑ *d* 30.96 V

6.46 In the diagram the time constant is

- ❑ *a* 0.25 s
- ❑ *b* 0.5 s
- ❑ *c* 1.0 s
- ❑ *d* 1.5 s

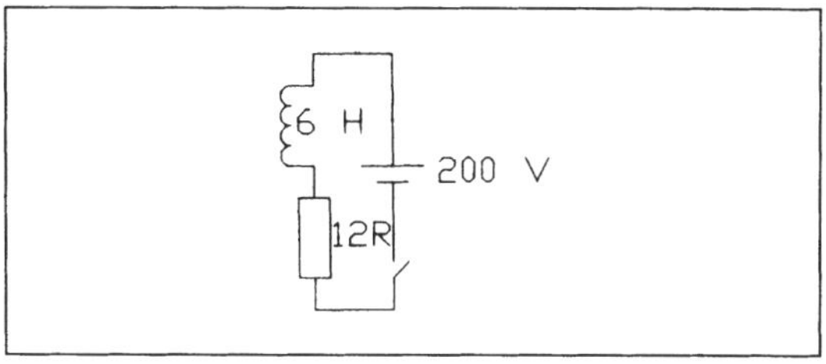

6.47 In Question 6.46 the current at $t = 1.0$ s is approximately

- ❑ *a* 2.6 A
- ❑ *b* 10.3 A
- ❑ *c* 14.4 A
- ❑ *d* 20.5 A

6.48 In Question 6.46 the current at $t = 2.0$ s is approximately

- ❑ *a* 25.22 A
- ❑ *b* 22.48 A
- ❑ *c* 19.41 A
- ❑ *d* 16.36 A

6.49 In Question 6.46 the voltage across the inductor at $t = 1.0$ s is approximately

- ❑ *a* 27 V
- ❑ *b* 123 V
- ❑ *c* 162 V
- ❑ *d* 193 V

6.50 In Question 6.46 the voltage across the resistor at $t = 1.0$ s is approximately

- ❑ *a* 102 V
- ❑ *b* 121 V
- ❑ *c* 173 V
- ❑ *d* 182 V

6.51 The diagram shows a circuit which has been connected to a 45-V DC supply until a steady state has been reached. The supply is disconnected and a short circuit is placed across the network as shown. The current at t = 0 s is

- ❑ *a* 10 A
- ❑ *b* 5 A
- ❑ *c* 4.5 A
- ❑ *d* 4.2 A

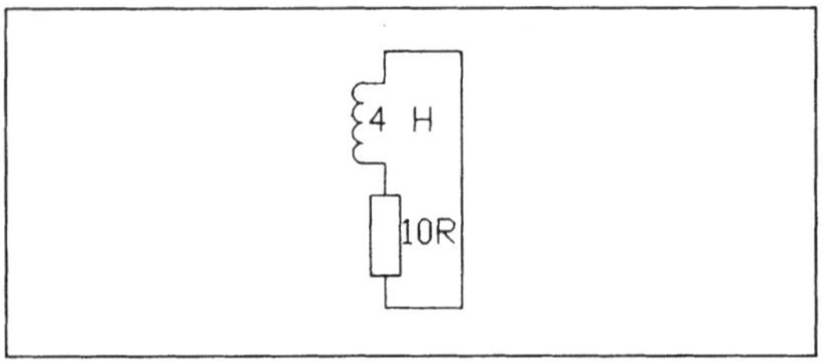

6.52 In Question 6.51 the current at t = 0.8 s is approximately

- ❑ *a* 0.2 A
- ❑ *b* 0.12 A
- ❑ *c* 0.08 A
- ❑ *d* 0.6 A

6.53 In Question 6.51 the voltage across the resistor at t = 0.4 s is approximately

- ❑ *a* 12.48 V
- ❑ *b* 16.55 V
- ❑ *c* 20.82 V
- ❑ *d* 40.16 V

6.54 In Question 6.51 the voltage across the inductor at t = 0.8 s is approximately

- ❑ *a* −6 V
- ❑ *b* −10 V
- ❑ *c* −15 V
- ❑ *d* −28 V

6.55 In Question 6.51 the voltage across the resistor at t = 2.0 s is approximately

- ❑ *a* 0.3 V
- ❑ *b* 1.3 V
- ❑ *c* 2.3 V
- ❑ *d* 6.6 V

6.56 The diagram shows a circuit which has been connected to a 200-V supply until a steady state has been reached. The supply is disconnected and a short circuit is placed across the network as shown. The current at t = 0 s is approximately

- ❑ *a* 10.44 A
- ❑ *b* 10.62 A
- ❑ *c* 14.32 A
- ❑ *d* 16.67 A

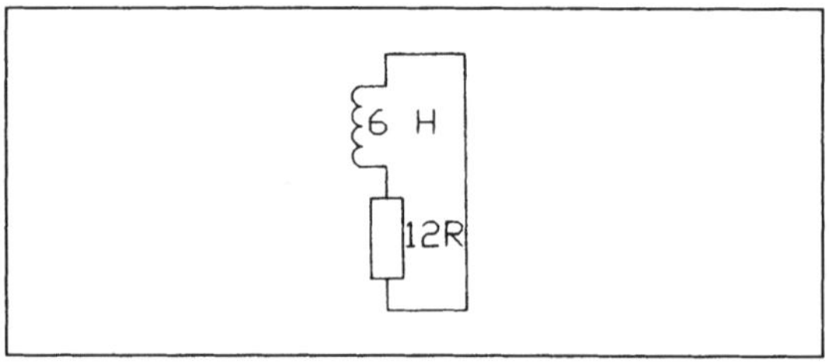

6.57 In Question 6.56 the current at t = 0.5 s is approximately

- ❑ *a* 2.62 A
- ❑ *b* 6.13 A
- ❑ *c* 10.44 A
- ❑ *d* 15.53 A

6.58 In Question 6.56 the voltage across the inductor at t = 1.0 s is approximately

- ❑ *a* −47 V
- ❑ *b* −40 V
- ❑ *c* −27 V
- ❑ *d* −52 V

6.59 In Question 6.56 the voltage across the resistor at t = 1.5 s is approximately

- ❑ *a* 9.96 V
- ❑ *b* 12.88 V
- ❑ *c* 86.67 V
- ❑ *d* 93.42 V

6.60 In Question 6.56 the voltage across the inductor at t = 2.0 s is approximately

- ❑ *a* − 3.66 V
- ❑ *b* − 4.02 V
- ❑ *c* − 8.06 V
- ❑ *d* − 9.33 V

6.61 The diagram shows a circuit which is suddenly connected to a 25-V supply. The time constant is

❑ *a* 0.02 s
❑ *b* 0.05 s
❑ *c* 0.1 s
❑ *d* 0.15 s

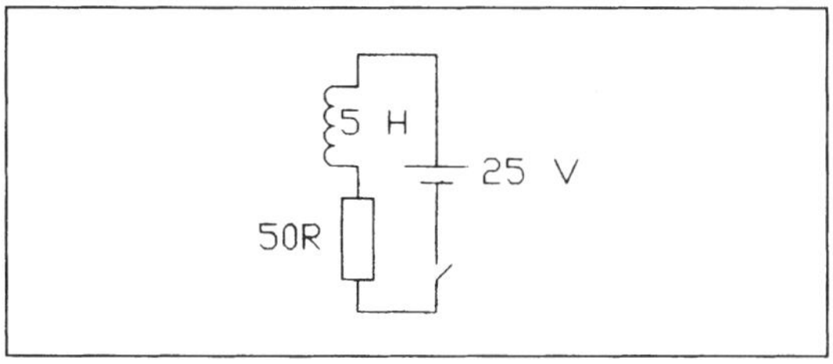

6.62 In Question 6.61 the time for the current to reach 0.5 A is approximately

❑ *a* 2.3 s
❑ *b* 4.1 s
❑ *c* 6.2 s
❑ *d* 8.8 s

6.63 In Question 6.61 the current after 0.2 s is approximately

❑ *a* 0.28 A
❑ *b* 0.43 A
❑ *c* 0.66 A
❑ *d* 0.92 A

6.64 The diagram shows a circuit which is suddenly connected to a 100-V supply. The time for the current to reach 2 A is approximately

❑ *a* 0.15 s
❑ *b* 0.32 s
❑ *c* 0.72 s
❑ *d* 1.45 s

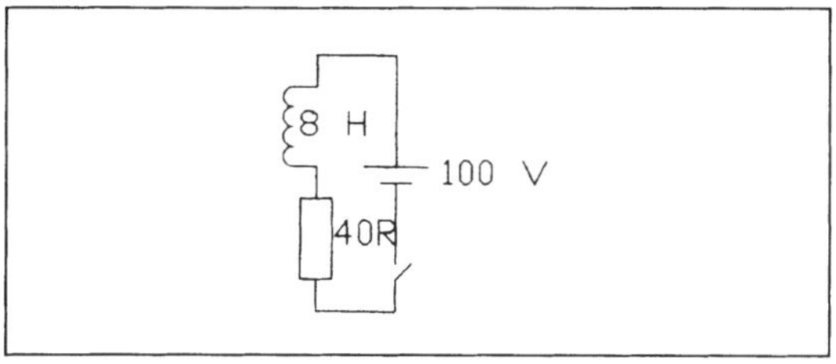

6.65 In Question 6.64 the current after 0.12 s is approximately

❑ *a* 0.33 A
❑ *b* 0.42 A
❑ *c* 0.81 A
❑ *d* 1.13 A

6.66 The diagram shows a circuit which is suddenly connected to a 100-V supply. The time for the capacitor voltage to reach 50 V is approximately

❑ *a* 0.69 s
❑ *b* 1.21 s
❑ *c* 1.86 s
❑ *d* 2.23 s

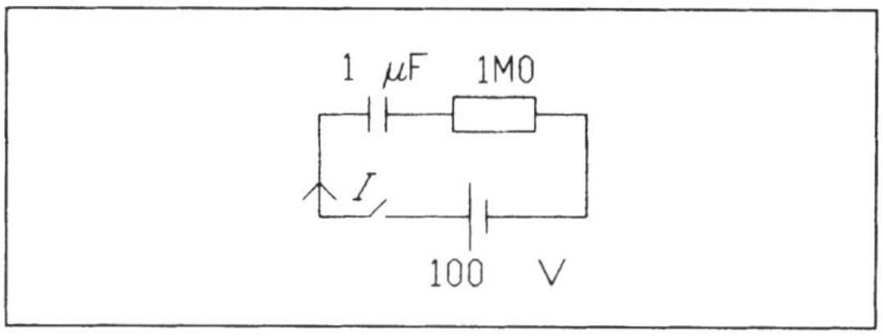

6.67 In Question 6.66 the current flowing when the capacitor voltage is 50 V is

❑ *a* 20 mA
❑ *b* 60 mA
❑ *c* 50 μA
❑ *d* 100 μA

6.68 If the capacitor in Question 6.66 is charged to 100 V, the supply is removed, and the capacitor voltage is discharged through the 1-MΩ resistor, the time taken for the voltage to drop to 20 V is

❑ *a* 1.2 s
❑ *b* 1.6 s
❑ *c* 2.3 s
❑ *d* 4.8 s

6.69 In Question 6.68 the current flowing when the capacitor voltage is 20 V is

❑ *a* 6 μA
❑ *b* 10 μA
❑ *c* 16 μA
❑ *d* 20 μA

6.70 The diagram shows a circuit which is suddenly connected to a 90-V supply. The time for the current to reach 3.75 A is approximately

❑ *a* 0.11 s
❑ *b* 0.23 s
❑ *c* 1.61 s
❑ *d* 2.62 s

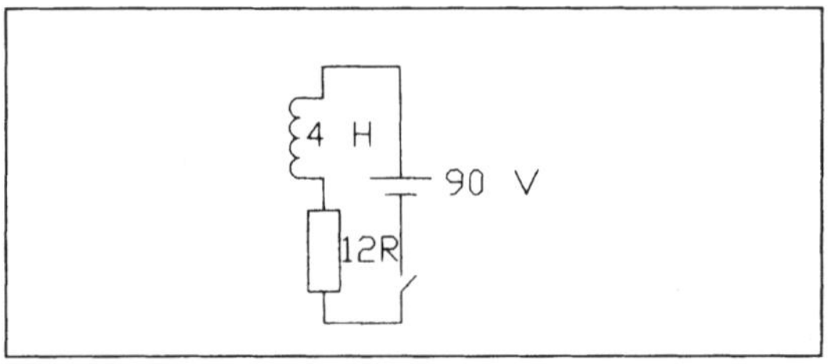

6.71 The voltage across the coil in Question 6.70 when the current is 3.75 A is approximately

❑ *a* –82 V
❑ *b* –80 V
❑ *c* –62 V
❑ *d* –45 V

6.72 The diagram shows a circuit which is suddenly connected to a 150-V supply. The time for the current to reach 4.69 A is approximately

❑ *a* 0.35 s
❑ *b* 0.92 s
❑ *c* 1.78 s
❑ *d* 2.42 s

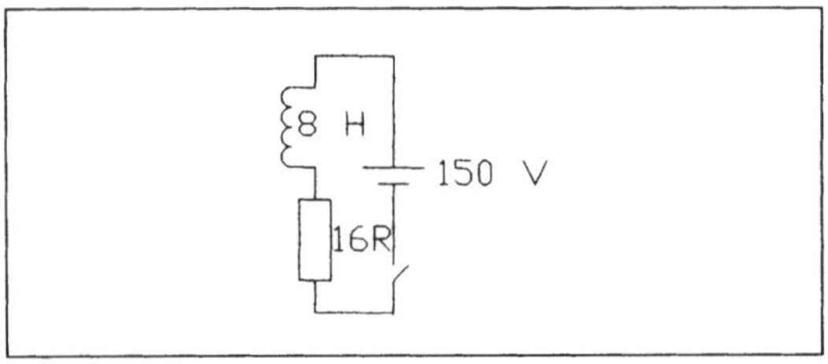

6.73 The voltage across the coil in Question 6.72 when the current flowing is 4.69 A is approximately

❑ *a* –35 V
❑ *b* –40 V
❑ *c* –75 V
❑ *d* –100 V

6.74 The diagram shows a circuit which is suddenly connected to a 12-V supply. The time for the current to reach 0.06 A is approximately

❑ *a* 2.9 s
❑ *b* 0.1 s
❑ *c* 0.015 s
❑ *d* 3.47×10^{-2} s

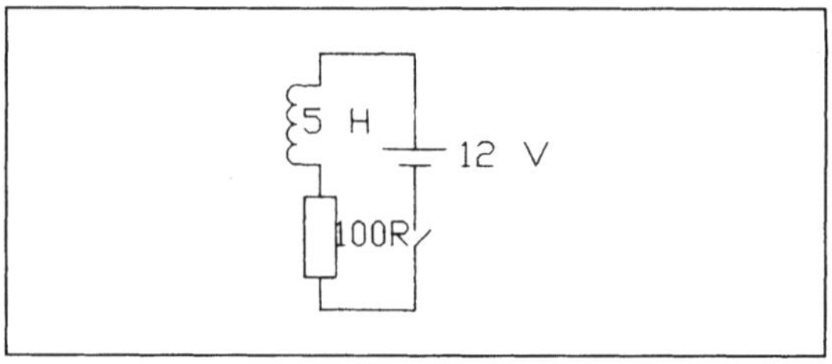

6.75 The coil voltage in Question 6.74 when the current flowing is 0.06 A is approximately

❑ *a* –4 V
❑ *b* –6 V
❑ *c* –8 V
❑ *d* –10 V

7 Electromagnetism

7.1 A conductor carrying 50 A is at a right angle to a magnetic field having a density of 0.5 T. If the conductor length is 1.0 m, the force on the conductor is

- ❑ *a* 25 N
- ❑ *b* 30 N
- ❑ *c* 50 N
- ❑ *d* 70 N

7.2 If the conductor in Question 7.1 is 0.2 m long, the current 24 A and the flux density 1.2 T, the force on the conductor is

- ❑ *a* 2.67 N
- ❑ *b* 3.41 N
- ❑ *c* 4.32 N
- ❑ *d* 5.76 N

7.3 If the conductor in Question 7.1 is 0.4 m long, the current 16 A and the flux density 60 mT, the force on the conductor is

- ❑ *a* 0.268 N
- ❑ *b* 0.384 N
- ❑ *c* 0.967 N
- ❑ *d* 1.324 N

7.4 If the conductor in Question 7.1 is 800 mm long, the current 14 A and the flux density 5 T, the force on the conductor is

- ❑ *a* 12 N
- ❑ *b* 24 N
- ❑ *c* 56 N
- ❑ *d* 100 N

7.5 If the force on a conductor 100 mm long is 1.0 N and the current in the conductor is 0.5 A, the flux density is

- ❑ *a* 1.2 T
- ❑ *b* 5 T
- ❑ *c* 10 T
- ❑ *d* 20 T

7.6 If the force on a conductor 242 mm long is 0.75 N and the current is 640 mA, the flux density is approximately

- ❑ *a* 1.67 T
- ❑ *b* 2.32 T
- ❑ *c* 4.84 T
- ❑ *d* 9.33 T

7.7 If the force on a conductor 0.242 m long is 1.5 N and the current is 0.64 A, the flux density is approximately

- ❑ *a* 2.86 T
- ❑ *b* 3.49 T
- ❑ *c* 4.68 T
- ❑ *d* 9.68 T

7.8 If the force on a conductor 55 mm long is 240 mN and the current is 1000 mA, the flux density is approximately

- ❑ *a* 4.36 T
- ❑ *b* 7.88 T
- ❑ *c* 9.02 T
- ❑ *d* 9.98 T

7.9 If a conductor has a force of 1.0 N exerted on it when carrying a current of 8 A in a magnetic field of 5 T, the conductor length is

- ❑ *a* 2.5 mm
- ❑ *b* 25 mm
- ❑ *c* 250 mm
- ❑ *d* 500 mm

7.10 If in Question 7.9 the force is 2 N, the current 4 A and the flux density 0.2 T, the conductor length is

- ❑ *a* 125 mm
- ❑ *b* 250 mm
- ❑ *c* 2.5 m
- ❑ *d* 25 m

7.11 If in Question 7.9 the force is 1.6 N, the current 6 A and the flux density 1.5 T, the conductor length is approximately

- ❑ *a* 122 mm
- ❑ *b* 178 mm
- ❑ *c* 603 mm
- ❑ *d* 804 mm

7.12 If in Question 7.9 the force is 1.2 N, the current 6 A and the flux density 200 mT, the conductor length is

- ❑ *a* 100 mm
- ❑ *b* 10 mm
- ❑ *c* 10 m
- ❑ *d* 1 m

7.13 If a conductor 5 m long moving in a magnetic field of 3 T has a force of 12 N exerted on it, the current in the conductor is

- ❑ *a* 800 mA
- ❑ *b* 60 mA
- ❑ *c* 32 mA
- ❑ *d* 1.4 mA

7.14 If in Question 7.13 the conductor is 200 mm long, the flux density 500 mT and the force 3.9 N, the current is

- ❑ *a* 10 A
- ❑ *b* 12 A
- ❑ *c* 39 A
- ❑ *d* 69 A

7.15 If in Question 7.13 the conductor is 0.2 m long, the flux density 200 mT and the force 4 N, the current is

- ❑ *a* 1 A
- ❑ *b* 10 A
- ❑ *c* 50 A
- ❑ *d* 100 A

7.16 If in Question 7.13 the conductor length is 5 m, the flux density 400 m T and the force 20 N, the current is

- ❑ *a* 10 mA
- ❑ *b* 100 mA
- ❑ *c* 1 A
- ❑ *d* 10 A

7.17 A conductor 100 mm long is moving with a velocity of 10 $m\,s^{-1}$ at right angles to a magnetic field having a flux density of 0.5 T. The EMF induced in the conductor is

- ❑ *a* 0.1 V
- ❑ *b* 0.5 V
- ❑ *c* 1.5 V
- ❑ *d* 2.5 V

7.18 If in Question 7.17 the magnetic field strength is doubled, the induced EMF is

- ❑ *a* 0.7 V
- ❑ *b* 0.8 V
- ❑ *c* 1.0 V
- ❑ *d* 2.0 V

7.19 If in Question 7.17 the conductor length is 0.45 m, the velocity 12 $m\,s^{-1}$ and the flux density 2 T, the induced EMF is

- ❑ *a* 1.9 V
- ❑ *b* 4.4 V
- ❑ *c* 6.2 V
- ❑ *d* 10.8 V

7.20 If in Question 7.19 the conductor length is doubled, the induced EMF is

- ❑ *a* 3.8 V
- ❑ *b* 8.8 V
- ❑ *c* 12.4 V
- ❑ *d* 21.6 V

7.21 If 2 V is induced in a conductor 50 mm long moving at a velocity of 5 $m\,s^{-1}$ at right angles to a magnetic field, the flux density is

- ❑ *a* 8 T
- ❑ *b* 10 T
- ❑ *c* 12 T
- ❑ *d* 50 T

7.22 If in Question 7.21 the EMF induced is 24 V, the conductor length 680 mm and the velocity 10 $m\,s^{-1}$, the flux density is approximately

- ❑ *a* 1.22 T
- ❑ *b* 3.53 T
- ❑ *c* 8.63 T
- ❑ *d* 9.17 T

7.23 If in Question 7.21 the EMF induced is 16 V, the conductor length 400 mm and the velocity 4 $m\,s^{-1}$, the flux density is

- ❑ *a* 0.1 T
- ❑ *b* 1 T
- ❑ *c* 10 T
- ❑ *d* 100 T

7.24 If the conductor length in Question 7.23 is increased by a factor of 4, the induced EMF and the velocity remaining the same, the flux density is

- ❑ *a* 0.4 T
- ❑ *b* 1.0 T
- ❑ *c* 1.25 T
- ❑ *d* 2.5 T

7.25 A voltage of 480 mV is induced in a conductor moving at a velocity of 6 $m\,s^{-1}$ in a magnetic field of 0.8 T. The conductor length is

- ❑ *a* 10 mm
- ❑ *b* 100 mm
- ❑ *c* 1 m
- ❑ *d* 10 m

7.26 If in Question 7.25 the voltage induced is 500 mV, the velocity 4 $m\,s^{-1}$ and the flux density 1.25 T, the conductor length is

- ❑ *a* 50 mm
- ❑ *b* 500 mm
- ❑ *c* 0.1 m
- ❑ *d* 1.0 m

7.27 If in Question 7.25 the voltage induced is 800 mV, the velocity 8 $m\,s^{-1}$ and the flux density 0.1 T, the conductor length is

- ❑ *a* 1.0 m
- ❑ *b* 2.0 m
- ❑ *c* 10 mm
- ❑ *d* 5 mm

7.28 If in Question 7.25 the voltage induced is 50 mV, the velocity 10 $m\,s^{-1}$ and the flux density 2 T, the conductor length is

- ❑ *a* 0.1 m
- ❑ *b* 0.2 m
- ❑ *c* 0.02 m
- ❑ *d* 0.0025 m

7.29 6 V is induced in a conductor 2 m long moving in a magnetic field of 6 T. The conductor velocity is

- ❑ *a* 0.1 $m\,s^{-1}$
- ❑ *b* 0.5 $m\,s^{-1}$
- ❑ *c* 0.7 $m\,s^{-1}$
- ❑ *d* 0.8 $m\,s^{-1}$

7.30 If in Question 7.29 the conductor length is 300 mm, the induced voltage 9 V and the flux density 3 T, the conductor velocity is

- ❑ *a* 3 $m\,s^{-1}$
- ❑ *b* 9 $m\,s^{-1}$
- ❑ *c* 10 $m\,s^{-1}$
- ❑ *d* 15 $m\,s^{-1}$

7.31 If in Question 7.30 the induced voltage is doubled, the new conductor velocity is

- ❑ *a* 6 $m\,s^{-1}$
- ❑ *b* 20 $m\,s^{-1}$
- ❑ *c* 18 $m\,s^{-1}$
- ❑ *d* 5 $m\,s^{-1}$

7.32 If in Question 7.29 the conductor length is 400 mm, the induced voltage 500 mV and the flux density 5 T, the conductor velocity is

- ❑ *a* 250 $mm\,s^{-1}$
- ❑ *b* 2.5 $m\,s^{-1}$
- ❑ *c* 25 $m\,s^{-1}$
- ❑ *d* 50 $m\,s^{-1}$

7.33 A magnetic flux of 600 μWb acts through a coil of 1500 turns and is reversed in 0.5 s. The average value of the induced EMF is

- ❑ *a* 1.6 V
- ❑ *b* 2.35 V
- ❑ *c* 3.6 V
- ❑ *d* 4.9 V

7.34 If the reversal time in Question 7.33 is doubled, the induced EMF is

- ❑ *a* 3.2 V
- ❑ *b* 1.8 V
- ❑ *c* 7.2 V
- ❑ *d* 9.8 V

7.35 If in Question 7.33 the flux is 460 μWb, the number of turns 450 and the reversal time 100 ms, the induced EMF is

- ❑ *a* 4.14 V
- ❑ *b* 2.36 V
- ❑ *c* 6.82 V
- ❑ *d* 1.22 V

7.36 If the reversal time in Question 7.35 is halved, the induced EMF is

- ❑ *a* 1.68 V
- ❑ *b* 2.07 V
- ❑ *c* 8.28 V
- ❑ *d* 2.44 V

7.37 A coil of 500 turns is wound on a wooden ring having a mean circumference of 200 mm. If the current in the coil is 2 A, the magnetic field strength is

- ❑ *a* 1000 $A\,m^{-1}$
- ❑ *b* 1500 $A\,m^{-1}$
- ❑ *c* 2500 $A\,m^{-1}$
- ❑ *d* 5000 $A\,m^{-1}$

7.38 If in Question 7.37 the number of turns is 400, the circumference 800 mm and the current 4 A, the field strength is

- ❑ *a* 1000 $A\,m^{-1}$
- ❑ *b* 2000 $A\,m^{-1}$
- ❑ *c* 3000 $A\,m^{-1}$
- ❑ *d* 4000 $A\,m^{-1}$

7.39 If the number of turns in Question 7.38 is doubled, the field strength is

- ❑ *a* 2000 $A\,m^{-1}$
- ❑ *b* 1000 $A\,m^{-1}$
- ❑ *c* 4000 $A\,m^{-1}$
- ❑ *d* 1500 $A\,m^{-1}$

7.40 If in Question 7.37 the number of turns is 150, the circumference 900 mm and the current 6 A, the field strength is

- ❑ *a* 1000 $A\,m^{-1}$
- ❑ *b* 2000 $A\,m^{-1}$
- ❑ *c* 4000 $A\,m^{-1}$
- ❑ *d* 6000 $A\,m^{-1}$

7.41 A coil 150 mm in length has a field strength of 1000 $A\,m^{-1}$ when the current flowing is 15 A. The number of turns on the coil is

- ❑ *a* 5
- ❑ *b* 10
- ❑ *c* 25
- ❑ *d* 65

7.42 If in Question 7.41 the length is 250 mm, the field strength 500 $A\,m^{-1}$ and the current 2.5 A, the number of turns is

- ❑ *a* 12
- ❑ *b* 20
- ❑ *c* 50
- ❑ *d* 100

7.43 If in Question 7.42 the current is doubled, the number of turns required to give the same field strength is

- ❑ *a* 24
- ❑ *b* 25
- ❑ *c* 100
- ❑ *d* 200

7.44 If in Question 7.41 the length is 200 mm, the field strength 1100 $A\,m^{-1}$ and the current 5.5 A, the number of turns is

- ❑ *a* 14
- ❑ *b* 15
- ❑ *c* 28
- ❑ *d* 40

7.45 The current required to produce a field strength of 1200 $A\,m^{-1}$ in a coil of 600 turns having a mean length of 140 mm is

- ❑ *a* 0.28 A
- ❑ *b* 0.92 A
- ❑ *c* 1.86 A
- ❑ *d* 2.44 A

7.46 If in Question 7.45 the field strength is 1800 $A\,m^{-1}$, the number of turns 90 and the length 100 mm, the current is

- ❑ *a* 1.6 A
- ❑ *b* 1.8 A
- ❑ *c* 2 A
- ❑ *d* 2.2 A

7.47 If in Question 7.45 the field strength is 1600 $A\,m^{-1}$, the number of turns 160 and the length 0.25 m, the current is
- ❑ *a* 1.11 A
- ❑ *b* 1.53 A
- ❑ *c* 1.92 A
- ❑ *d* 2.5 A

7.48 If in Question 7.45 the field strength is 750 $A\,m^{-1}$, the number of turns 2500 and the length 340 mm, the current is
- ❑ *a* 102 mA
- ❑ *b* 900 mA
- ❑ *c* 1.62 A
- ❑ *d* 9.22 A

7.49 The circumference of a ring having 150 turns and a current of 5 A producing a field strength of 2500 $A\,m^{-1}$ is
- ❑ *a* 150 mm
- ❑ *b* 300 mm
- ❑ *c* 600 mm
- ❑ *d* 900 mm

7.50 If in Question 7.49 the number of turns is 200, the current 8 A and the field strength 1600 $A\,m^{-1}$, the circumference is
- ❑ *a* 280 mm
- ❑ *b* 640 mm
- ❑ *c* 1000 mm
- ❑ *d* 1200 mm

7.51 A coil has 100 turns wound on a non-magnetic former with a circumference of 500 mm and a cross-sectional area of 600 mm^2. If the current in the coil is 10 A, the magnetic field strength is
- ❑ *a* 2000 $A\,m^{-1}$
- ❑ *b* 2500 $A\,m^{-1}$
- ❑ *c* 6000 $A\,m^{-1}$
- ❑ *d* 7500 $A\,m^{-1}$

7.52 The flux density in Question 7.51, given that the permeability of free space is $4\pi \times 10^{-7}$ $H\,m^{-1}$, is approximately
- ❑ *a* 1.33 mT
- ❑ *b* 1.62 mT
- ❑ *c* 1.92 mT
- ❑ *d* 2.51 mT

7.53 In Question 7.51 the total flux is approximately
- ❑ *a* 1.2 μWb
- ❑ *b* 1.5 μWb
- ❑ *c* 6.1 μWb
- ❑ *d* 7.7 μWb

7.54 If in Question 7.51 the number of turns is 200, the circumference 400 mm, the cross-sectional area 200 mm^2 and the current 20 A, the magnetic field strength is
- ❑ *a* 3350 $A\,m^{-1}$
- ❑ *b* 6200 $A\,m^{-1}$
- ❑ *c* 8500 $A\,m^{-1}$
- ❑ *d* 10 000 $A\,m^{-1}$

7.55 The flux density in Question 7.54 is approximately
- ❑ *a* 6.39 mT
- ❑ *b* 12.57 mT
- ❑ *c* 15.62 mT
- ❑ *d* 16.44 mT

7.56 The total flux in Question 7.54 is approximately
- ❑ *a* 2.47 μWb
- ❑ *b* 4.11 μWb
- ❑ *c* 2.51 μWb
- ❑ *d* 6.32 μWb

7.57 A coil wound on a wooden bobbin has a magnetic field strength of 2500 $A\,m^{-1}$. If the total flux is 25 μWb and the permeability of free space $4\pi \times 10^{-7}$ $H\,m^{-1}$, the cross-sectional area of the bobbin is approximately
- ❑ *a* 3261 mm^2
- ❑ *b* 4126 mm^2
- ❑ *c* 7958 mm^2
- ❑ *d* 8600 mm^2

7.58 If in Question 7.57 the field strength is 2200 $A\,m^{-1}$ and the total flux 30 μWb, the cross-sectional area is approximately
- ❑ *a* 4622 mm^2
- ❑ *b* 5831 mm^2
- ❑ *c* 6800 mm^2
- ❑ *d* 10 851 mm^2

7.59 If in Question 7.57 the field strength is 2000 A m^{-1} and the total flux 2 μWb, the cross-sectional area is approximately

- ❑ *a* 240 mm^2
- ❑ *b* 796 mm^2
- ❑ *c* 850 mm^2
- ❑ *d* 900 mm^2

7.60 A coil wound on a plastic former has a total flux of 20 μWb and a cross-sectional area of 4500 mm^2. Given that the permeability of free space is $4\pi \times 10^{-7}$ H m^{-1}, the field strength is approximately

- ❑ *a* 3537 A m^{-1}
- ❑ *b* 4520 A m^{-1}
- ❑ *c* 5522 A m^{-1}
- ❑ *d* 6010 A m^{-1}

7.61 If in Question 7.60 the total flux is 40 μWb and the area is 4500 mm^2, the field strength is approximately

- ❑ *a* 2681 A m^{-1}
- ❑ *b* 3082 A m^{-1}
- ❑ *c* 5200 A m^{-1}
- ❑ *d* 7074 A m^{-1}

7.62 If in Question 7.60 the total flux is 2.3 μWb and the area 240 mm^2, the field strength is approximately

- ❑ *a* 4800 A m^{-1}
- ❑ *b* 7626 A m^{-1}
- ❑ *c* 8100 A m^{-1}
- ❑ *d* 9216 A m^{-1}

7.63 If a coil has a field strength of 2500 A m^{-1}, and the permeability of free space is $4\pi \times 10^{-7}$ H m^{-1}, the flux density is approximately

- ❑ *a* 1.28 mT
- ❑ *b* 2.61 mT
- ❑ *c* 3.14 mT
- ❑ *d* 5.33 mT

7.64 If in Question 7.63 the field strength is 450 A m^{-1}, the flux density is approximately

- ❑ *a* 0.57 mT
- ❑ *b* 1.12 mT
- ❑ *c* 2.88 mT
- ❑ *d* 3.06 mT

7.65 If in Question 7.63 the field strength is 6000 A m^{-1}, the flux density is approximately

- ❑ *a* 7.54 mT
- ❑ *b* 8.62 mT
- ❑ *c* 9.05 mT
- ❑ *d* 9.94 mT

7.66 A coil wound uniformly on a non-magnetic ring has an air gap 4 mm long. The cross-sectional area is 20 cm^2. The number of ampere turns required to establish a total flux of 10 mWb in the air gap, given that the permeability of free space equals $4\pi \times 10^{-7}$ H m^{-1}, is approximately

- ❑ *a* 15 915 A T^{-1}
- ❑ *b* 17 622 A T^{-1}
- ❑ *c* 18 731 A T^{-1}
- ❑ *d* 19 600 A T^{-1}

7.67 If in Question 7.66 the air gap is 2 mm, the area 25 cm^2 and the total flux 15 mWb, the number of ampere turns required is approximately

- ❑ *a* 5120
- ❑ *b* 6233
- ❑ *c* 9549
- ❑ *d* 10 600

7.68 If in Question 7.66 the air gap is doubled, the number of ampere turns required is

- ❑ *a* Halved
- ❑ *b* Doubled
- ❑ *c* Increased by a factor of 4
- ❑ *d* Decreased by a factor of 4

7.69 An iron ring has a cross-sectional area of 500 mm^2, a mean length of 0.2 m and a field strength of 200 A m^{-1}. If the coil has 400 turns and the permeability of free space is taken as $4\pi \times 10^{-7}$ H m^{-1}, the current required to produce a magnetic flux of 550 μWb is

- ❑ *a* 0.06 A
- ❑ *b* 0.1 A
- ❑ *c* 0.8 A
- ❑ *d* 1.3 A

7.70 The flux density in Question 7.69 is

- ❑ *a* 0.9 T
- ❑ *b* 1.0 T
- ❑ *c* 1.1 T
- ❑ *d* 1.2 T

7.71 The relative permeability of the iron in Question 7.69 is approximately

- ❑ *a* 1553
- ❑ *b* 1982
- ❑ *c* 2602
- ❑ *d* 4377

7.72 If in Question 7.69 the area is 800 mm^2, the length 0.1 m, the field strength 300 $A\,m^{-1}$, the number of turns 500 and the total flux 400 μWb, the flux density is

- ❑ *a* 0.5 T
- ❑ *b* 1.5 T
- ❑ *c* 2.5 T
- ❑ *d* 4.5 T

7.73 The required current in Question 7.72 is

- ❑ *a* 3 mA
- ❑ *b* 6 mA
- ❑ *c* 60 mA
- ❑ *d* 600 mA

7.74 The relative permeability in Question 7.72 is approximately

- ❑ *a* 606
- ❑ *b* 732
- ❑ *c* 940
- ❑ *d* 1326

7.75 If the cross-sectional area in Question 7.72 was changed to 400 mm^2, the flux density would then be

- ❑ *a* 0.75 T
- ❑ *b* 0.85 T
- ❑ *c* 1.0 T
- ❑ *d* 1.5 T

8 AC circuits

8.1 In the diagram the current I is

- ❑ *a* 0.62 A
- ❑ *b* 1.23 A
- ❑ *c* 1.59 A
- ❑ *d* 2.31 A

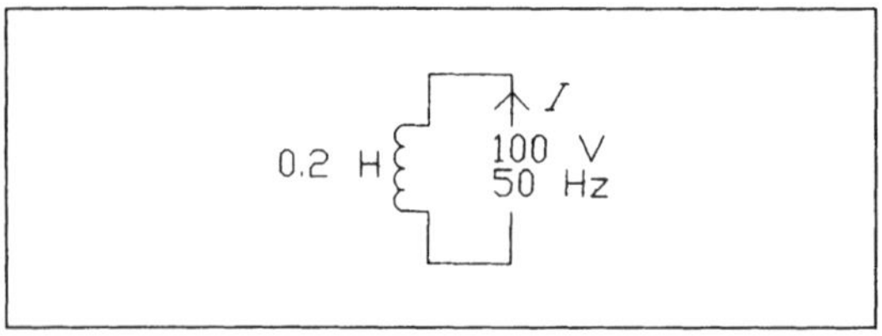

8.2 If in Question 8.1 the coil has an inductance of 0.5 H and the input voltage is 60 V with a frequency of 100 Hz, the current is

- ❑ *a* 0.19 A
- ❑ *b* 0.76 A
- ❑ *c* 1.34 A
- ❑ *d* 2.66 A

8.3 If in Question 8.1 the inductor is 1.5 H and the current I 4.5 A, the input voltage at a frequency of 20 Hz is approximately

- ❑ *a* 240 V
- ❑ *b* 410 V
- ❑ *c* 640 V
- ❑ *d* 848 V

8.4 If in Question 8.1 the inductor is 550 mH and the current I is 100 mA, the input voltage at a frequency of 1 kHz is approximately

- ❑ *a* 128 V
- ❑ *b* 346 V
- ❑ *c* 624 V
- ❑ *d* 860 V

8.5 If in Question 8.1 the voltage is 140 V 125 Hz and the current 3 A, the inductance of the coil is approximately

- ❑ *a* 0.032 H
- ❑ *b* 0.059 H
- ❑ *c* 0.62 H
- ❑ *d* 0.94 H

8.6 If in Question 8.1 the voltage is 50 V 60 Hz and the current 7.6 A, the inductance is

- ❑ *a* 17.45 mH
- ❑ *b* 240.6 mH
- ❑ *c* 6.2 H
- ❑ *d* 9.7 H

8.7 If in Question 8.1 the voltage is 12 V, the current 12.4 A and the inductance 240 μH, the frequency is approximately

- ❑ *a* 246 Hz
- ❑ *b* 390 Hz
- ❑ *c* 484 Hz
- ❑ *d* 642 Hz

8.8 If in Question 8.1 the voltage is 1.6 V, the current 100 mA and the inductance 100 μH, the frequency is

- ❑ *a* 16.823 kHz
- ❑ *b* 25.464 kHz
- ❑ *c* 40 kHz
- ❑ *d* 60 kHz

8.9 In the diagram the current I is approximately

- ❑ *a* 0.2 A
- ❑ *b* 0.8 A
- ❑ *c* 1.1 A
- ❑ *d* 2.6 A

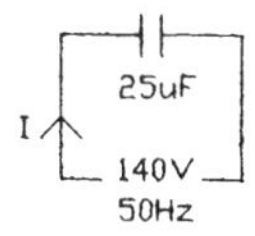

8.10 If in Question 8.9 the voltage is 240 V 200 Hz and the capacitance 0.5 μF, the current is

- ❑ *a* 0.64 A
- ❑ *b* 0.08 A
- ❑ *c* 0.11 A
- ❑ *d* 0.15 A

8.11 If in Question 8.9 the capacitance is 2 μF and the current 0.5 A, the voltage at a frequency of 60 Hz is approximately

- ❑ *a* 663 V
- ❑ *b* 748 V
- ❑ *c* 820 V
- ❑ *d* 915 V

8.12 If in Question 8.9 the capacitance is 3.5 μF and the current 1.8 A, the voltage at a frequency of 130 Hz is approximately

- ❑ *a* 336 V
- ❑ *b* 480 V
- ❑ *c* 630 V
- ❑ *d* 900 V

8.13 If in Question 8.9 the current is 4.2 A and the voltage 210 V 80 Hz, the capacitance is approximately

- ❑ *a* 39.8 μF
- ❑ *b* 45.2 μF
- ❑ *c* 50.3 μF
- ❑ *d* 70.7 μF

8.14 If in Question 8.9 the current is 1.6 A and the voltage 180 V 250 Hz, the capacitance is approximately

- ❑ *a* 4.3 μF
- ❑ *b* 5.7 μF
- ❑ *c* 10.8 μF
- ❑ *d* 12.6 μF

8.15 If in Question 8.9 the capacitance is 680 pF, the current 500 mA and the voltage 1 kV, the frequency is approximately

- ❑ *a* 47 kHz
- ❑ *b* 86 kHz
- ❑ *c* 93 kHz
- ❑ *d* 117 kHz

8.16 If in Question 8.9 the capacitance is 150 pF, the current 2 mA and the voltage 2.5 kV, the frequency is approximately

- ❑ *a* 630 Hz
- ❑ *b* 724 Hz
- ❑ *c* 849 Hz
- ❑ *d* 933 Hz

8.17 The diagram shows a series *R–L* circuit. The current I is approximately

- ❑ *a* 1.93 A
- ❑ *b* 2.16 A
- ❑ *c* 3.147 A
- ❑ *d* 4.231 A

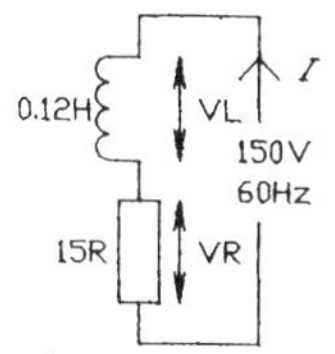

8.18 The voltage V_L in Question 8.17 is approximately

- ❑ *a* 88 V
- ❑ *b* 93 V
- ❑ *c* 106 V
- ❑ *d* 142 V

8.19 The voltage V_R in Question 8.17 is approximately
- ❑ *a* 47 V
- ❑ *b* 69 V
- ❑ *c* 83 V
- ❑ *d* 91 V

8.20 If in Question 8.17 the resistance is 10 Ω, the inductance 0.2 H and the voltage 120 V (50 Hz), the current is approximately
- ❑ *a* 1.22 A
- ❑ *b* 1.89 A
- ❑ *c* 2.68 A
- ❑ *d* 3.27 A

8.21 In Question 8.20 the true power is approximately
- ❑ *a* 16.8 W
- ❑ *b* 35.6 W
- ❑ *c* 47.8 W
- ❑ *d* 60.7 W

8.22 In Question 8.20 the circuit phase angle is approximately
- ❑ *a* 36°
- ❑ *b* 45°
- ❑ *c* 67°
- ❑ *d* 81°

8.23 If in Question 8.17 the resistance is 15 Ω, the inductance 0.12 H and the voltage 150 V (60 Hz), the circuit impedance is approximately
- ❑ *a* 19.34 Ω
- ❑ *b* 27.68 Ω
- ❑ *c* 39.33 Ω
- ❑ *d* 47.66 Ω

8.24 In Question 8.23 the power factor is approximately
- ❑ *a* 0.315
- ❑ *b* 0.464
- ❑ *c* 0.723
- ❑ *d* 0.814

8.25 In Question 8.23 the phase angle is approximately
- ❑ *a* 83.2°
- ❑ *b* 71.7°
- ❑ *c* 46.6°
- ❑ *d* 33.1°

8.26 The diagram shows a series *R–C* circuit. The reactive impedance is approximately
- ❑ *a* 260 Ω
- ❑ *b* 318 Ω
- ❑ *c* 473 Ω
- ❑ *d* 566 Ω

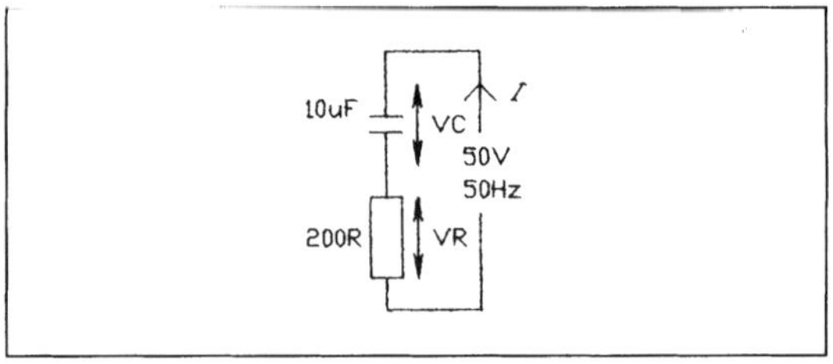

8.27 In Question 8.26 the circuit impedance is approximately
- ❑ *a* 120 Ω
- ❑ *b* 261 Ω
- ❑ *c* 376 Ω
- ❑ *d* 498 Ω

8.28 In Question 8.26 the current *I* is approximately
- ❑ *a* 103 mA
- ❑ *b* 119 mA
- ❑ *c* 121 mA
- ❑ *d* 133 mA

8.29 If in Question 8.26 the resistance is 100 Ω, the capacitance 12 μF and the voltage 60 V (120 Hz), the voltage across the resistor is approximately
- ❑ *a* 40 V
- ❑ *b* 45 V
- ❑ *c* 50 V
- ❑ *d* 55 V

8.30 In Question 8.29 the voltage across the capacitor is approximately
- ❑ *a* 19.8 V
- ❑ *b* 26.6 V
- ❑ *c* 30.2 V
- ❑ *d* 44.5 V

8.31 In Question 8.29 the true power is approximately
- ❑ *a* 16.2 W
- ❑ *b* 43.3 W
- ❑ *c* 46.6 W
- ❑ *d* 55.4 W

8.32 In Question 8.29 the apparent power is approximately
- ❑ *a* 20.6 VA
- ❑ *b* 24.2 VA
- ❑ *c* 28.2 VA
- ❑ *d* 30.4 VA

8.33 In Question 8.29 the power factor is
- ❑ *a* 0.22
- ❑ *b* 0.44
- ❑ *c* 0.67
- ❑ *d* 0.96

8.34 In Question 8.29 the circuit phase angle is approximately
- ❑ *a* 24.6°
- ❑ *b* 47.9°
- ❑ *c* 55.4°
- ❑ *d* 60.7°

8.35 The diagram shows a series *R–L–C* circuit. The circuit impedance is approximately
- ❑ *a* 67 Ω
- ❑ *b* 96 Ω
- ❑ *c* 100 Ω
- ❑ *d* 118 Ω

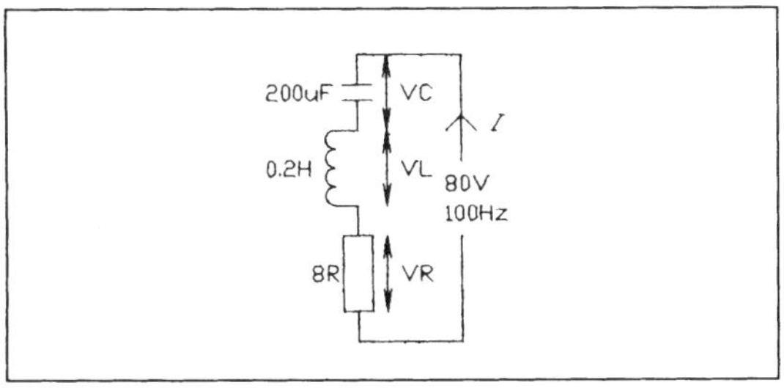

8.36 The current *I* in Question 8.35 is approximately
- ❑ *a* 430 mA
- ❑ *b* 678 mA
- ❑ *c* 782 mA
- ❑ *d* 906 mA

8.37 In Question 8.35 the true power is approximately
- ❑ *a* 1.39 W
- ❑ *b* 2.18 W
- ❑ *c* 2.67 W
- ❑ *d* 3.68 W

8.38 If in Question 8.35 the resistor value is 6 Ω, the capacitance 280 μF, the inductance 0.015 H and the voltage 110 V (60 Hz), the current *I* is approximately
- ❑ *a* 10.4 A
- ❑ *b* 12.6 A
- ❑ *c* 15.5 A
- ❑ *d* 19.8 A

8.39 In Question 8.38 the voltage across the resistor is approximately
- ❑ *a* 31.2 V
- ❑ *b* 40.8 V
- ❑ *c* 60.6 V
- ❑ *d* 92.8 V

8.40 In Question 8.38 the voltage across the inductor is approximately
- ❑ *a* 42.2 V
- ❑ *b* 70.7 V
- ❑ *c* 87.5 V
- ❑ *d* 99.9 V

8.41 In Question 8.38 the voltage across the capacitor is approximately
- ❑ *a* 150 V
- ❑ *b* 147 V
- ❑ *c* 110 V
- ❑ *d* 55 V

8.42 In Question 8.38 the true power is approximately
- ❑ *a* 664 W
- ❑ *b* 762 W
- ❑ *c* 900 W
- ❑ *d* 1435 W

8.43 In Question 8.38 the circuit phase angle is approximately
❑ *a* 24.1°
❑ *b* 32.5°
❑ *c* 40.8°
❑ *d* 70.7°

8.44 In a series circuit with an inductance of 0.0318 H and a capacitance of 318 μF, the resonant frequency is
❑ *a* 50 Hz
❑ *b* 120 Hz
❑ *c* 160 Hz
❑ *d* 350 Hz

8.45 If a series circuit has an inductance of 1.4 H and a capacitance of 14 μF, the resonant frequency is
❑ *a* 36 Hz
❑ *b* 98 Hz
❑ *c* 164 Hz
❑ *d* 408 Hz

8.46 The diagram shows a parallel *R–L* circuit. The current I_R is approximately
❑ *a* 1.22 A
❑ *b* 1.48 A
❑ *c* 1.64 A
❑ *d* 1.79 A

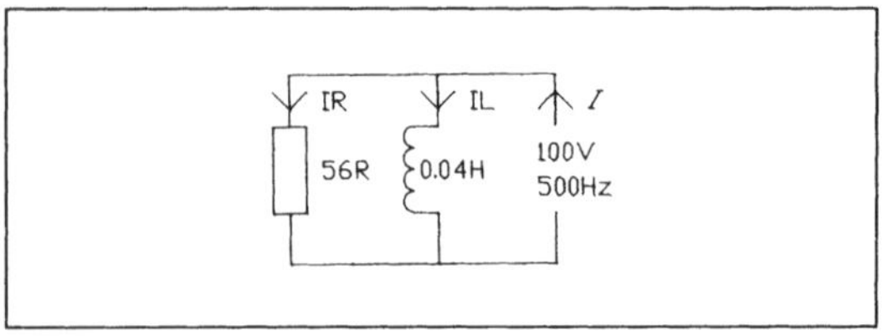

8.47 In Question 8.46 the current I_L is approximately
❑ *a* 0.1 A
❑ *b* 0.5 A
❑ *c* 0.8 A
❑ *d* 1.2 A

8.48 In Question 8.46 the total current *I* is approximately
❑ *a* 1.42 A
❑ *b* 1.96 A
❑ *c* 2.87 A
❑ *d* 3.49 A

8.49 If the parameters in Question 8.46 are R = 100 Ω, L = 0.06 H and V = 120 V (400 Hz), the total circuit impedance is approximately
❑ *a* 83.34 Ω
❑ *b* 90.66 Ω
❑ *c* 95.82 Ω
❑ *d* 99.21 Ω

8.50 The circuit phase angle in Question 8.49 is approximately
❑ *a* 20.71°
❑ *b* 26.42°
❑ *c* 33.55°
❑ *d* 45.18°

8.51 The true power in Question 8.49 is
❑ *a* 106 W
❑ *b* 144 W
❑ *c* 162 W
❑ *d* 198 W

8.52 The total current *I* in Question 8.49 is approximately
❑ *a* 1.16 A
❑ *b* 1.32 A
❑ *c* 1.44 A
❑ *d* 1.68 A

8.53 The reactive power in Question 8.49 is approximately
❑ *a* 20.6 VA
❑ *b* 50.4 VA
❑ *c* 60.2 VA
❑ *d* 95.5 VA

8.54 The apparent power in Question 8.49 is approximately
❑ *a* 120 VA
❑ *b* 139 VA
❑ *c* 160 VA
❑ *d* 173 VA

8.55 The diagram shows a parallel R–C circuit. The current I_R is approximately

- ❑ *a* 4.2 A
- ❑ *b* 5.3 A
- ❑ *c* 6.8 A
- ❑ *d* 7.6 A

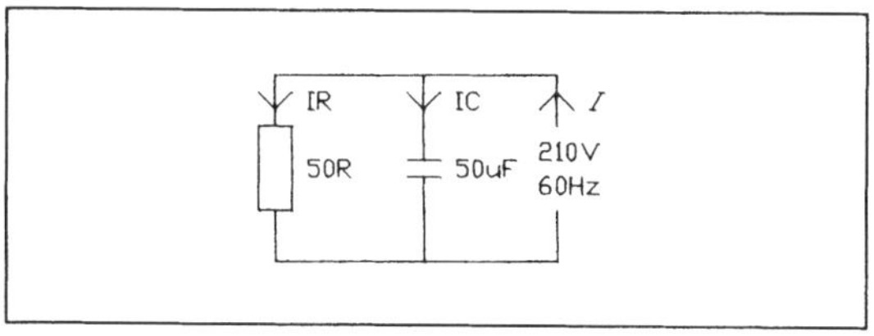

8.56 In Question 8.55 the current I_C is approximately

- ❑ *a* 1.67 A
- ❑ *b* 2.48 A
- ❑ *c* 3.96 A
- ❑ *d* 4.88 A

8.57 In Question 8.55 the total circuit impedance is approximately

- ❑ *a* 30.42 Ω
- ❑ *b* 36.39 Ω
- ❑ *c* 50.81 Ω
- ❑ *d* 62.38 Ω

8.58 In Question 8.55 the total current is approximately

- ❑ *a* 3.88 A
- ❑ *b* 4.21 A
- ❑ *c* 5.77 A
- ❑ *d* 6.31 A

8.59 In Question 8.55 the circuit phase angle is

- ❑ *a* 43.3°
- ❑ *b* 60.5°
- ❑ *c* 68.7°
- ❑ *d* 75.4°

8.60 In Question 8.55 the true power is

- ❑ *a* 555 W
- ❑ *b* 624 W
- ❑ *c* 882 W
- ❑ *d* 900 W

8.61 If in Question 8.55 the parameters are R = 68 Ω, C = 45 μF, V = 180 V (40 Hz), the total current I is approximately

- ❑ *a* 2.67 A
- ❑ *b* 2.98 A
- ❑ *c* 3.02 A
- ❑ *d* 3.34 A

8.62 In Question 8.61 the phase angle is approximately

- ❑ *a* 25.82°
- ❑ *b* 31.22°
- ❑ *c* 37.56°
- ❑ *d* 60.68°

8.63 In Question 8.61 the apparent power is

- ❑ *a* 482 VA
- ❑ *b* 601 VA
- ❑ *c* 830 VA
- ❑ *d* 962 VA

8.64 The diagram shows a parallel L–C circuit. The total current I is approximately

- ❑ *a* 515 mA
- ❑ *b* 603 mA
- ❑ *c* 728 mA
- ❑ *d* 770 mA

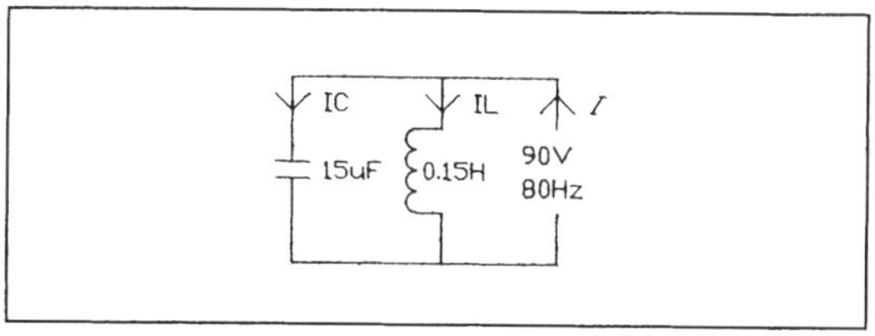

8.65 In Question 8.64 the circuit phase angle is

- ❑ *a* 30°
- ❑ *b* 45°
- ❑ *c* 55°
- ❑ *d* 90°

8.66 In Question 8.64 the true power consumed is

- ❑ *a* 0 W
- ❑ *b* 20 W
- ❑ *c* 90 W
- ❑ *d* 95 W

8.67 The diagram shows a parallel *R–L–C* circuit. The current I_{LR} is approximately

- ❑ *a* 1.25 A
- ❑ *b* 1.88 A
- ❑ *c* 2.05 A
- ❑ *d* 2.62 A

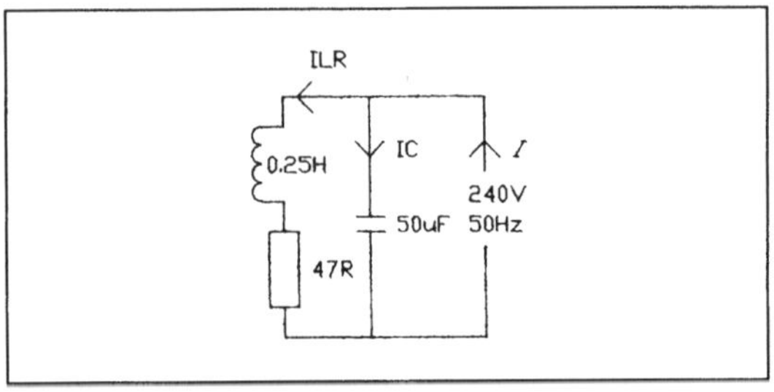

8.68 In Question 8.67 the current I_C is approximately

- ❑ *a* 2.09 A
- ❑ *b* 3.77 A
- ❑ *c* 6.81 A
- ❑ *d* 7.45 A

8.69 In Question 8.67 the total current *I* is approximately

- ❑ *a* 1.46 A
- ❑ *b* 2.03 A
- ❑ *c* 3.44 A
- ❑ *d* 6.81 A

8.70 In Question 8.67 the circuit phase angle is approximately

- ❑ *a* 48.46 °
- ❑ *b* 55.13 °
- ❑ *c* 65.83 °
- ❑ *d* 75.86 °

8.71 In Question 8.67 the true power is

- ❑ *a* 110 W
- ❑ *b* 200 W
- ❑ *c* 323 W
- ❑ *d* 639 W

8.72 In Question 8.67 the apparent power is

- ❑ *a* 361 VA
- ❑ *b* 487 VA
- ❑ *c* 705 VA
- ❑ *d* 758 VA

8.73 The diagram shows a parallel *R–L–C* circuit. The resonant frequency is

- ❑ *a* 11.26 Hz
- ❑ *b* 14.8 Hz
- ❑ *c* 20 Hz
- ❑ *d* 68 Hz

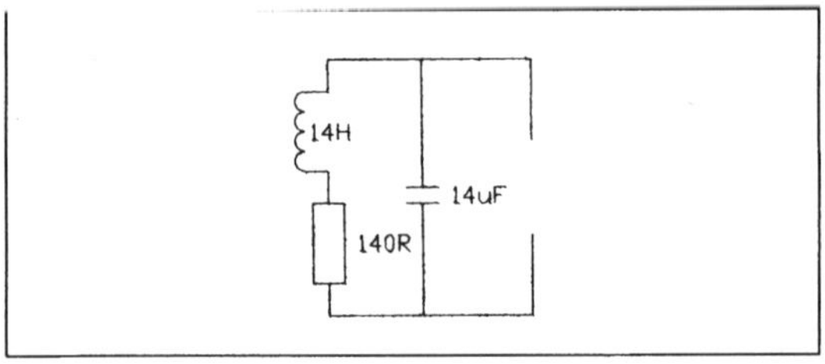

8.74 If the component values in Question 8.73 are $R = 2\ \Omega$, $L = 1.6$ H and $C = 3.3\ \mu$F, the resonant frequency is

- ❑ *a* 17 Hz
- ❑ *b* 32 Hz
- ❑ *c* 44 Hz
- ❑ *d* 69 Hz

8.75 If the component values in Question 8.73 are $R = 4\ \Omega$, $L = 1$ mH and $C = 0.02\ \mu$F, the resonant frequency is

- ❑ *a* 16.341 kHz
- ❑ *b* 35.582 kHz
- ❑ *c* 47.881 kHz
- ❑ *d* 58.403 kHz

9 Phasors and transformers

9.1 $x + jy$ is called
- ❑ *a* An exponential number
- ❑ *b* An odd number
- ❑ *c* An even number
- ❑ *d* A complex number

9.2 In the expression $x + jy$, x, the horizontal component, is
- ❑ *a* A complex number
- ❑ *b* A real number
- ❑ *c* An imaginary number
- ❑ *d* An odd number

9.3 In the expression $x + jy$, jy, the vertical component, is
- ❑ *a* A real number
- ❑ *b* An imaginary number
- ❑ *c* An even number
- ❑ *d* A complex number

9.4 An expression such as $OA = OB + jOC$ is called
- ❑ *a* A rectangular or cartesian notation
- ❑ *b* A trigonometric notation
- ❑ *c* An argument
- ❑ *d* A polar notation

9.5 An expression such as $OA\,(\cos\theta + j\sin\theta)$ is called
- ❑ *a* A modulus
- ❑ *b* A trigonometric notation
- ❑ *c* A polar notation
- ❑ *d* An argument

9.6 A phasor written as $OA\,\underline{/\theta}$ is called
- ❑ *a* A polar notation
- ❑ *b* A trigonometric notation
- ❑ *c* A rectangular notation
- ❑ *d* An exponential

9.7 The operator j rotates a phasor
- ❑ *a* 180° clockwise
- ❑ *b* 180° anticlockwise
- ❑ *c* 90° clockwise
- ❑ *d* 90° anticlockwise

9.8 The operator j is equal to
- ❑ *a* -1
- ❑ *b* $\sqrt{-1}$
- ❑ *c* $+1$
- ❑ *d* x

9.9 The expression j^2 is equal to
- ❑ *a* 0
- ❑ *b* y
- ❑ *c* x
- ❑ *d* -1

9.10 The expression j^2x is equal to
- ❑ *a* $+1$
- ❑ *b* -1
- ❑ *c* $-x$
- ❑ *d* $-y$

9.11 The expression j^3 is equal to
- ❑ *a* j
- ❑ *b* $-j$
- ❑ *c* 1
- ❑ *d* -1

9.12 The expression j^4 is equal to
- ❑ *a* 1
- ❑ *b* -1
- ❑ *c* j
- ❑ *d* -1

9.13 The expression j^5 is equal to
- ☐ *a* 1
- ☐ *b* −1
- ☐ *c* j
- ☐ *d* $-j$

9.14 $13 + j23$ expressed in polar form is
- ☐ *a* $18.67\angle 45.32^\circ$
- ☐ *b* $22.34\angle 50.48^\circ$
- ☐ *c* $26.42\angle 60.52^\circ$
- ☐ *d* $45\angle 70^\circ$

9.15 $32 + j34$ expressed in polar form is
- ☐ *a* $46.69\angle 46.74^\circ$
- ☐ *b* $50.21\angle 50.41^\circ$
- ☐ *c* $60.66\angle 20.32^\circ$
- ☐ *d* $70.71\angle 45^\circ$

9.16 $200 + j150$ expressed in polar form is
- ☐ *a* $190\angle 30^\circ$
- ☐ *b* $200\angle 60^\circ$
- ☐ *c* $220\angle 45.62^\circ$
- ☐ *d* $250\angle 36.87^\circ$

9.17 $-4 + j3$ expressed in polar form is
- ☐ *a* $5\angle -143.13^\circ$
- ☐ *b* $5\angle 143.13^\circ$
- ☐ *c* $4.5\angle 143.13^\circ$
- ☐ *d* $4.5\angle -143.13^\circ$

9.18 $-18 + j12$ expressed in polar form is
- ☐ *a* $21.63\angle 146.3^\circ$
- ☐ *b* $33.42\angle -50^\circ$
- ☐ *c* $48.69\angle 120.4^\circ$
- ☐ *d* $63.33\angle -116^\circ$

9.19 $-2 + j$ expressed in polar form is
- ☐ *a* $2.12\angle 62^\circ$
- ☐ *b* $2.18\angle 73.42^\circ$
- ☐ *c* $2.21\angle 120.31^\circ$
- ☐ *d* $2.24\angle 153.44^\circ$

9.20 $5 - j3$ expressed in polar form is
- ☐ *a* $4.61\angle 30^\circ$
- ☐ *b* $5.83\angle -30.96^\circ$
- ☐ *c* $6.26\angle 60.33^\circ$
- ☐ *d* $6.26\angle -60^\circ$

9.21 $-4 - j3$ expressed in polar form is
- ☐ *a* $5\angle 143.13^\circ$
- ☐ *b* $5\angle -36.87^\circ$
- ☐ *c* $5\angle -143.13^\circ$
- ☐ *d* $5\angle 36.87^\circ$

9.22 $5\angle 36.87$ expressed in rectangular form is
- ☐ *a* $4 + 4j$
- ☐ *b* $4 - j3$
- ☐ *c* $3 + j3$
- ☐ *d* $4 + j3$

9.23 $5\angle 143.13^\circ$ expressed in rectangular form is
- ☐ *a* $-4 - j3$
- ☐ *b* $4 + j3$
- ☐ *c* $-4 + j3$
- ☐ *d* $-4 + j5$

9.24 $5\angle -143.13^\circ$ expressed in rectangular form is
- ☐ *a* $-4 - j3$
- ☐ *b* $4 - j3$
- ☐ *c* $-4 + j3$
- ☐ *d* $4 + j3$

9.25 $5\angle -36.87^\circ$ expressed in rectangular form is
- ☐ *a* $4 + j3$
- ☐ *b* $4 - j3$
- ☐ *c* $-4 + j3$
- ☐ *d* $-4 - j3$

9.26 When $2 + j3$ is added to $4 - j$ the answer in polar form is
- ☐ *a* $6.32\angle 18.43^\circ$
- ☐ *b* $7.11\angle 25.18^\circ$
- ☐ *c* $8.46\angle 60^\circ$
- ☐ *d* $9.22\angle 65.4^\circ$

9.27 When $4 - j$ is added to $6 + j4$ the answer in polar form is
- ☐ *a* $1.24\angle 45^\circ$
- ☐ *b* $3.31\angle 63.2^\circ$
- ☐ *c* $7.44\angle 12.2^\circ$
- ☐ *d* $10.44\angle 16.7^\circ$

9.28 When $-2 - j4$ is added to $-3 - j9$ the answer in polar form is
- ❑ *a* $14.6\angle -20^\circ$
- ❑ *b* $15.5\angle 101^\circ$
- ❑ *c* $13.9\angle -111^\circ$
- ❑ *d* $21.2\angle 120^\circ$

9.29 When $40 - j60$ is added to $25 - j50$ the answer in polar form is
- ❑ *a* $104.63\angle -30^\circ$
- ❑ *b* $127.77\angle -59.42^\circ$
- ❑ *c* $130.94\angle 60^\circ$
- ❑ *d* $145.22\angle 45^\circ$

9.30 When $20 + j30$ is added to $40 + j60$ the answer in polar form is
- ❑ *a* $108.17\angle 56.3^\circ$
- ❑ *b* $62.33\angle 46.2^\circ$
- ❑ *c* $70.61\angle 27^\circ$
- ❑ *d* $81.12\angle 33.2^\circ$

9.31 When $12 - j45$ is added to $18 + j60$ the answer in polar form is
- ❑ *a* $12.68\angle 60.3^\circ$
- ❑ *b* $15.672\angle 43^\circ$
- ❑ *c* $33.54\angle 26.57^\circ$
- ❑ *d* $25.3\angle 30^\circ$

9.32 When $4.5 - j30$ is added to $5 - j60$ the answer in polar form is
- ❑ *a* $90.5\angle -83.97^\circ$
- ❑ *b* $62.38\angle 27.1^\circ$
- ❑ *c* $75.66\angle -18.7^\circ$
- ❑ *d* $82.34\angle 16^\circ$

9.33 When $8 + j60$ is added to $7 - j25$ the answer in polar form is
- ❑ *a* $6\angle -32^\circ$
- ❑ *b* $38.08\angle 66.8^\circ$
- ❑ *c* $18\angle 42^\circ$
- ❑ *d* $27\angle -20^\circ$

9.34 When $6 + j4$ is subtracted from $4 - j2$ the answer in polar form is
- ❑ *a* $62.3\angle -110^\circ$
- ❑ *b* $53\angle 120.68^\circ$
- ❑ *c* $18.54\angle 107^\circ$
- ❑ *d* $6.32\angle -108.44^\circ$

9.35 When $4 + j2$ is subtracted from $6 + j3$ the answer in polar form is
- ❑ *a* $1.23\angle 68.28^\circ$
- ❑ *b* $2.24\angle 26.57^\circ$
- ❑ *c* $4.44\angle 82.87^\circ$
- ❑ *d* $5.15\angle -32.2^\circ$

9.36 When $3 + j5$ is subtracted from $4 - j2$ the answer in polar form is
- ❑ *a* $5.41\angle 35.51^\circ$
- ❑ *b* $6.32\angle -61.48^\circ$
- ❑ *c* $7.07\angle -81.87^\circ$
- ❑ *d* $8.08\angle 62.84^\circ$

9.37 When $10\angle 30^\circ$ is subtracted from $10\angle 60^\circ$ the answer is
- ❑ *a* $5.17\angle 135^\circ$
- ❑ *b* $6.81\angle 120^\circ$
- ❑ *c* $7.66\angle 45^\circ$
- ❑ *d* $9.92\angle 120^\circ$

9.38 When $2 + j3$ is multiplied by $4 - j$ the answer is
- ❑ *a* $6 + j6$
- ❑ *b* $3 - j2$
- ❑ *c* $6 - j4$
- ❑ *d* $11 + j10$

9.39 When $6 + j6$ is multiplied by $5 + j5$ the answer is
- ❑ *a* $j30$
- ❑ *b* $j60$
- ❑ *c* $6 + j30$
- ❑ *d* $6 + j60$

9.40 When $2 + j9$ is divided by $5 - j2$ the answer is
- ❑ *a* $-0.15 + j4.2$
- ❑ *b* $0.81 + j6.3$
- ❑ *c* $0.96 + j1.73$
- ❑ *d* $-0.276 + j1.69$

9.41 When $1 - j3$ is divided by $7 - j5$ the answer is
- ❑ *a* $0.297 - j0.216$
- ❑ *b* $0.622 + j0.8$
- ❑ *c* $1.6 + j1.8$
- ❑ *d* $2.42 + j1.89$

9.42 The diagram shows an ideal transformer. If the primary voltage V_p is 36 V, the number of turns on the primary N_p is 240, and the number of turns on the secondary N_s is 2400, the output voltage from the secondary V_s is

❑ *a* 3.6 V
❑ *b* 36 V
❑ *c* 360 V
❑ *d* 240 V

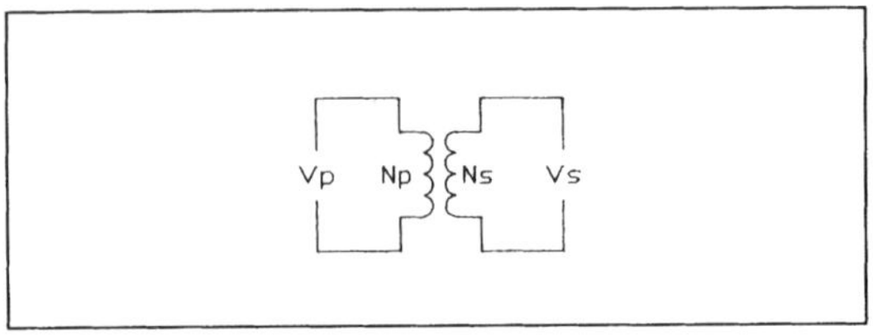

9.43 If in Question 9.42 $V_p = 2.5$ V, $N_p = 45$ and $N_s = 22.5$, the value of V_s is

❑ *a* 1.25 V
❑ *b* 6.25 V
❑ *c* 8.25 V
❑ *d* 9.25 V

9.44 If in Question 9.42 $V_p = 1600$ V, $N_p = 2000$ and $N_s = 80$, the value of V_s is

❑ *a* 640 V
❑ *b* 320 V
❑ *c* 32 V
❑ *d* 64 V

9.45 If in Question 9.42 $V_s = 72$ mV, $N_p = 145$ and $N_s = 290$, the input voltage V_p is

❑ *a* 14 mV
❑ *b* 36 mV
❑ *c* 100 mV
❑ *d* 144 mV

9.46 If in Question 9.42 $V_s = 50$ mV, $N_p = 2500$ and $N_s = 125$, the input voltage V_p is

❑ *a* 0.1 V
❑ *b* 1.0 V
❑ *c* 2.0 V
❑ *d* 10.0 V

9.47 If in Question 9.42 $V_s = 400$ V, $V_p = 20$ V and $N_s = 300$, the number of primary turns N_p is

❑ *a* 10
❑ *b* 300
❑ *c* 15
❑ *d* 6000

9.48 If in Question 9.42 $V_s = 12$, $V_p = 240$ and $N_p = 300$, the number of secondary turns N_s is

❑ *a* 15
❑ *b* 150
❑ *c* 30
❑ *d* 300

9.49 The turns ratio of a transformer that transforms 1500 V to 30 V is

❑ *a* 25 : 1
❑ *b* 1 : 25
❑ *c* 50 : 1
❑ *d* 1 : 50

9.50 The diagram shows an ideal transformer with a resistive load. If the primary has 20 turns, the number of turns on the secondary is

❑ *a* 2
❑ *b* 20
❑ *c* 200
❑ *d* 2000

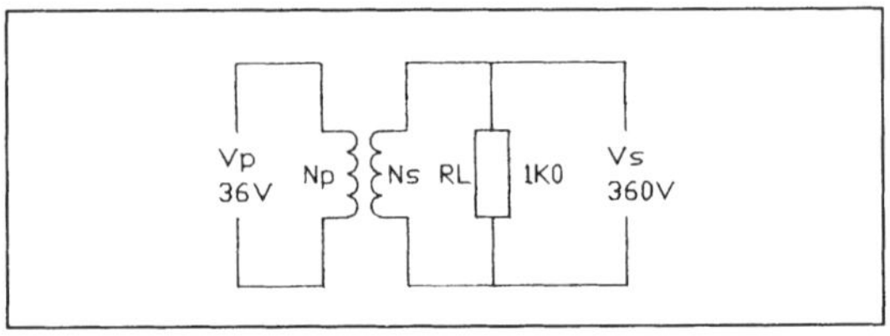

9.51 In Question 9.50 the reflected resistance is

❑ *a* 10 Ω
❑ *b* 20 Ω
❑ *c* 100 Ω
❑ *d* 2000 Ω

9.52 In Question 9.50 the primary current is
- ❑ *a* 1.8 A
- ❑ *b* 3.6 A
- ❑ *c* 7.2 A
- ❑ *d* 8.4 A

9.53 In Question 9.50 the secondary current is
- ❑ *a* 0.12 A
- ❑ *b* 0.24 A
- ❑ *c* 0.36 A
- ❑ *d* 6.4 A

9.54 If the circuit in Question 9.50 had the parameters $V_p = 240$ V, $V_s = 12$ V, $N_p = 360$ and $R_L = 50\ \Omega$, the number of secondary turns is
- ❑ *a* 9
- ❑ *b* 90
- ❑ *c* 180
- ❑ *d* 18

9.55 In Question 9.54 the reflected resistance is
- ❑ *a* 10 kΩ
- ❑ *b* 20 kΩ
- ❑ *c* 100 Ω
- ❑ *d* 200 Ω

9.56 In Question 9.54 the primary current is
- ❑ *a* 12 mA
- ❑ *b* 120 mA
- ❑ *c* 1.2 A
- ❑ *d* 12 A

9.57 In Question 9.54 the secondary current is
- ❑ *a* 120 mA
- ❑ *b* 240 mA
- ❑ *c* 480 mA
- ❑ *d* 960 mA

9.58 The diagram shows a transformer with a source resistance R_s and a load resistance R_L. The turns ratio required to match the load to the source is
- ❑ *a* 2
- ❑ *b* 4
- ❑ *c* 8
- ❑ *d* 20

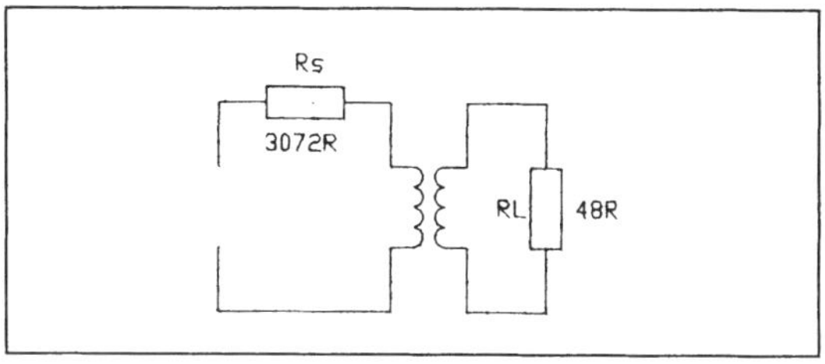

9.59 If in Question 9.58 the source resistance is 500 Ω and the load resistance 5 Ω, the turns ratio required for matching is
- ❑ *a* 10
- ❑ *b* 12
- ❑ *c* 14
- ❑ *d* 24

9.60 A transformer with a source resistance of 200 Ω is matched to a 20-Ω load. If the number of secondary turns is 400, the number of primary turns is
- ❑ *a* 750
- ❑ *b* 850
- ❑ *c* 1100
- ❑ *d* 1265

9.61 In the diagram the primary current is
- ❑ *a* 1.28 A
- ❑ *b* 1.94 A
- ❑ *c* 2.76 A
- ❑ *d* 4.43 A

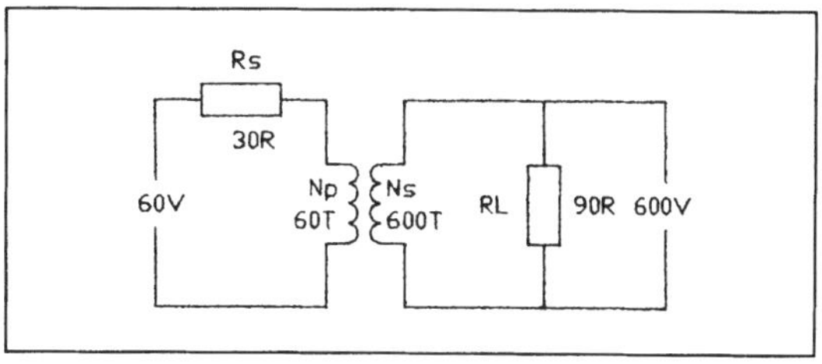

9.62 In Question 9.61 the secondary current is
- ❑ *a* 0.194 A
- ❑ *b* 1.26 A
- ❑ *c* 2.831 A
- ❑ *d* 3.32 A

9.63 In Question 9.61 the power dissipated in R_s is
- ❑ *a* 88 W
- ❑ *b* 100 W
- ❑ *c* 106.2 W
- ❑ *d* 113.1 W

9.64 In Question 9.61 the power dissipated in R_L is
- ❑ *a* 1.2 W
- ❑ *b* 2.7 W
- ❑ *c* 3.4 W
- ❑ *d* 3.9 W

9.65 A 400-kVA, 8000-V/200-V transformer has 100 turns on the secondary winding. Assuming an ideal transformer, the maximum permissible primary current is
- ❑ *a* 10 A
- ❑ *b* 40 A
- ❑ *c* 50 A
- ❑ *d* 80 A

9.66 In Question 9.65 the maximum permissible secondary current is
- ❑ *a* 1000 A
- ❑ *b* 2000 A
- ❑ *c* 1500 A
- ❑ *d* 200 A

9.67 In Question 9.65 the number of primary turns is
- ❑ *a* 500
- ❑ *b* 1000
- ❑ *c* 2000
- ❑ *d* 4000

9.68 A transformer is connected to a 140-V (100-Hz) supply and the output voltage is 20 V. The transformer core has a cross-sectional area of 5 cm^2 and the maximum flux density is 0.1 T. The number of primary turns is
- ❑ *a* 2800
- ❑ *b* 3800
- ❑ *c* 4700
- ❑ *d* 6300

9.69 In Question 9.68 the number of secondary turns is
- ❑ *a* 900
- ❑ *b* 400
- ❑ *c* 300
- ❑ *d* 100

9.70 If in Question 9.68 the parameters are input voltage = 100 V (50 Hz), output voltage = 500 V, area = 12 cm^2 and flux density = 0.4 T, the number of primary turns is
- ❑ *a* 640
- ❑ *b* 938
- ❑ *c* 1120
- ❑ *d* 1438

9.71 In Question 9.68 the number of secondary turns is
- ❑ *a* 2481
- ❑ *b* 3122
- ❑ *c* 4690
- ❑ *d* 6300

9.72 If in Question 9.68 the parameters are input voltage = 240 V (150 Hz), output voltage = 24 V, area = 8 cm^2 and flux density = 0.04 T, the number of primary turns is
- ❑ *a* 10 000
- ❑ *b* 10 400
- ❑ *c* 11 261
- ❑ *d* 14 623

9.73 In Question 9.72 the number of secondary turns is
- ❑ *a* 820
- ❑ *b* 940
- ❑ *c* 1000
- ❑ *d* 1126

9.74 A transformer has an input voltage of 15 V (200 Hz) and the output voltage is 750 V. The area of the core is 6 cm^2 and the flux density is 0.3 T. The number of secondary turns is

- ❑ *a* 3822
- ❑ *b* 4692
- ❑ *c* 5500
- ❑ *d* 6700

9.75 A transformer has a 200-turn primary winding. If 240 V is applied to the winding and the maximum flux is 5 mWb, the frequency of the applied voltage is

- ❑ *a* 54 Hz
- ❑ *b* 78 Hz
- ❑ *c* 84 Hz
- ❑ *d* 91 Hz

10 DC supplies, batteries and battery chargers

10.1 The diagram shows a battery with an internal resistance *r* and a load resistance *R*. The terminal voltage between A and B is

❑ *a* 8.5 V
❑ *b* 9.2 V
❑ *c* 9.5 V
❑ *d* 9.95 V

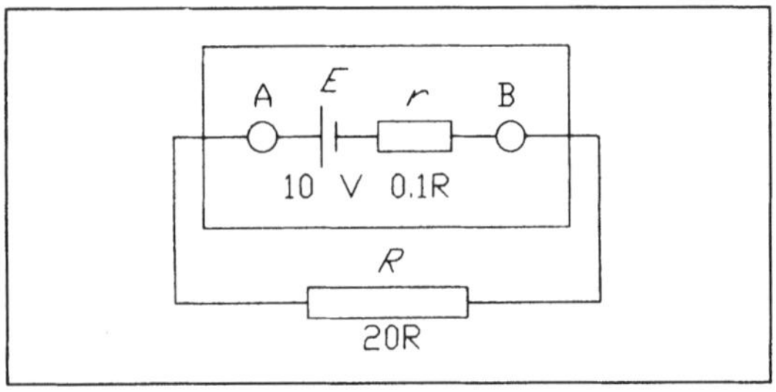

10.2 If in Question 10.1 $E = 12$ V, $r = 0.2\ \Omega$ and $R = 36\ \Omega$, the terminal voltage is

❑ *a* 11.05 V
❑ *b* 11.93 V
❑ *c* 11.96 V
❑ *d* 11.99 V

10.3 If in Question 10.1 $E = 100$ V, $r = 2\ \Omega$ and $R = 200\ \Omega$, the terminal voltage is

❑ *a* 90 V
❑ *b* 90.5 V
❑ *c* 99 V
❑ *d* 99.5 V

10.4 The diagram shows a battery similar to the one in Question 10.1. The value of the internal resistance *r* is

❑ *a* 0.12 Ω
❑ *b* 0.66 Ω
❑ *c* 0.88 Ω
❑ *d* 0.94 Ω

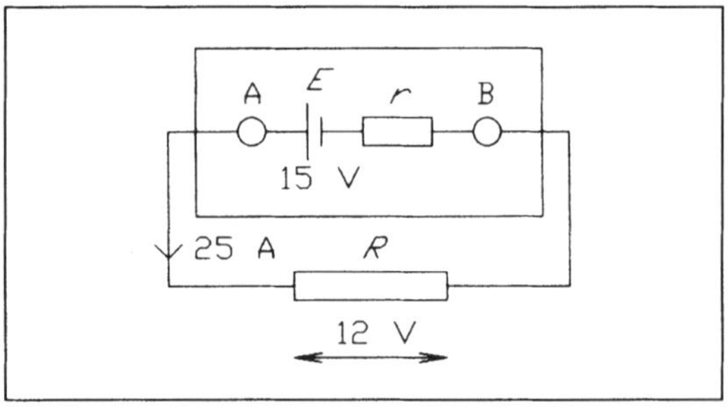

10.5 If in Question 10.4 $E = 240$ V and the current and the terminal voltage respectively are 10 A and 220 V, the internal resistance is

❑ *a* 0.5 Ω
❑ *b* 1 Ω
❑ *c* 2 Ω
❑ *d* 3 Ω

10.6 If in Question 10.4 $E = 120$ V and the current and the terminal voltage respectively are 30 A and 105 V, the internal resistance is

❑ *a* 0.5 Ω
❑ *b* 0.8 Ω
❑ *c* 0.9 Ω
❑ *d* 1.0 Ω

10.7 If a battery with an internal resistance of 0.4 Ω has a terminal voltage of 16 V when delivering 4 A, the EMF is

- ❑ *a* 17.4 V
- ❑ *b* 17.6 V
- ❑ *c* 18.2 V
- ❑ *d* 18.4 V

10.8 If in Question 10.7 the internal resistance is 1.2 Ω, the terminal voltage 48 V and the current 12 A, the EMF is

- ❑ *a* 53.7 V
- ❑ *b* 56.5 V
- ❑ *c* 60.8 V
- ❑ *d* 62.4 V

10.9 If in Question 10.7 the internal resistance is 0.8 Ω, the terminal voltage 36 V and the current 8 A, the EMF is

- ❑ *a* 40.3 V
- ❑ *b* 42.4 V
- ❑ *c* 44.1 V
- ❑ *d* 46.2 V

10.10 Measurements of the terminal voltages and the currents drawn, during an experiment on a car battery, gave the following data:

V	*I* (A)
8	6
6	10

The internal resistance of the battery is

- ❑ *a* 0.2 Ω
- ❑ *b* 0.4 Ω
- ❑ *c* 0.5 Ω
- ❑ *d* 0.9 Ω

10.11 The battery EMF in Question 10.10 is

- ❑ *a* 9.5 V
- ❑ *b* 10 V
- ❑ *c* 10.5 V
- ❑ *d* 11 V

10.12 An experiment similar to the one in Question 10.10 gave the following data:

V	*I* (A)
96	24
92	28

The internal resistance of the battery is

- ❑ *a* 1.0 Ω
- ❑ *b* 2.5 Ω
- ❑ *c* 2.6 Ω
- ❑ *d* 3.0 Ω

10.13 The battery EMF in Question 10.12 is

- ❑ *a* 100 V
- ❑ *b* 120 V
- ❑ *c* 125 V
- ❑ *d* 130 V

10.14 The diagram shows a battery with four cells. The terminal voltage is

- ❑ *a* 3.7 V
- ❑ *b* 3.8 V
- ❑ *c* 4.0 V
- ❑ *d* 4.2 V

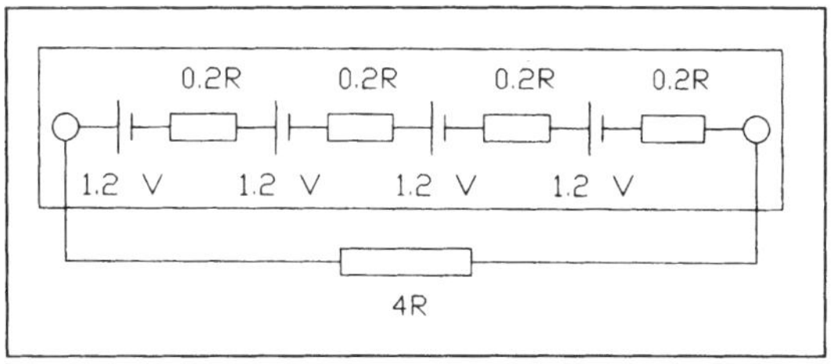

10.15 If a battery similar to the one in Question 10.14 has six cells, each with an EMF of 2.4 V and internal resistance of 0.04 Ω, connected to a 12-Ω load, the terminal voltage is

- ❑ *a* 12.06 V
- ❑ *b* 12.12 V
- ❑ *c* 13.64 V
- ❑ *d* 14.12 V

10.16 If a battery similar to the one in Question 10.14 has five cells, each with an EMF of 2 V and internal resistance of 0.1 Ω, connected to a 16-Ω load, the terminal voltage is

- ❑ *a* 9.6 V
- ❑ *b* 9.7 V
- ❑ *c* 9.8 V
- ❑ *d* 9.9 V

10.17 The diagram shows three cells connected in parallel to a load R_L. The terminal voltage is

- ❑ *a* 10.55 V
- ❑ *b* 10.98 V
- ❑ *c* 11.25 V
- ❑ *d* 11.85 V

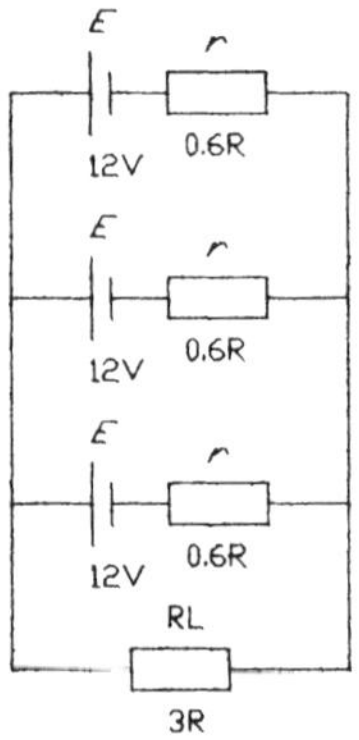

10.18 If in Question 10.17 there are four 6-V cells, each with a resistance *r* of 0.4 Ω, connected to a 10-Ω load, the terminal voltage is

- ❑ *a* 4.62 V
- ❑ *b* 5.94 V
- ❑ *c* 4.87 V
- ❑ *d* 5.26 V

10.19 If in Question 10.17 there are five 50-V cells, each with an internal resistance *r* of 0.5 Ω, connected to a 10-Ω load, the terminal voltage is

- ❑ *a* 40.5 V
- ❑ *b* 43.5 V
- ❑ *c* 45.5 V
- ❑ *d* 49.5 V

10.20 The diagram shows a series-parallel arrangement of battery cells. The current *I* is

- ❑ *a* 2.67 A
- ❑ *b* 3.33 A
- ❑ *c* 3.74 A
- ❑ *d* 4.33 A

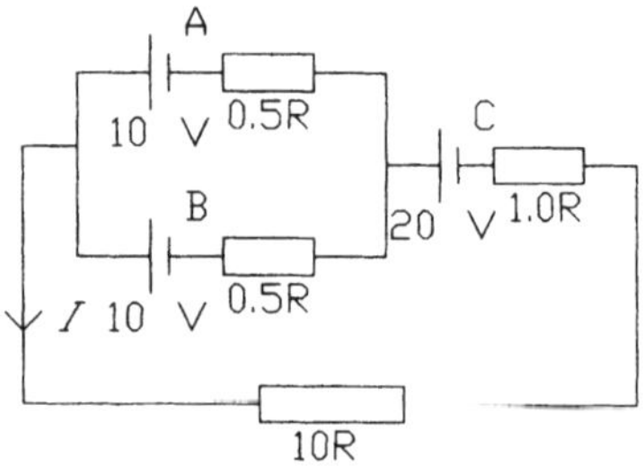

10.21 In Question 10.20 the terminal voltage of batteries A and B is

- ❑ *a* 7.82 V
- ❑ *b* 8.86 V
- ❑ *c* 9.33 V
- ❑ *d* 9.93 V

10.22 In Question 10.20 the terminal voltage of battery C is

- ❑ *a* 15.98 V
- ❑ *b* 16.22 V
- ❑ *c* 16.51 V
- ❑ *d* 17.33 V

10.23 The diagram shows a series-parallel arrangement of battery cells. The current *I* is

- ❑ *a* 4.41 A
- ❑ *b* 5.63 A
- ❑ *c* 6.82 A
- ❑ *d* 8.44 A

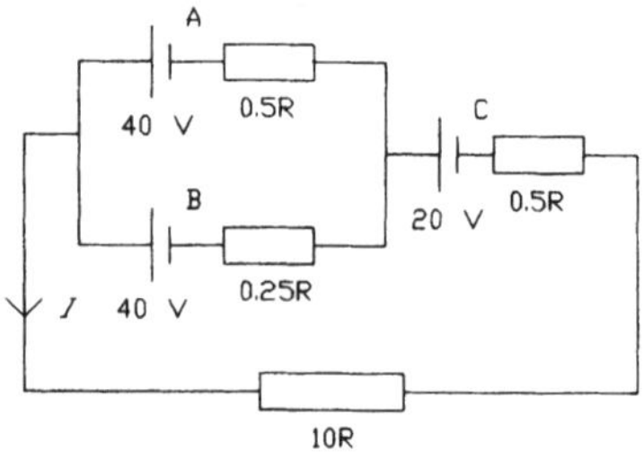

10.24 In Question 10.23 the terminal voltage of batteries A and B is

❑ *a* 30.16 V
❑ *b* 33.28 V
❑ *c* 35.16 V
❑ *d* 39.06 V

10.25 In Question 10.23 the terminal voltage of battery C is

❑ *a* 16.12 V
❑ *b* 17.19 V
❑ *c* 19.12 V
❑ *d* 19.87 V

10.26 The diagram shows a battery charger recharging a secondary cell. At switch-on, the initial charging current is

❑ *a* 10 A
❑ *b* 11 A
❑ *c* 12 A
❑ *d* 13 A

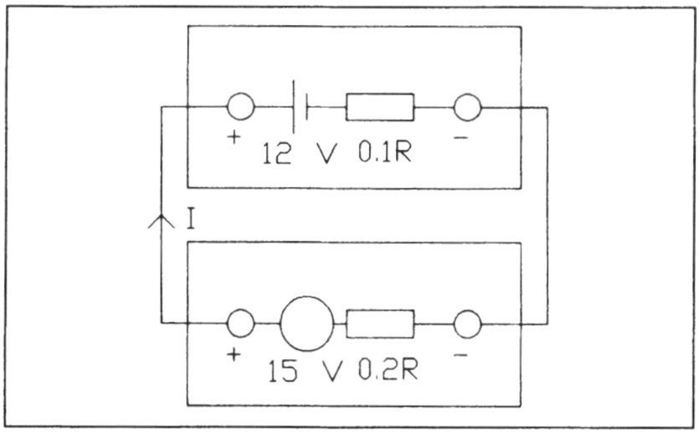

10.27 In Question 10.26 the charging current when the battery voltage has risen to 13.5 V is

❑ *a* 3 A
❑ *b* 4 A
❑ *c* 5 A
❑ *d* 6 A

10.28 If in Question 10.26 a 30 V charger is used with an internal resistance of 0.15 Ω and the battery has an internal resistance of 0.3 Ω and a voltage of 24 V, the initial charging current is

❑ *a* 10.67 A
❑ *b* 13.33 A
❑ *c* 15.67 A
❑ *d* 17.33 A

10.29 In Question 10.28 the charging current when the battery voltage has risen to 27 V is

❑ *a* 4.96 A
❑ *b* 5.16 A
❑ *c* 5.55 A
❑ *d* 6.67 A

10.30 A 24-V battery with an internal resistance of 0.3 Ω is connected to a 30-V charger with an internal resistance of 0.15 Ω. It is a requirement to limit the initial charging current to 10 A by including a series resistor in the charger. The value of this resistor is

❑ *a* 0.06 Ω
❑ *b* 0.15 Ω
❑ *c* 0.85 Ω
❑ *d* 1.25 Ω

10.31 In Question 10.30 using the series resistance the value of the charging current when the battery voltage has risen to 28 V is

❑ *a* 2.16 A
❑ *b* 2.42 A
❑ *c* 2.67 A
❑ *d* 3.33 A

10.32 A battery consists of 50 cells, each in series with an EMF of 2.5 V when fully charged. The internal resistance of each cell is 0.01 Ω. If the voltage of each cell has fallen to 2.0 V, the charger voltage needed to fully charge the battery is

❑ *a* 115 V
❑ *b* 125 V
❑ *c* 130 V
❑ *d* 135 V

10.33 In Question 10.32, given that the internal resistance of the charger is 1.0 Ω, the initial charging current is

❑ *a* 13.33 A
❑ *b* 14.85 A
❑ *c* 16.67 A
❑ *d* 20.45 A

10.34 In Question 10.32 the charging current when the EMF of each cell has risen to 2.4 V is
- ❑ *a* 3.3 A
- ❑ *b* 4.6 A
- ❑ *c* 6.1 A
- ❑ *d* 6.5 A

10.35 A battery consists of 50 1.5-V cells, each with a resistance of 0.01 Ω. The voltage of each cell has fallen to 1.25 V. It is a requirement to charge the battery at a constant current of 2 A from a charger with an internal resistance of 1.5 Ω. The applied voltage at the commencement of charge is
- ❑ *a* 60.5 V
- ❑ *b* 63.8 V
- ❑ *c* 66.5 V
- ❑ *d* 70.4 V

10.36 In Question 10.35 the applied voltage at the end of charge is
- ❑ *a* 70 V
- ❑ *b* 71 V
- ❑ *c* 74 V
- ❑ *d* 79 V

10.37 If in Question 10.35 the battery consists of 44 2-V cells, each with an internal resistance of 0.12 Ω, and the voltage of each cell has fallen to 1.8 V, and it is a requirement to charge the battery at a constant current of 4 A from a charger having an internal resistance of 2 Ω, the applied voltage at the commencement of charge is
- ❑ *a* 108.32 V
- ❑ *b* 120.67 V
- ❑ *c* 130.33 V
- ❑ *d* 135.15 V

10.38 In Question 10.37 the applied voltage at the end of charge is
- ❑ *a* 101.69 V
- ❑ *b* 103.18 V
- ❑ *c* 105.16 V
- ❑ *d* 117.12 V

10.39 A battery is charged with a constant current of 14 A for 10 hours. It is discharged at the rate of 12 A for 10 hours. The ampere-hour efficiency is
- ❑ *a* 82.3%
- ❑ *b* 85.7%
- ❑ *c* 90.1%
- ❑ *d* 92.6%

10.40 If in Question 10.39 the charging current is 18 A over 20 hours and the discharge current is 12 A over 19 hours, the ampere-hour efficiency is
- ❑ *a* 60.13%
- ❑ *b* 61.67%
- ❑ *c* 63.33%
- ❑ *d* 70.13%

10.41 A battery is fully charged by a constant current of 5 A flowing for 15 hours. The average charging voltage is 1.6 V. It is discharged at a constant current of 4 A flowing for 20 hours. If the average terminal voltage during discharge is 1.4 V, the watt-hour efficiency is
- ❑ *a* 79.68%
- ❑ *b* 82.44%
- ❑ *c* 87.67%
- ❑ *d* 93.33%

10.42 The diagram shows a half-wave rectifier circuit connected to a 240-V mains supply. If the peak voltage output from the rectifier is 50 V, the peak current is
- ❑ *a* 0.1 A
- ❑ *b* 0.5 A
- ❑ *c* 0.6 A
- ❑ *d* 0.8 A

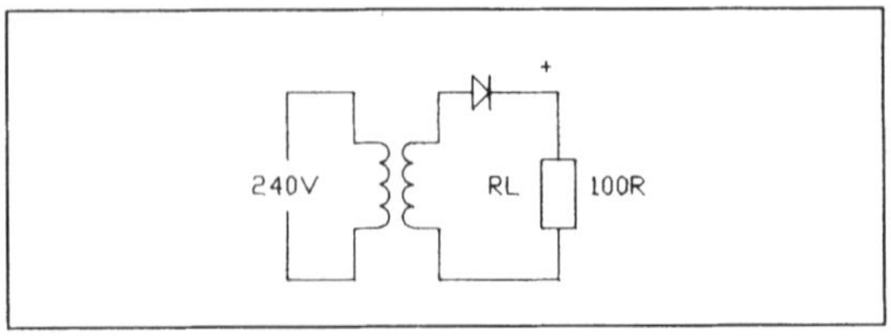

10.43 In Question 10.42 the DC current is approximately
- ❑ *a* 159 mA
- ❑ *b* 172 mA
- ❑ *c* 188 mA
- ❑ *d* 200 mA

10.44 In Question 10.42 the r.m.s. current is
- ❑ *a* 170 mA
- ❑ *b* 200 mA
- ❑ *c* 250 mA
- ❑ *d* 260 mA

10.45 In Question 10.42 the DC voltage across R_L is approximately
- ❑ *a* 15.9 V
- ❑ *b* 16.8 V
- ❑ *c* 17.3 V
- ❑ *d* 18.6 V

10.46 In Question 10.42 the r.m.s. voltage across R_L is
- ❑ *a* 18 V
- ❑ *b* 20 V
- ❑ *c* 22 V
- ❑ *d* 25 V

10.47 If in a circuit identical to the one in Question 10.42 the peak voltage is 200 V and R_L is 600 Ω, the peak current is
- ❑ *a* 293 mA
- ❑ *b* 333 mA
- ❑ *c* 400 mA
- ❑ *d* 500 mA

10.48 In Question 10.47 the DC current is
- ❑ *a* 94 mA
- ❑ *b* 98 mA
- ❑ *c* 106 mA
- ❑ *d* 121 mA

10.49 In Question 10.47 the r.m.s. current is
- ❑ *a* 167 mA
- ❑ *b* 183 mA
- ❑ *c* 192 mA
- ❑ *d* 220 mA

10.50 In Question 10.47 the DC voltage across R_L is approximately
- ❑ *a* 50.14 V
- ❑ *b* 63.66 V
- ❑ *c* 69.84 V
- ❑ *d* 75.33 V

10.51 In Question 10.47 the r.m.s. voltage across R_L is
- ❑ *a* 70 V
- ❑ *b* 85 V
- ❑ *c* 90 V
- ❑ *d* 100 V

10.52 The diagram shows a full-wave rectifier circuit connected to a 240-V mains supply. If the peak output voltage from the rectifier is 50 V, the peak current is
- ❑ *a* 0.5 A
- ❑ *b* 1.0 A
- ❑ *c* 1.5 A
- ❑ *d* 2.0 A

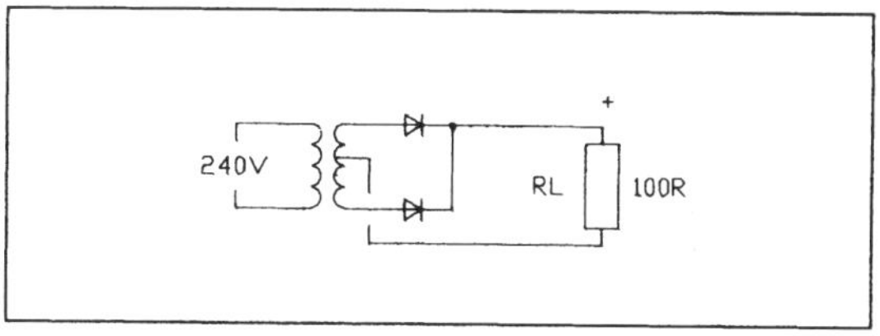

10.53 In Question 10.52 the DC current is approximately
- ❑ *a* 210 mA
- ❑ *b* 260 mA
- ❑ *c* 318 mA
- ❑ *d* 350 mA

10.54 In Question 10.52 the r.m.s. current is approximately
- ❑ *a* 300 mA
- ❑ *b* 354 mA
- ❑ *c* 386 mA
- ❑ *d* 400 mA

10.55 In Question 10.52 the DC voltage across R_L is
- ❑ *a* 31.83 V
- ❑ *b* 35.63 V
- ❑ *c* 39.12 V
- ❑ *d* 40.16 V

10.56 In Question 10.52 the r.m.s. voltage across R_L is

- ❑ *a* 30.22 V
- ❑ *b* 35.36 V
- ❑ *c* 38.12 V
- ❑ *d* 40.61 V

10.57 If in Question 10.52 the peak voltage is 64 V and R_L is 320 Ω, the DC current is approximately

- ❑ *a* 105 mA
- ❑ *b* 110 mA
- ❑ *c* 127 mA
- ❑ *d* 160 mA

10.58 In Question 10.57 the r.m.s. voltage across R_L is approximately

- ❑ *a* 37.64 V
- ❑ *b* 38.34 V
- ❑ *c* 40.15 V
- ❑ *d* 45.25 V

10.59 The diagram shows a voltage regulator. The value of the series resistor R_s is approximately

- ❑ *a* 8.61 Ω
- ❑ *b* 9.52 Ω
- ❑ *c* 9.78 Ω
- ❑ *d* 9.93 Ω

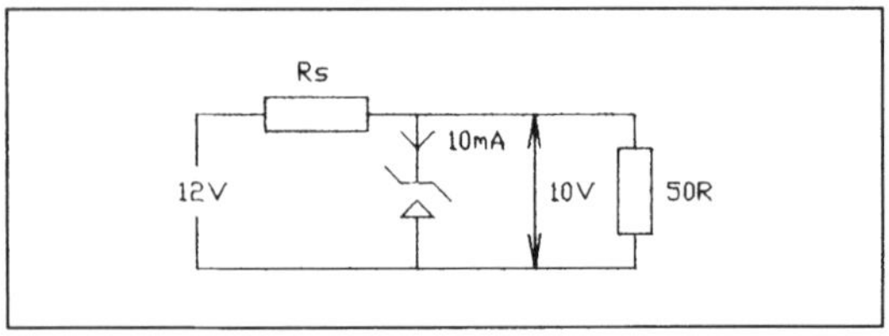

10.60 In Question 10.59 the power dissipated in R_s is

- ❑ *a* 0.24 W
- ❑ *b* 0.33 W
- ❑ *c* 0.42 W
- ❑ *d* 0.81 W

10.61 If the parameters of the circuit in Question 10.59 are input voltage = 240 V, output voltage = 200 V, load resistance = 1200 Ω and the zener diode current = 15 mA, the value of the series resistor R_s is

- ❑ *a* 220 Ω
- ❑ *b* 270 Ω
- ❑ *c* 330 Ω
- ❑ *d* 390 Ω

10.62 In Question 10.61 the power dissipated in the series resistor is approximately

- ❑ *a* 6.31 W
- ❑ *b* 6.82 W
- ❑ *c* 7.27 W
- ❑ *d* 7.94 W

10.63 If the load resistance in Question 10.61 is doubled and the diode current is 20 mA, the value of the series resistor R_s is

- ❑ *a* 231 Ω
- ❑ *b* 246 Ω
- ❑ *c* 260 Ω
- ❑ *d* 387 Ω

10.64 In Question 10.63 the power dissipated in R_s is approximately

- ❑ *a* 3.91 W
- ❑ *b* 4.13 W
- ❑ *c* 4.67 W
- ❑ *d* 5.12 W

10.65 The diagram shows a voltage regulator. The diode current I_z is

- ❑ *a* 61.67 mA
- ❑ *b* 14.33 mA
- ❑ *c* 58.62 mA
- ❑ *d* 70.31 mA

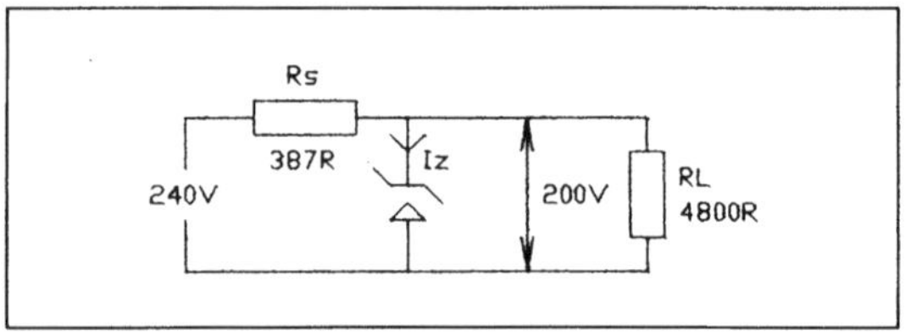

10.66 If the circuit in Question 10.65 has input voltage = 28 V, output voltage = 18 V, R_L = 2400 Ω and R_s = 200 Ω, the diode current is

- ❑ *a* 41.2 mA
- ❑ *b* 42.5 mA
- ❑ *c* 46.8 mA
- ❑ *d* 48.2 mA

10.67 If the circuit in Question 10.65 has an input voltage of 34 V, an output voltage of 30 V, a load resistance of 4700 Ω and a series resistance of 180 Ω, the diode current is approximately

- ❑ *a* 11.21 mA
- ❑ *b* 12.48 mA
- ❑ *c* 14.67 mA
- ❑ *d* 15.84 mA

10.68 A zener diode stabilizing circuit is operated with an input voltage of 24 V and a diode current of 10 mA, to provide 12 V across a load of 1800 Ω. The value of the series resistor is

- ❑ *a* 720 Ω
- ❑ *b* 780 Ω
- ❑ *c* 810 Ω
- ❑ *d* 900 Ω

10.69 If in Question 10.68 the input voltage rises to 25 V, the new diode current is approximately

- ❑ *a* 10.06 mA
- ❑ *b* 11.39 mA
- ❑ *c* 14.48 mA
- ❑ *d* 16.67 mA

10.70 The diagram shows a voltage regulator. The value of R_s is approximately

- ❑ *a* 105.38 Ω
- ❑ *b* 110.14 Ω
- ❑ *c* 115.38 Ω
- ❑ *d* 121.63 Ω

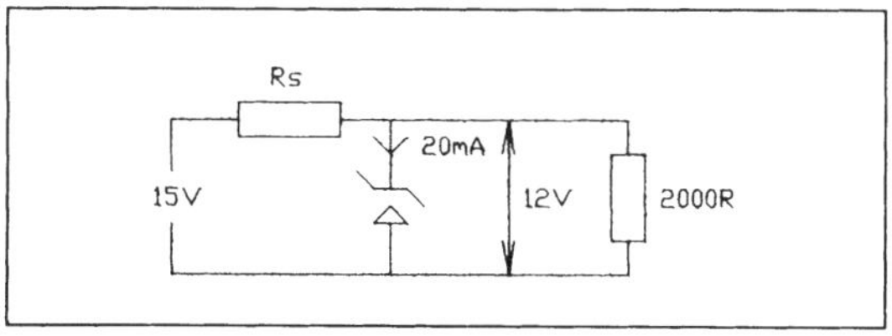

10.71 If the input voltage in Question 10.70 increases to 16 V, the diode current is

- ❑ *a* 19.18 mA
- ❑ *b* 21.63 mA
- ❑ *c* 23.41 mA
- ❑ *d* 28.67 mA

10.72 If the parameters of the circuit in Question 10.70 are input voltage 110 V, output voltage 100 V, load resistance 4000 Ω and diode current 15 mA, the value of R_s is

- ❑ *a* 250 Ω
- ❑ *b* 350 Ω
- ❑ *c* 400 Ω
- ❑ *d* 480 Ω

10.73 If the input voltage in Question 10.72 increases to 120 V, the diode current is

- ❑ *a* 48 mA
- ❑ *b* 51 mA
- ❑ *c* 55 mA
- ❑ *d* 62 mA

10.74 The diagram shows a half-wave rectifier circuit using a 10:1 step-down transformer. If the input voltage is 240 V r.m.s., and assuming no losses, the peak voltage across the load is approximately

- ❑ *a* 28 V
- ❑ *b* 33 V
- ❑ *c* 40 V
- ❑ *d* 45 V

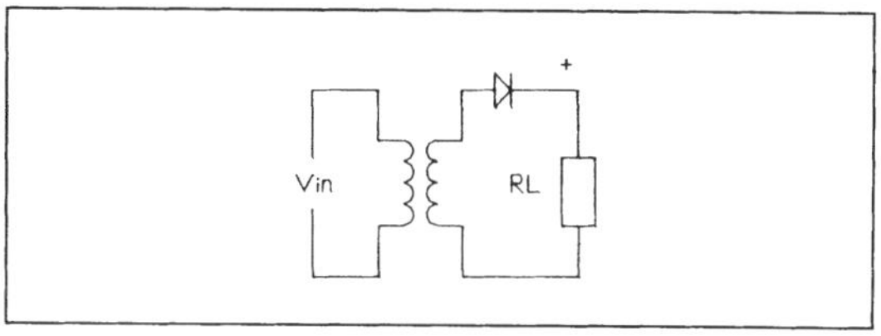

10.75 If in Question 10.74 the input voltage is 415 V r.m.s., the peak voltage across the load is approximately

- ❑ *a* 51.32 V
- ❑ *b* 58.68 V
- ❑ *c* 66.75 V
- ❑ *d* 70.68 V

11 Amplifiers

11.1 The diagram shows the simplest practical n-p-n transistor amplifier. The emitter current I_E is

❑ *a* 10.06 mA
❑ *b* 10.6 mA
❑ *c* 10.0 mA
❑ *d* 16.0 mA

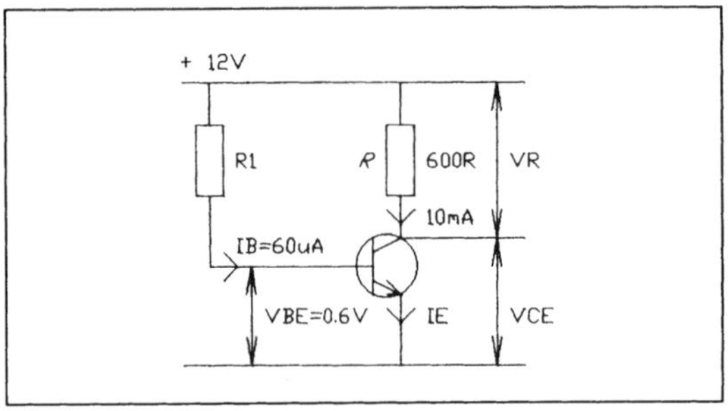

11.2 In Question 11.1 the collector to emitter voltage V_{CE} is

❑ *a* 4 V
❑ *b* 6 V
❑ *c* 8 V
❑ *d* 10 V

11.3 In Question 11.1 the value of the bias resistor R_1 is

❑ *a* 120 kΩ
❑ *b* 190 kΩ
❑ *c* 220 kΩ
❑ *d* 470 kΩ

11.4 The DC input resistance in Question 11.1 is

❑ *a* 4 kΩ
❑ *b* 6 kΩ
❑ *c* 8 kΩ
❑ *d* 10 kΩ

11.5 The DC output resistance in Question 11.1 is

❑ *a* 200 Ω
❑ *b* 400 Ω
❑ *c* 600 Ω
❑ *d* 800 Ω

11.6 The DC current gain in Question 11.1 is

❑ *a* 122
❑ *b* 167
❑ *c* 178
❑ *d* 200

11.7 The diagram shows a transistor amplifier identical to the one in Question 11.1 but with different parameters. The emitter current is

❑ *a* 12.1 mA
❑ *b* 12.11 mA
❑ *c* 12.01 mA
❑ *d* 12.001 mA

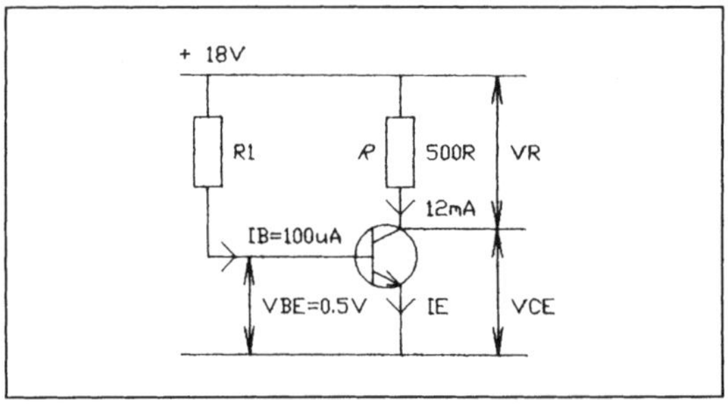

11.8 In Question 11.7 the collector to emitter voltage is

❑ *a* 8 V
❑ *b* 10 V
❑ *c* 11 V
❑ *d* 12 V

11.9 In Question 11.7 the value of the bias resistor is
- ❑ *a* 160 kΩ
- ❑ *b* 175 kΩ
- ❑ *c* 210 kΩ
- ❑ *d* 230 kΩ

11.10 In Question 11.7 the DC input resistance is
- ❑ *a* 3 kΩ
- ❑ *b* 4 kΩ
- ❑ *c* 5 kΩ
- ❑ *d* 6 kΩ

11.11 In Question 11.7 the DC output resistance is
- ❑ *a* 1 kΩ
- ❑ *b* 2 kΩ
- ❑ *c* 3 kΩ
- ❑ *d* 4 kΩ

11.12 In Question 11.7 the DC current gain is
- ❑ *a* 85
- ❑ *b* 90
- ❑ *c* 100
- ❑ *d* 120

11.13 The diagram shows a transistor amplifier with an alternative biasing arrangement. The voltage V_{CE} is
- ❑ *a* 6.95 V
- ❑ *b* 7.88 V
- ❑ *c* 8.64 V
- ❑ *d* 9.38 V

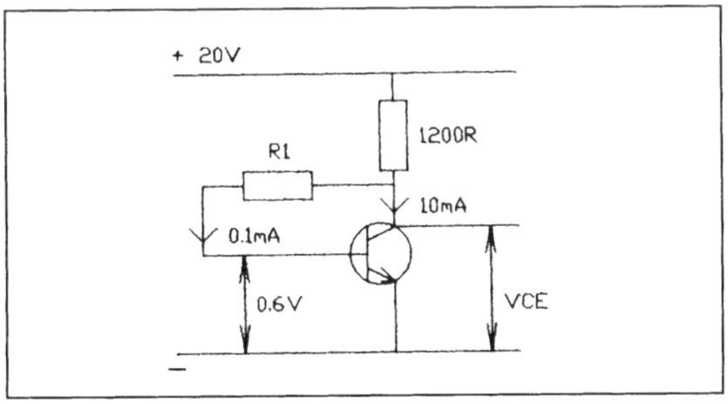

11.14 In Question 11.13 the value of the bias resistor R_1 is
- ❑ *a* 60.6 kΩ
- ❑ *b* 67.5 kΩ
- ❑ *c* 72.8 kΩ
- ❑ *d* 84.1 kΩ

11.15 In Question 11.13 the DC output resistance is
- ❑ *a* 588 Ω
- ❑ *b* 610 Ω
- ❑ *c* 640 Ω
- ❑ *d* 788 Ω

11.16 The diagram shows an amplifier similar to the one in Question 11.13. The voltage V_{CE} is
- ❑ *a* 5.97 V
- ❑ *b* 6.9 V
- ❑ *c* 9.0 V
- ❑ *d* 9.5 V

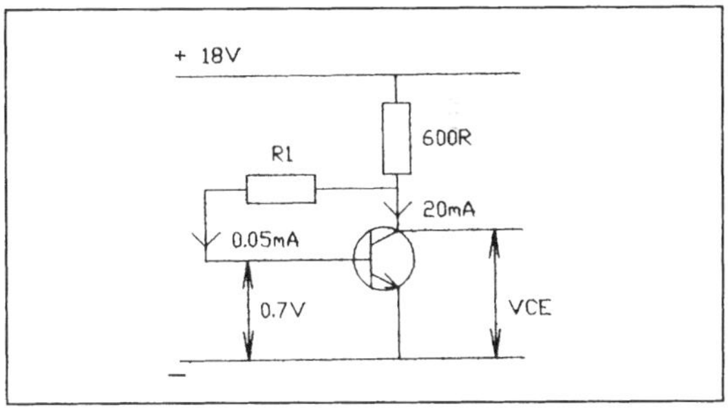

11.17 In Question 11.16 the value of the bias resistor is
- ❑ *a* 105.4 kΩ
- ❑ *b* 120 kΩ
- ❑ *c* 160 kΩ
- ❑ *d* 470 kΩ

11.18 In Question 11.16 the DC output resistance is
- ❑ *a* 200 Ω
- ❑ *b* 298.5 Ω
- ❑ *c* 680 Ω
- ❑ *d* 970 Ω

11.19 The diagram shows another method of biasing an amplifier. If I_1 is equal to $5I_B$ and $I_B = 100\ \mu A$, the value of R_1 is

❑ *a* 16 kΩ
❑ *b* 18 kΩ
❑ *c* 20 kΩ
❑ *d* 25 kΩ

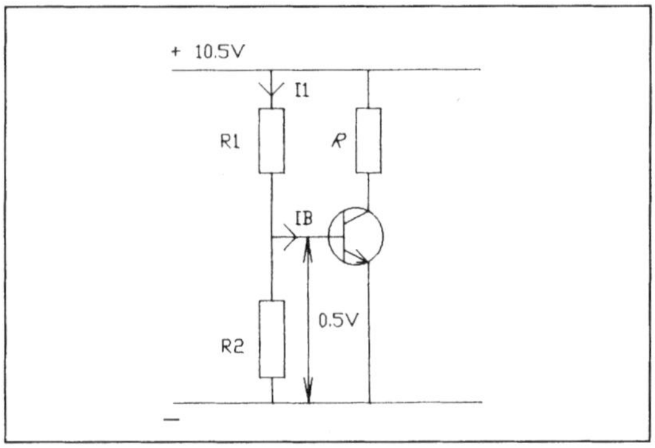

11.20 In Question 11.19 the value of R_2 is

❑ *a* 1250 Ω
❑ *b* 1500 Ω
❑ *c* 1800 Ω
❑ *d* 1900 Ω

11.21 The diagram shows a biasing circuit similar to the one in Question 11.19, where $I_B = 50\ \mu A$, VBE = 0.5 V, $I_1 = 5I_B$ and the applied voltage is 12 V. The value of R_1 is

❑ *a* 21 kΩ
❑ *b* 23 kΩ
❑ *c* 40 kΩ
❑ *d* 46 kΩ

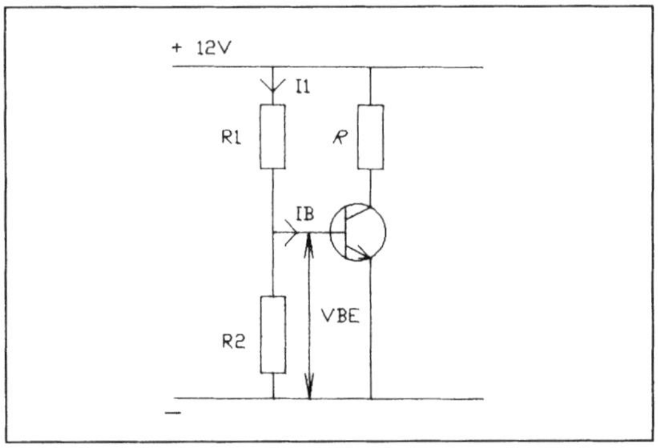

11.22 The value of R_2 in Question 11.21 is

❑ *a* 2.5 kΩ
❑ *b* 4.5 kΩ
❑ *c* 6.2 kΩ
❑ *d* 7.3 kΩ

11.23 The diagram shows a transistor amplifier where $I_C = 8$ mA, $R_L = 800\ \Omega$, $V_{BE} = 0.5$ V and $I_B = 40\ \mu A$. The voltage drop across R_L is

❑ *a* 6.0 V
❑ *b* 6.2 V
❑ *c* 6.4 V
❑ *d* 7.0 V

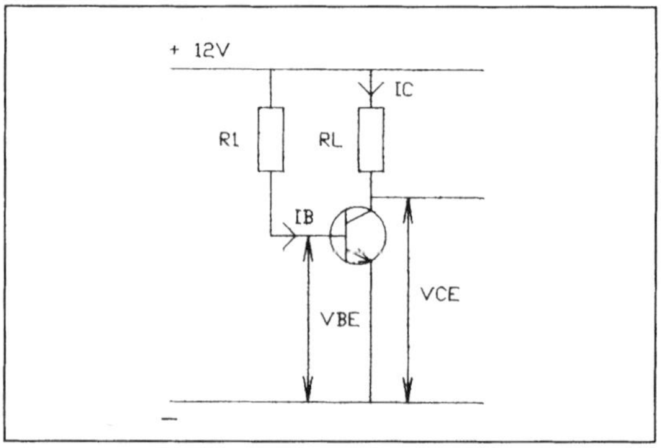

11.24 In Question 11.23 the value of V_{CE} is

❑ *a* 5.6 V
❑ *b* 5.8 V
❑ *c* 6.4 V
❑ *d* 6.5 V

11.25 In Question 11.23 the DC input resistance is

❑ *a* 10.2 kΩ
❑ *b* 12.5 kΩ
❑ *c* 14.3 kΩ
❑ *d* 15.6 kΩ

11.26 In Question 11.23 the DC output resistance is

❑ *a* 460 Ω
❑ *b* 500 Ω
❑ *c* 560 Ω
❑ *d* 700 Ω

11.27 In Question 11.23 the DC current gain is
- ❑ *a* 200
- ❑ *b* 215
- ❑ *c* 285
- ❑ *d* 310

11.28 In Question 11.23 the emitter current is
- ❑ *a* 8.004 mA
- ❑ *b* 8.04 mA
- ❑ *c* 8.4 mA
- ❑ *d* 8.2 mA

11.29 The diagram shows an inverting operational amplifier. If R_1 = 50 Ω and R_2 = 50 Ω, the voltage gain is
- ❑ *a* −1
- ❑ *b* −5
- ❑ *c* −10
- ❑ *d* −50

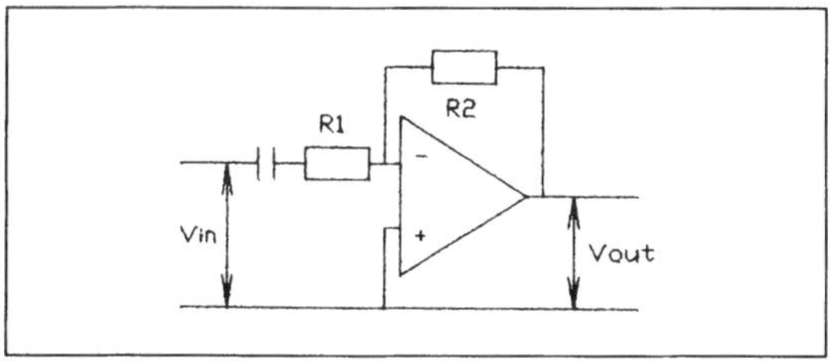

11.30 If in Question 11.29 R_1 = 600 Ω and R_2 = 6000 Ω, the voltage gain is
- ❑ *a* −1
- ❑ *b* −5
- ❑ *c* −10
- ❑ *d* −100

11.31 If in Question 11.29 R_1 = 10 kΩ and R_2 = 20 kΩ, the voltage gain is
- ❑ *a* −0.5
- ❑ *b* −1
- ❑ *c* −2
- ❑ *d* −20

11.32 The diagram shows a non-inverting amplifier. If R_1 = 1500 Ω and R_2 = 15 kΩ, the voltage gain is
- ❑ *a* 8
- ❑ *b* 9
- ❑ *c* 10
- ❑ *d* 11

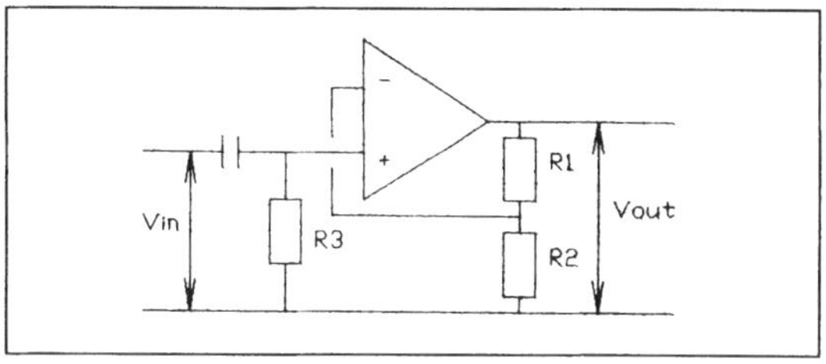

11.33 In Question 11.32, if R_1 = 60 Ω and R_2 = 300 Ω, the voltage gain is
- ❑ *a* 6
- ❑ *b* 10
- ❑ *c* 16
- ❑ *d* 24

11.34 In Question 11.32, if R_1 and R_2 are both 450 Ω, the voltage gain is
- ❑ *a* 1
- ❑ *b* 2
- ❑ *c* 6
- ❑ *d* 8

11.35 A non-inverting amplifier identical to the one shown in Question 11.32 has a gain of 21. If R_1 is 2700 Ω, the value of R_2 is
- ❑ *a* 11 kΩ
- ❑ *b* 27 kΩ
- ❑ *c* 46 kΩ
- ❑ *d* 54 kΩ

11.36 If in Question 11.35 the gain is 11 and R_2 = 47 kΩ, the value of R_1 is
- ❑ *a* 3300 Ω
- ❑ *b* 4700 Ω
- ❑ *c* 6800 Ω
- ❑ *d* 8200 Ω

11.37 If in Question 11.35 the gain is 20 and the value of R_2 is 228 kΩ, the value of R_1 is

❑ *a* 12 kΩ
❑ *b* 14 kΩ
❑ *c* 28 kΩ
❑ *d* 68 kΩ

11.38 The diagram shows a two-input summing amplifier. If R_1 = 120 Ω, R_2 = 220 Ω, R_3 = 330 Ω, V_1 = 1.5 V and V_2 = 2.5 V, the value of the output voltage V_3 is

❑ *a* –6.255 V
❑ *b* –7.875 V
❑ *c* –8.864 V
❑ *d* –9.862 V

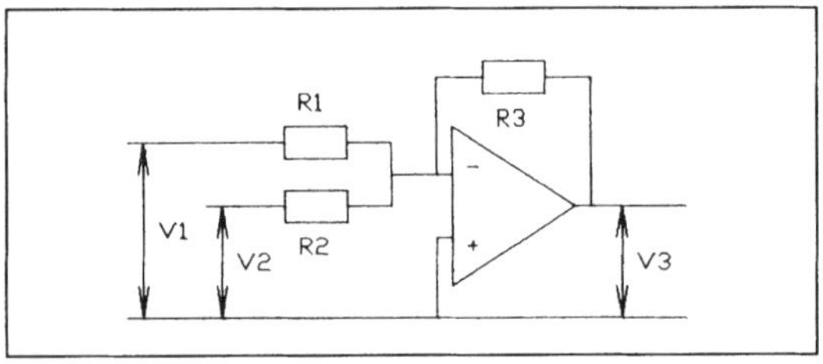

11.39 In Question 11.38, if R_1 = 2000 Ω, R_2 = 1500 Ω, R_3 = 1200 Ω, V_1 = 10 V and V_2 = 12 V, the value of V_3 is

❑ *a* –10.3 V
❑ *b* –11.6 V
❑ *c* –12.1 V
❑ *d* –15.6 V

11.40 In Question 11.38, if all three resistors are 2400 Ω and V_1 is 5 V and V_2 is 15 V, the output voltage V_3 is

❑ *a* –6.8 V
❑ *b* –9.7 V
❑ *c* –20 V
❑ *d* –24 V

11.41 A two-input summing amplifier identical to the one in Question 11.38 has the following parameters: R_1 = 100 kΩ, R_2 = 120 kΩ, V_1 = 2 V, V_2 = 4 V and V_3 = –24 V. The value of R_3 is

❑ *a* 270 kΩ
❑ *b* 450 kΩ
❑ *c* 510 kΩ
❑ *d* 600 kΩ

11.42 If a summing amplifier has R_1 = 3900 Ω, R_2 = 3900 Ω, V_1 = 4.5 V, V_2 = 6.5 V and V_3 = –12 V, the value of R_3 is approximately

❑ *a* 1861 Ω
❑ *b* 3622 Ω
❑ *c* 4255 Ω
❑ *d* 6800 Ω

11.43 If a summing amplifier has R_1 = 1 kΩ, R_2 = 2 kΩ, V_1 = 12 V , V_2 = 12 V and V_3 = –48 V, the value of R_3 is approximately

❑ *a* 2667 Ω
❑ *b* 3300 Ω
❑ *c* 3900 Ω
❑ *d* 4137 Ω

11.44 If a summing amplifier has R_1 = 8 kΩ, R_3 = 60 kΩ, V_1 = 2 V, V_2 = 10 V and V_3 = –30 V, the value of R_2 is

❑ *a* 22 kΩ
❑ *b* 30 kΩ
❑ *c* 36 kΩ
❑ *d* 40 kΩ

11.45 If a summing amplifier has R_1 = R_3 = 470 Ω, V_1 = 5 V, V_2 = 9 V and V_3 = –21 V, the approximate value of R_2 is

❑ *a* 264 Ω
❑ *b* 461 Ω
❑ *c* 688 Ω
❑ *d* 750 Ω

11.46 A summing amplifier has R_1 = 1400 Ω, R_3 = 2400 Ω, V_1 = V_2 = 3.5 V and V_3 = –10 V. The value of R_2 is

❑ *a* 1400 Ω
❑ *b* 2100 Ω
❑ *c* 2400 Ω
❑ *d* 3000 Ω

11.47 The diagram shows a three-input summing amplifier. If $R_1 = R_2 = R_3 = R_4 = 3300\ \Omega$, $V_1 = 2$ V, $V_2 = 4$ V and $V_3 = 8$ V, the value of the output voltage V_4 is

- ❑ *a* −10 V
- ❑ *b* −11 V
- ❑ *c* −14 V
- ❑ *d* −18 V

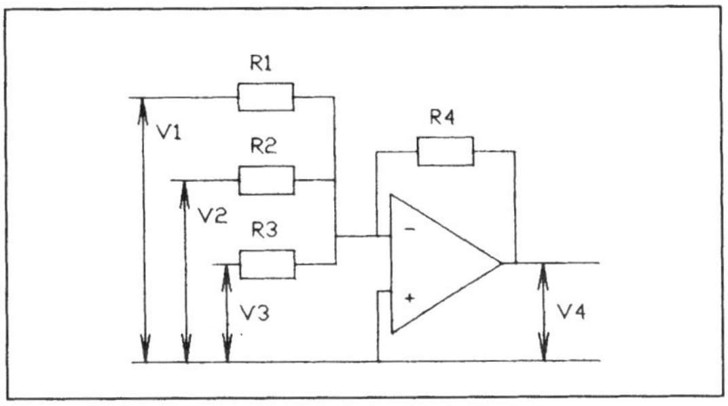

11.48 If in Question 11.47 $V_1 = V_2 = V_3 = 5$ V, the output voltage V_4 is

- ❑ *a* −10 V
- ❑ *b* −11 V
- ❑ *c* −12 V
- ❑ *d* −15 V

11.49 A three-input summing amplifier has the following parameters: $R_1 = 120\ \Omega$, $R_2 = 150\ \Omega$, $R_3 = 180\ \Omega$, $R_4 = 210\ \Omega$, $V_1 = 2$ V, $V_2 = 4$ V and $V_3 = 8$ V. The value of V_4 is

- ❑ *a* −10.88 V
- ❑ *b* −15.52 V
- ❑ *c* −16.26 V
- ❑ *d* −18.43 V

11.50 In a three-input summing amplifier, $R_1 = 400\ \Omega$, $R_3 = 500\ \Omega$, $R_4 = 600\ \Omega$, $V_1 = 3$ V, $V_2 = 2$ V, $V_3 = 7$ V and $V_4 = -24$ V. The value of R_2 is approximately

- ❑ *a* 100 Ω
- ❑ *b* 108 Ω
- ❑ *c* 166 Ω
- ❑ *d* 240 Ω

11.51 In a three-input summing amplifier, $R_1 = R_3 = R_4 = 1$ kΩ. If $V_1 = 4.6$ V, $V_2 = 3.4$ V, $V_3 = 2.8$ V and $V_4 = -32$ V, R_2 is approximately

- ❑ *a* 124 Ω
- ❑ *b* 138 Ω
- ❑ *c* 190 Ω
- ❑ *d* 264 Ω

11.52 In a three-input summing amplifier, $R_1 = R_3 = R_4 = 1$ kΩ, and $V_1 = V_2 = V_3 = 3$ V. If $V_4 = -32$ V, the approximate value of R_2 is

- ❑ *a* 102 Ω
- ❑ *b* 107 Ω
- ❑ *c* 115 Ω
- ❑ *d* 167 Ω

11.53 The diagram shows a subtracting amplifier where $R_1 = R_2$ and $R_3 = R_4$. If $V_1 = 3.6$ V and $V_2 = 4$ V and the gain is 10, the output voltage V_3 is

- ❑ *a* 4 V
- ❑ *b* 5 V
- ❑ *c* 6 V
- ❑ *d* 7 V

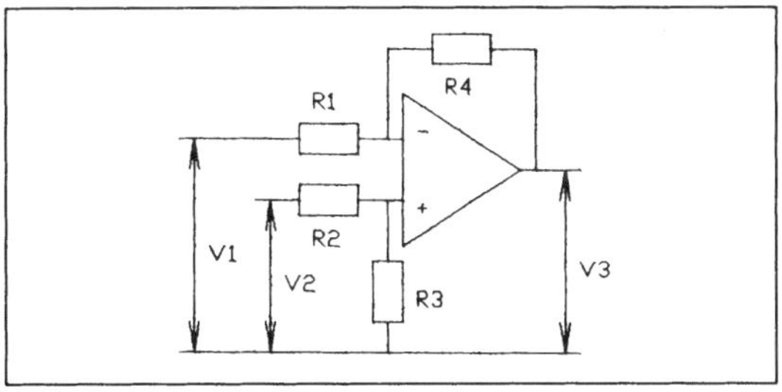

11.54 If in Question 11.53 the gain is 20, the output voltage V_4 is

- ❑ *a* 4 V
- ❑ *b* 6 V
- ❑ *c* 8 V
- ❑ *d* 10 V

11.55 If in Question 11.53 $V_1 = 3$ V, $V_2 = 5$ V and the gain is 24, the output voltage V_3 is

- ❑ *a* 12 V
- ❑ *b* 24 V
- ❑ *c* 36 V
- ❑ *d* 48 V

11.56 If in Question 11.53 $V_1 = 3$ V, $V_3 = 4$ V and the gain is 12, V_2 is

❑ *a* 2.12 V
❑ *b* 3.33 V
❑ *c* 4.61 V
❑ *d* 5.08 V

11.57 If in Question 11.53 V_1 is 5 V, V_3 is 5 V and the gain is 10, V_2 is

❑ *a* 4.0 V
❑ *b* 5.5 V
❑ *c* 6.2 V
❑ *d* 8.0 V

11.58 If in Question 11.53 V_1 is 5 V, V_3 is 10 V and the gain is unity, V_2 is

❑ *a* 10 V
❑ *b* 11 V
❑ *c* 12 V
❑ *d* 15 V

11.59 The diagram shows a simple voltage amplifier. An alternating voltage V_{be} of 40 mV is applied to the base which causes a change in the base current I_b of 60 μA. The changes in the collector current I_c, and the collector to emitter voltage V_{ce}, are 12 mA and 6 V respectively. The AC input resistance is

❑ *a* 333.3 Ω
❑ *b* 666.7 Ω
❑ *c* 774.6 Ω
❑ *d* 813.8 Ω

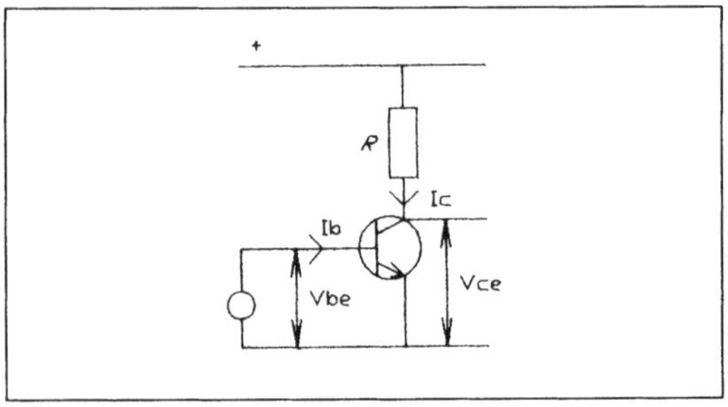

11.60 In Question 11.59 the AC output resistance is

❑ *a* 300 Ω
❑ *b* 450 Ω
❑ *c* 500 Ω
❑ *d* 750 Ω

11.61 In Question 11.59 the AC current gain is

❑ *a* 200
❑ *b* 240
❑ *c* 265
❑ *d* 270

11.62 In Question 11.59 the AC voltage gain is

❑ *a* 105
❑ *b* 120
❑ *c* 150
❑ *d* 160

11.63 In Question 11.59 the AC power gain is

❑ *a* 10 000
❑ *b* 12 000
❑ *c* 24 000
❑ *d* 30 000

11.64 The circuit in Question 11.59 has the following data: Input voltage V_{be} of 80 mV, change in base current I_b of 20 μA, change in collector current I_c of 10 mA, change in collector to emitter voltage V_{ce} of 8 V. The AC input resistance is

❑ *a* 2000 Ω
❑ *b* 4000 Ω
❑ *c* 5500 Ω
❑ *d* 6200 Ω

11.65 In Question 11.64 the AC output resistance is

❑ *a* 600 Ω
❑ *b* 620 Ω
❑ *c* 750 Ω
❑ *d* 800 Ω

11.66 In Question 11.64 the AC current gain is

❑ *a* 500
❑ *b* 350
❑ *c* 220
❑ *d* 450

11.67 In Question 11.64 the AC voltage gain is

❑ *a* 200
❑ *b* 150
❑ *c* 100
❑ *d* 250

11.68 In Question 11.64 the AC power gain is
- ❑ *a* 1.8×10^2
- ❑ *b* 2.6×10^4
- ❑ *c* 5×10^4
- ❑ *d* 4.2×10^3

11.69 A study of the transistor amplifier characteristics gives the following data:
(i) $V_{be} = 75$ mV
(ii) $\delta I_b = 25$ μA
(iii) $\delta I_c = 15$ mA
(iv) $\delta V_{ce} = 10$ V
The AC input resistance is
- ❑ *a* 1500 Ω
- ❑ *b* 2200 Ω
- ❑ *c* 2500 Ω
- ❑ *d* 3000 Ω

11.70 In Question 11.69 the AC output resistance is
- ❑ *a* 667 Ω
- ❑ *b* 700 Ω
- ❑ *c* 821 Ω
- ❑ *d* 851 Ω

11.71 In Question 11.69 the AC current gain is
- ❑ *a* 700
- ❑ *b* 600
- ❑ *c* 500
- ❑ *d* 200

11.72 In Question 11.69 the AC voltage gain is
- ❑ *a* 133
- ❑ *b* 200
- ❑ *c* 223
- ❑ *d* 300

11.73 In Question 11.69 the AC power gain is
- ❑ *a* 36 100
- ❑ *b* 45 000
- ❑ *c* 47 600
- ❑ *d* 79 800

11.74 The diagram shows an emitter follower. The AC current gain is
- ❑ *a* 180
- ❑ *b* 200
- ❑ *c* 220
- ❑ *d* 230

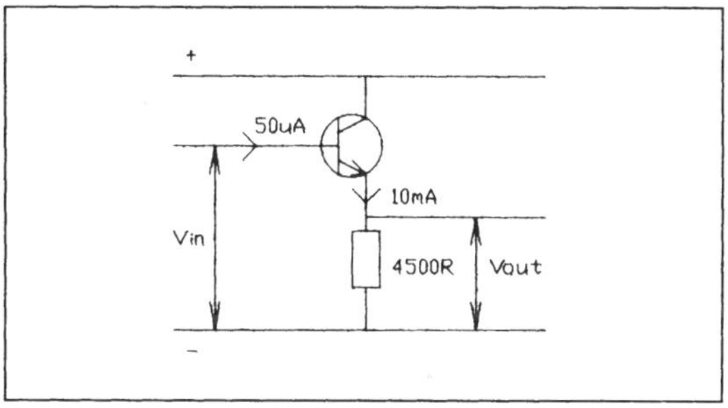

11.75 If the input resistance in Question 11.74 is 1 MΩ, the AC voltage gain is
- ❑ *a* 0.9
- ❑ *b* 1.2
- ❑ *c* 1.4
- ❑ *d* 1.9

12 Oscillators

12.1 The natural frequency of a sinusoidal oscillator where the capacitance is 2.5 μF, the inductance 800 μH and the resistance 2 Ω is

- ❑ *a* 3553 Hz
- ❑ *b* 5561 Hz
- ❑ *c* 6670 Hz
- ❑ *d* 8875 Hz

12.2 If in Question 12.1 $C = 10$ μF, $L = 0.9$ mH and $R = 4$ Ω, the natural frequency is

- ❑ *a* 1111 Hz
- ❑ *b* 1320 Hz
- ❑ *c* 1640 Hz
- ❑ *d* 1832 Hz

12.3 If in Question 12.1 $C = 20$ μF, $L = 0.5$ mH and $R = 2$ Ω, the natural frequency is

- ❑ *a* 1050 Hz
- ❑ *b* 1321 Hz
- ❑ *c* 1428 Hz
- ❑ *d* 1559 Hz

12.4 If in Question 12.1 $C = 0.4$ μF, $L = 12$ μH and $R = 0.5$ Ω, the natural frequency is

- ❑ *a* 47.161 kHz
- ❑ *b* 50.323 kHz
- ❑ *c* 60.667 kHz
- ❑ *d* 72.568 kHz

12.5 A sinusoidal oscillator comprising an inductance and capacitance has a resonant frequency. If the circuit has $C = 0.4$ μF and $L = 12$ μH, the resonant frequency is

- ❑ *a* 67.331 kHz
- ❑ *b* 72.643 kHz
- ❑ *c* 80.875 kHz
- ❑ *d* 90.162 kHz

12.6 In Question 12.5, if the capacitance is 16 μF and the inductance 0.7 mH, the resonant frequency is

- ❑ *a* 1060 Hz
- ❑ *b* 1212 Hz
- ❑ *c* 1306 Hz
- ❑ *d* 1504 Hz

12.7 In Question 12.5, if the capacitance is 160 μF and the inductance is 0.7 mH, the resonant frequency is

- ❑ *a* 476 Hz
- ❑ *b* 680 Hz
- ❑ *c* 910 Hz
- ❑ *d* 927 Hz

12.8 If in Question 12.5 the capacitance is 16 μF and the inductance 350 μH, the resonant frequency is

- ❑ *a* 1668 Hz
- ❑ *b* 2010 Hz
- ❑ *c* 2127 Hz
- ❑ *d* 3020 Hz

12.9 The periodic time of a sinusoidal oscillator having a capacitance of 0.35 μF and an inductance of 0.4 H is approximately

- ❑ *a* 2.35 ms
- ❑ *b* 4.82 ms
- ❑ *c* 6.91 ms
- ❑ *d* 7.74 ms

12.10 The periodic time of a sinusoidal oscillator with a capacitance of 21 μF and an inductance of 13 H is approximately

- ❑ *a* 88 ms
- ❑ *b* 96 ms
- ❑ *c* 104 ms
- ❑ *d* 150 ms

12.11 The periodic time of a sinusoidal oscillator with a capacitance of 10 μF and an inductance of 2 H is

❑ *a* 14 ms
❑ *b* 28 ms
❑ *c* 56 ms
❑ *d* 84 ms

12.12 The periodic time of a sinusoidal oscillator having a capacitance of 105 μF and an inductance of 13 H is approximately

❑ *a* 104 ms
❑ *b* 196 ms
❑ *c* 204 ms
❑ *d* 232 ms

12.13 An oscillator is required to produce a 1-kHz sine wave. If the value of the capacitance is 0.12 μF, the value of the inductance is approximately

❑ *a* 132 mH
❑ *b* 150 mH
❑ *c* 211 mH
❑ *d* 260 mH

12.14 In Question 12.13, if C = 3.4 μF and f = 2400 Hz, the value of L is approximately

❑ *a* 1.02 mH
❑ *b* 1.09 mH
❑ *c* 1.16 mH
❑ *d* 1.29 mH

12.15 In Question 12.13, if C = 0.2 μF and f = 500 Hz, the value of the inductance is

❑ *a* 0.1 H
❑ *b* 0.5 H
❑ *c* 1.5 H
❑ *d* 2.6 H

12.16 In Question 12.13, if the capacitance is 0.55 μF and the frequency 120 Hz, the value of the inductance is approximately

❑ *a* 3.2 H
❑ *b* 4.4 H
❑ *c* 6.4 H
❑ *d* 7.7 H

12.17 Given that an oscillator has a resonant frequency of 1 MHz and that the inductance is 20 μH, the value of the capacitance is

❑ *a* 1.16 μF
❑ *b* 1.24 μF
❑ *c* 1.27 nF
❑ *d* 1.86 pF

12.18 If an oscillator has a resonant frequency of 800 kHz and the inductance is 14 μH, the value of the capacitance is approximately

❑ *a* 2.83 nF
❑ *b* 4.91 nF
❑ *c* 6.62 nF
❑ *d* 8.87 nF

12.19 Given that an oscillator has a resonant frequency of 1.5 MHz and that the inductance is 4 μH, the value of the capacitance is approximately

❑ *a* 1.16 nF
❑ *b* 1.42 nF
❑ *c* 1.61 nF
❑ *d* 2.81 nF

12.20 If an oscillator has a resonant frequency of 900 kHz and the inductance is 10 μH, the value of the capacitance is approximately

❑ *a* 1.06 nF
❑ *b* 3.13 nF
❑ *c* 1.06 pF
❑ *d* 3.13 pF

12.21 The diagram shows a typical phase-shift oscillator. If $C = C_1 = C_2 = C_3$ = 2.8 μF and $R = R_1 = R_2 = R_{in} = R_L$ = 1800 Ω,

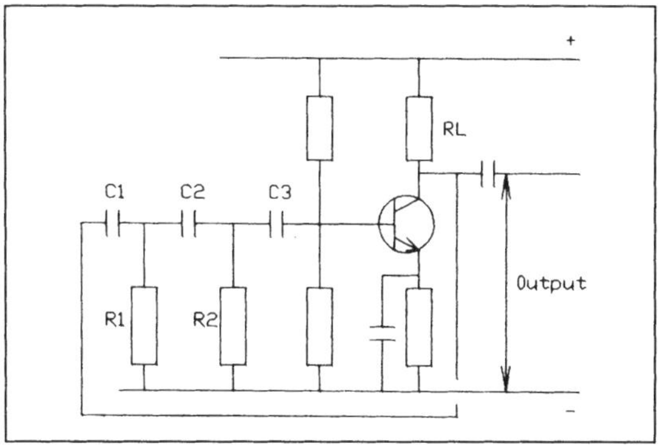

where R_{in} is the input resistance, the frequency of oscillation is approximately

❑ *a* 13 Hz
❑ *b* 65 Hz
❑ *c* 88 Hz
❑ *d* 99 Hz

12.22 If in Question 12.21 $C = 0.1$ μF and $R = 2$ kΩ, the frequency of oscillation is approximately

❑ *a* 109 Hz
❑ *b* 165 Hz
❑ *c* 325 Hz
❑ *d* 491 Hz

12.23 If in Question 12.21 $C = 0.75$ μF and $R = 1500$ Ω, the frequency of oscillation is approximately

❑ *a* 30 Hz
❑ *b* 34 Hz
❑ *c* 49 Hz
❑ *d* 58 Hz

12.24 If in Question 12.21 $C = 0.02$ μF and $R = 120$ Ω, the oscillation frequency is

❑ *a* 14.312 kHz
❑ *b* 27.072 kHz
❑ *c* 43.604 kHz
❑ *d* 48.122 kHz

12.25 In a phase shift oscillator identical to the one shown in Question 12.21 all the capacitors are equal to 15 μF. If the oscillation frequency is 100 Hz, the value of R is approximately

❑ *a* 39.6 Ω
❑ *b* 43.3 Ω
❑ *c* 51.7 Ω
❑ *d* 82.6 Ω

12.26 If in Question 12.25 $C = 0.6$ μF and $f = 200$ Hz, the value of R is approximately

❑ *a* 345.6 Ω
❑ *b* 471.8 Ω
❑ *c* 541.5 Ω
❑ *d* 667.6 Ω

12.27 If in Question 12.25 the capacitance is 0.6 μF and the frequency 2 kHz, the resistance is approximately

❑ *a* 20.48 Ω
❑ *b* 21.22 Ω
❑ *c* 47.66 Ω
❑ *d* 54.15 Ω

12.28 If in Question 12.25 the value of C is 1.0 μF and the frequency is 50 Hz, the value of R is approximately

❑ *a* 1000 Ω
❑ *b* 1299 Ω
❑ *c* 1687 Ω
❑ *d* 2051 Ω

12.29 If the frequency of oscillation in an R–C oscillator similar to the one in Question 12.21 is 6912 Hz and the value of R is 4.7 kΩ the value of C is

❑ *a* 2 nF
❑ *b* 5 nF
❑ *c* 7 nF
❑ *d* 9 nF

12.30 If in Question 12.29 the frequency is 5 kHz and the resistance 3.3 kΩ, the capacitance is approximately

❑ *a* 2.21 nF
❑ *b* 2.86 nF
❑ *c* 3.94 nF
❑ *d* 5.62 nF

12.31 If in Question 1 2.29 the frequency is 200 Hz and the resistance 400 Ω, the capacitance is approximately

❑ *a* 460 nF
❑ *b* 580 nF
❑ *c* 720 nF
❑ *d* 812 nF

12.32 If in Question 12.31 the frequency is increased by a factor of 4, the capacitance value is then

❑ *a* 162 nF
❑ *b* 203 nF
❑ *c* 230 nF
❑ *d* 360 nF

12.33 The diagram shows a Wien bridge oscillator. If the capacitor C is 10 μF and the resistance R is 120 Ω, the frequency of oscillation is

❑ *a* 100 Hz
❑ *b* 116 Hz
❑ *c* 133 Hz
❑ *d* 144 Hz

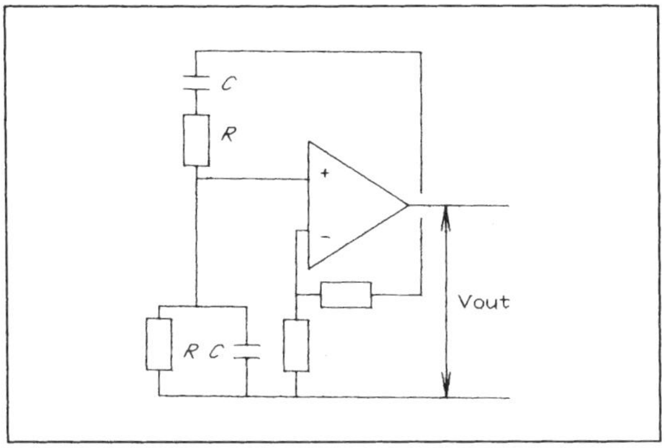

12.34 If in Question 12.33 C = 47 μF and R = 100 Ω, the oscillating frequency is approximately

❑ *a* 34 Hz
❑ *b* 68 Hz
❑ *c* 72 Hz
❑ *d* 97 Hz

12.35 If in Question 12.33 C = 4.4 μF and R = 220 Ω, the oscillating frequency is approximately

❑ *a* 128 Hz
❑ *b* 164 Hz
❑ *c* 175 Hz
❑ *d* 183 Hz

12.36 If in Question 12.33 the capacitance is 6 μF and the resistance 1.5 kΩ, the frequency is

❑ *a* 15.33 Hz
❑ *b* 17.68 Hz
❑ *c* 21.62 Hz
❑ *d* 38.44 Hz

12.37 A Wien bridge oscillator oscillates at a frequency of 4 kHz. If the resistance is 330 Ω, the value of the capacitance is

❑ *a* 0.06 μF
❑ *b* 0.12 μF
❑ *c* 1.21 μF
❑ *d* 2.21 μF

12.38 If a Wien bridge oscillator has a frequency of 20 kHz and the resistance R is 1200 Ω, the value of C is approximately

❑ *a* 1.14×10^{-2} μF
❑ *b* 2.88×10^{-2} μF
❑ *c* 4.61×10^{-3} μF
❑ *d* 6.63×10^{-3} μF

12.39 If a Wien bridge oscillator has a frequency of 120 Hz and a resistance of 820 Ω, the value of C is approximately

❑ *a* 1.33 μF
❑ *b* 1.62 μF
❑ *c* 3.48 μF
❑ *d* 5.72 μF

12.40 If a Wien bridge oscillator has a frequency of 240 Hz and the value of R is 820 Ω, the value of C is

❑ *a* 0.81 μF
❑ *b* 0.98 μF
❑ *c* 1.67 μF
❑ *d* 2.27 μF

12.41 A Wien bridge oscillator oscillates at a frequency of 500 Hz. If the value of C is 3.183 μF, the value of R is

❑ *a* 75 Ω
❑ *b* 90 Ω
❑ *c* 100 Ω
❑ *d* 250 Ω

12.42 If a Wien bridge oscillator has a frequency of 600 Hz and the capacitance is 10 μF, the value of R is approximately

❑ *a* 11.12 Ω
❑ *b* 15.61 Ω
❑ *c* 26.53 Ω
❑ *d* 78.68 Ω

12.43 If a Wien bridge oscillator has a frequency of 640 Hz and $C = 0.8\ \mu F$, the value of R is approximately
- ❑ *a* 120 Ω
- ❑ *b* 167 Ω
- ❑ *c* 258 Ω
- ❑ *d* 311 Ω

12.44 If a Wien bridge oscillator has a frequency of 1280 Hz and the capacitance is 0.8 μF, the value of R is approximately
- ❑ *a* 140 Ω
- ❑ *b* 155 Ω
- ❑ *c* 181 Ω
- ❑ *d* 271 Ω

12.45 The diagram shows a twin-T oscillator. If the capacitance C is 2 nF and the resistance R is 2 MΩ, the oscillating frequency is
- ❑ *a* 20 Hz
- ❑ *b* 25 Hz
- ❑ *c* 40 Hz
- ❑ *d* 80 Hz

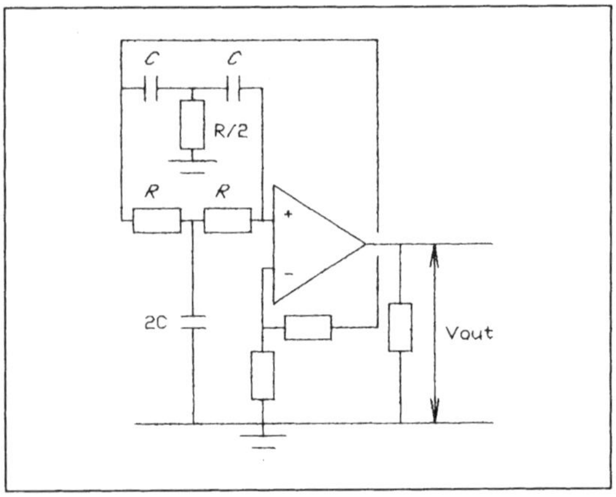

12.46 If a twin-T oscillator has a capacitance of 1.2 nF and a resistance of 1 MΩ, the frequency is approximately
- ❑ *a* 109 Hz
- ❑ *b* 116 Hz
- ❑ *c* 124 Hz
- ❑ *d* 133 Hz

12.47 If a twin-T oscillator has a capacitance of 500 pF and a resistance of 1 MΩ, the frequency is approximately
- ❑ *a* 318 Hz
- ❑ *b* 628 Hz
- ❑ *c* 780 Hz
- ❑ *d* 825 Hz

12.48 If a twin-T oscillator has a capacitance of 40 pF and the resistance is 2 MΩ, the frequency is
- ❑ *a* 1.989 kHz
- ❑ *b* 2.631 kHz
- ❑ *c* 4.828 kHz
- ❑ *d* 5.711 kHz

12.49 The diagram shows a typical multivibrator circuit. If $C = C_1 = C_2 = 0.02\ \mu F$ and $R = R_1 = R_2 = 22\ k\Omega$, the oscillating frequency is
- ❑ *a* 1.623 kHz
- ❑ *b* 1.813 kHz
- ❑ *c* 2.416 kHz
- ❑ *d* 4.421 kHz

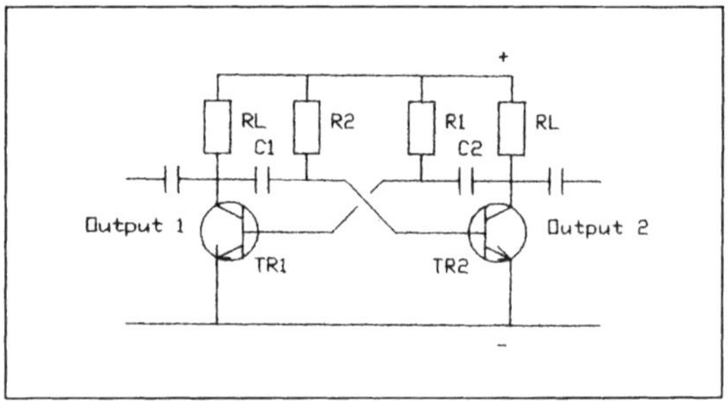

12.50 If in Question 12.49 $C = 0.024\ \mu F$ and $R = 4700\ \Omega$, the frequency is
- ❑ *a* 3.331 kHz
- ❑ *b* 4.161 kHz
- ❑ *c* 6.332 kHz
- ❑ *d* 8.128 kHz

12.51 If in Question 12.49 the capacitance C is 0.5 μF and the resistance R is 150 Ω, the frequency is
- ❑ *a* 9.524 kHz
- ❑ *b* 10.621 kHz
- ❑ *c* 11.822 kHz
- ❑ *d* 16.667 kHz

12.52 If in Question 12.49 the capacitance is 1.0 nF and the resistance 33 kΩ, the frequency is
- ❑ *a* 16.432 kHz
- ❑ *b* 18.516 kHz
- ❑ *c* 19.128 kHz
- ❑ *d* 21.645 kHz

12.53 If a multivibrator is running at 50 Hz and $C = C_1 = C_2 = 14.286$ µF, the value of R where $R = R_1 = R_2$ is
- ❑ *a* 1600 Ω
- ❑ *b* 1000 Ω
- ❑ *c* 1800 Ω
- ❑ *d* 1220 Ω

12.54 If in Question 12.53 the frequency is 1 kHz and C is 13 µF, the value of R is approximately
- ❑ *a* 23.37 Ω
- ❑ *b* 47.16 Ω
- ❑ *c* 54.95 Ω
- ❑ *d* 66.62 Ω

12.55 If in Question 12.53 the frequency is 5 kHz and the capacitance 60 nF, the value of R is
- ❑ *a* 1330 Ω
- ❑ *b* 1662 Ω
- ❑ *c* 1771 Ω
- ❑ *d* 2381 Ω

12.56 If in Question 12.53 the frequency is 10 kHz and the capacitance is 4 nF, the resistance R is
- ❑ *a* 10.062 kΩ
- ❑ *b* 13.128 kΩ
- ❑ *c* 14.604 kΩ
- ❑ *d* 17.857 kΩ

12.57 If a multivibrator is running at 952 Hz and $R = R_1 = R_2 = 500$ Ω, the value of C where $C = C_1 = C_2$ is
- ❑ *a* 1500 nF
- ❑ *b* 1700 nF
- ❑ *c* 1850 nF
- ❑ *d* 1910 nF

12.58 If in Question 12.57 the frequency is 250 Hz and R is 680 Ω, the value of C is
- ❑ *a* 2.9 µF
- ❑ *b* 3.6 µF
- ❑ *c* 4.2 µF
- ❑ *d* 7.6 µF

12.59 If in Question 12.57 the frequency is 100 kHz and R is 2 kΩ, the value of C is approximately
- ❑ *a* 1.61 nF
- ❑ *b* 3.57 nF
- ❑ *c* 4.82 nF
- ❑ *d* 9.06 nF

12.60 If in Question 12.57 the frequency is 200 kHz and R is 2 kΩ, the value of C is approximately
- ❑ *a* 1.08 nF
- ❑ *b* 1.18 nF
- ❑ *c* 1.43 nF
- ❑ *d* 1.79 nF

12.61 The diagram shows a free-running multivibrator. If $R_1 = 10$ kΩ, $R_2 = 20$ kΩ, $C_1 = 0.5$ µF and $C_2 = 0.1$ µF, the frequency of oscillation is
- ❑ *a* 104 Hz
- ❑ *b* 106 Hz
- ❑ *c* 130 Hz
- ❑ *d* 237 Hz

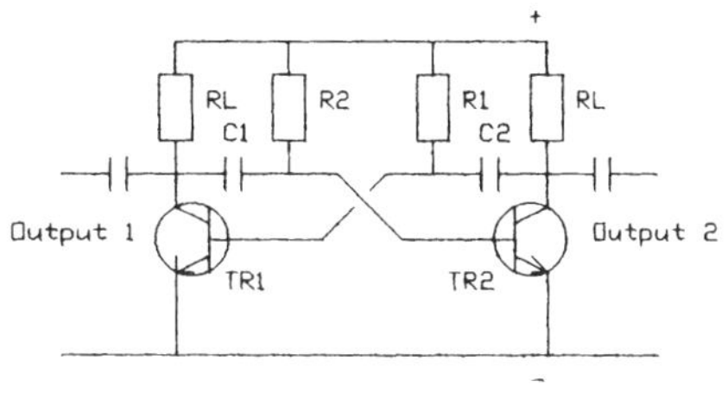

12.62 If in Question 12.61 $R_1 = 8.2$ kΩ, $R_2 = 15$ kΩ, $C_1 = 1.2$ µF and $C_2 = 1.5$ µF, the frequency is
- ❑ *a* 47 Hz
- ❑ *b* 75 Hz
- ❑ *c* 90 Hz
- ❑ *d* 98 Hz

12.63 If in Question 12.61 $R_1 = 6.8$ kΩ, $R_2 = 6.8$ kΩ, $C_1 = 0.02$ μF and $C_2 = 0.02$ μF, the frequency is

- ❑ *a* 4.621 kHz
- ❑ *b* 5.252 kHz
- ❑ *c* 6.383 kHz
- ❑ *d* 9.311 kHz

12.64 If in Question 12.61 $C_1 = 0.02$ μF, $C_2 = 0.2$ μF and $R_1 = R_2 = 6800$ Ω, the frequency is approximately

- ❑ *a* 450 Hz
- ❑ *b* 630 Hz
- ❑ *c* 955 Hz
- ❑ *d* 990 Hz

12.65 If in the circuit of Question 12.61 the frequency is 170 Hz, $C_1 = 1.5$ μF, $C_2 = 2.5$ μF and $R_1 = 1400$ Ω, the value of R_2 is approximately

- ❑ *a* 1445 Ω
- ❑ *b* 1800 Ω
- ❑ *c* 2461 Ω
- ❑ *d* 3269 Ω

12.66 If in the circuit of Question 12.61 the frequency is 170 Hz, $C_1 = 3$ μF, $C_2 = 5$ μF and $R_1 = 1400$ Ω, the value of R_2 is approximately

- ❑ *a* 232 Ω
- ❑ *b* 468 Ω
- ❑ *c* 711 Ω
- ❑ *d* 809 Ω

12.67 If in the circuit of Question 12.61 the frequency is 230 Hz, $C_1 = 0.9$ μF, $C_2 = 0.7$ μF and $R_1 = 560$ Ω, the value of R_2 is

- ❑ *a* 3.234 k Ω
- ❑ *b* 4.318 k Ω
- ❑ *c* 6.466 k Ω
- ❑ *d* 7.702 k Ω

12.68 If in the circuit of Question 12.61 the frequency is 330 Hz, $C_1 = C_2 = 0.02$ μF and $R_1 = 1400$ Ω, the value of R_2 is approximately

- ❑ *a* 215 kΩ
- ❑ *b* 416 kΩ
- ❑ *c* 720 kΩ
- ❑ *d* 805 kΩ

12.69 If in the circuit of Question 12.61 the frequency is 15 kHz, $R_1 = 100$ Ω, $R_2 = 150$ Ω and $C_2 = 0.2$ μF, the value of C_1 is

- ❑ *a* 0.5 μF
- ❑ *b* 1.0 μF
- ❑ *c* 2.0 μF
- ❑ *d* 5.5 μF

12.70 If in Question 12.61 the frequency is 1 kHz, $R_1 = R_2 = 120$ Ω and $C_2 = 0.5$ μF, the value of C_1 is

- ❑ *a* 2.8 μF
- ❑ *b* 9.3 μF
- ❑ *c* 10.6 μF
- ❑ *d* 11.4 μF

12.71 If in Question 12.61 the frequency is 800 Hz, $R_1 = 80$ Ω, $R_2 = 60$ Ω and $C_2 = 1.4$ μF, the value of C_1 is

- ❑ *a* 16.3 μF
- ❑ *b* 27.9 μF
- ❑ *c* 46.3 μF
- ❑ *d* 48.6 μF

12.72 The diagram shows a triangular waveform generator using an operational amplifier. Given that the input voltage is 10 V, the output voltage is 2 V, $R = 10$ kΩ and $C = 0.5$ μF, the frequency of the input waveform is

- ❑ *a* 120 Hz
- ❑ *b* 150 Hz
- ❑ *c* 250 Hz
- ❑ *d* 350 Hz

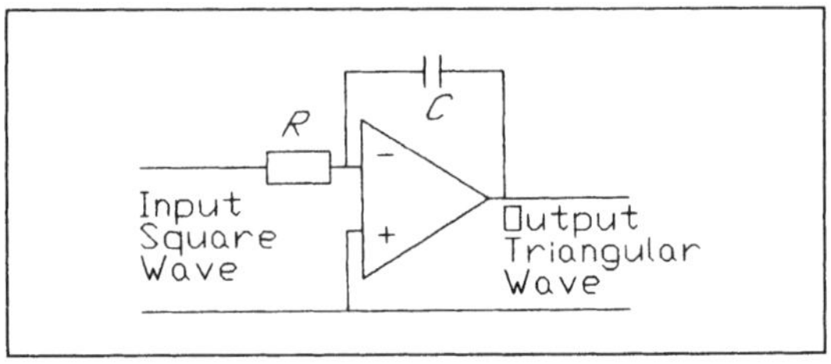

12.73 If in Question 12.72 $V_{in} = 8$ V, $V_{out} = 1.0$ V, $R = 15$ kΩ and $C = 1.0$ μF, the frequency is

- ❑ *a* 133.33 Hz
- ❑ *b* 166.66 Hz
- ❑ *c* 203.43 Hz
- ❑ *d* 233.33 Hz

12.74 If in Question 12.72 V_{in} = 6.6 V, V_{out} = 2.4 V, R = 20 kΩ and C = 1.5 μF, the frequency is

❑ *a* 15 Hz
❑ *b* 23 Hz
❑ *c* 67 Hz
❑ *d* 69 Hz

12.75 If in Question 12.72 V_{in} = 7 V, V_{out} = 3.5 V, R = 820 Ω and C = 0.6 μF, the frequency is

❑ *a* 864 Hz
❑ *b* 970 Hz
❑ *c* 1016 Hz
❑ *d* 1808 Hz

13 Radio

13.1 Radio waves are propagated in space at

- ❑ *a* $3 \times 10^5\ \mathrm{m\,s^{-1}}$
- ❑ *b* $3 \times 10^6\ \mathrm{m\,s^{-1}}$
- ❑ *c* $3 \times 10^7\ \mathrm{m\,s^{-1}}$
- ❑ *d* $3 \times 10^8\ \mathrm{m\,s^{-1}}$

13.2 The long wave band extends from approximately

- ❑ *a* 50 to 100 kHz
- ❑ *b* 60 to 110 kHz
- ❑ *c* 150 to 350 kHz
- ❑ *d* 150 to 300 MHz

13.3 The medium wave band extends from approximately

- ❑ *a* 530 to 1600 kHz
- ❑ *b* 660 to 1000 kHz
- ❑ *c* 800 to 1100 kHz
- ❑ *d* 950 to 1500 kHz

13.4 The FM radio band extends from approximately

- ❑ *a* 16 to 60 MHz
- ❑ *b* 20 to 65 MHz
- ❑ *c* 88 to 108 MHz
- ❑ *d* 120 to 400 MHz

13.5 A radio station has a wavelength of 1500 m. The frequency is

- ❑ *a* 150 kHz
- ❑ *b* 200 kHz
- ❑ *c* 600 kHz
- ❑ *d* 901 kHz

13.6 A local station has a wavelength of 362 m. The frequency is

- ❑ *a* 440 kHz
- ❑ *b* 382 kHz
- ❑ *c* 540 kHz
- ❑ *d* 828 kHz

13.7 A radio station has a frequency of 100 MHz. The wavelength is

- ❑ *a* 3 m
- ❑ *b* 3.3 m
- ❑ *c* 30 m
- ❑ *d* 300 m

13.8 A radio station has a frequency of 648 kHz. The wavelength is

- ❑ *a* 160 m
- ❑ *b* 240 m
- ❑ *c* 463 m
- ❑ *d* 581 m

13.9 An aerial is required to be 0.25 λ in length. If such an aerial is used to transmit a signal with a frequency of 100 MHz, the length of the aerial is

- ❑ *a* 0.25 m
- ❑ *b* 0.5 m
- ❑ *c* 0.75 m
- ❑ *d* 1.5 m

13.10 A frequency of 200 kHz is in the

- ❑ *a* Low frequency band (LF)
- ❑ *b* Super high frequency band (SHF)
- ❑ *c* Very high frequency band (VHF)
- ❑ *d* High frequency band (HF)

13.11 A frequency of 2500 MHz is in the

- ❑ *a* Very low frequency band (VLF)
- ❑ *b* Medium frequency band (MF)
- ❑ *c* Ultra high frequency band (UHF)
- ❑ *d* Extremely high frequency band (EHF)

13.12 A frequency below 30 kHz is in the
- ❑ *a* VLF band
- ❑ *b* LF band
- ❑ *c* MF band
- ❑ *d* HF band

13.13 A frequency of 40 000 MHz is in the
- ❑ *a* VHF band
- ❑ *b* SHF band
- ❑ *c* EHF band
- ❑ *d* HF band

13.14 A frequency of 100 MHz is in the
- ❑ *a* LF band
- ❑ *b* MF band
- ❑ *c* UHF band
- ❑ *d* VHF band

13.15 A frequency of 4000 MHz is in the
- ❑ *a* EHF band
- ❑ *b* SHF band
- ❑ *c* UHF band
- ❑ *d* VHF band

13.16 A frequency of 4000 kHz is in the
- ❑ *a* HF band
- ❑ *b* MF band
- ❑ *c* LF band
- ❑ *d* SHF band

13.17 A frequency of 1000 kHz is in the
- ❑ *a* HF band
- ❑ *b* MF band
- ❑ *c* LF band
- ❑ *d* VHF band

13.18 In amplitude modulation, if the peak signal is 0.5 V and the carrier peak is 1.0 V, the depth of modulation is
- ❑ *a* 25%
- ❑ *b* 50%
- ❑ *c* 75%
- ❑ *d* 100%

13.19 A modulated carrier has a peak signal of 1.6 V. If the depth of modulation is 80%, the peak carrier is
- ❑ *a* 1.2 V
- ❑ *b* 1.8 V
- ❑ *c* 2.0 V
- ❑ *d* 2.2 V

13.20 A carrier has a peak value of 2.0 V. If the depth of modulation is 20%, the peak signal is
- ❑ *a* 0.4 V
- ❑ *b* 0.8 V
- ❑ *c* 1.2 V
- ❑ *d* 1.6 V

13.21 The maximum percentage modulation without distortion in an amplitude modulation system is
- ❑ *a* 20%
- ❑ *b* 120%
- ❑ *c* 110%
- ❑ *d* 100%

13.22 If a carrier with a frequency F_c is modulated by a sinusoidal voltage with a frequency F_m, the number of frequencies produced are
- ❑ *a* $F_c - F_m$
- ❑ *b* F_c, $F_c + F_m$, $F_c - F_m$
- ❑ *c* F_c
- ❑ *d* $F_c + F_m$, $F_c - F_m$

13.23 In Question 13.22 the frequency $F_c - F_m$ is called the
- ❑ *a* Modulator
- ❑ *b* Modulating index
- ❑ *c* Upper sideband
- ❑ *d* Lower sideband

13.24 If a carrier is amplitude modulated by a sine wave, the magnitude of the sideband voltages are half the magnitude of the carrier if the modulation depth is
- ❑ *a* 20%
- ❑ *b* 50%
- ❑ *c* 60%
- ❑ *d* 100%

13.25 A 1-MHz carrier is amplitude modulated by a 20-kHz sinusoidal voltage. The frequency of the lower sideband is
- ❑ *a* 0.089 MHz
- ❑ *b* 0.89 MHz
- ❑ *c* 0.098 MHz
- ❑ *d* 0.98 MHz

13.26 The frequency of the upper sideband in Question 13.25 is
- ❑ *a* 1.2 MHz
- ❑ *b* 1.02 MHz
- ❑ *c* 1.002 MHz
- ❑ *d* 1.22 MHz

13.27 A 200-kHz carrier is modulated by a 10-kHz sinusoidal voltage. The value of the upper sideband is
- ❑ *a* 205 000 Hz
- ❑ *b* 210 000 Hz
- ❑ *c* 220 000 Hz
- ❑ *d* 225 000 Hz

13.28 The value of the lower sideband in Question 13.27 is
- ❑ *a* 0.17 MHz
- ❑ *b* 0.18 MHz
- ❑ *c* 0.19 MHz
- ❑ *d* 0.21 MHz

13.29 If a 1-MHz carrier is modulated with a range of frequencies from 100 to 10 000 Hz, the lower sideband will contain frequencies ranging from
- ❑ *a* 989 to 990.9 kHz
- ❑ *b* 989 to 990 kHz
- ❑ *c* 999 to 999.9 kHz
- ❑ *d* 990 to 999.9 kHz

13.30 In Question 13.29 the upper sideband contains a range of frequencies ranging from
- ❑ *a* 1.01 to 1.1 MHz
- ❑ *b* 1.001 to 1.1 MHz
- ❑ *c* 1.0001 to 1.01 MHz
- ❑ *d* 1.001 to 1.01 MHz

13.31 For the transmission of speech and music the sidebands should ideally extend
- ❑ *a* 7 kHz each side of the carrier
- ❑ *b* 10 kHz each side of the carrier
- ❑ *c* 15 kHz each side of the carrier
- ❑ *d* 20 kHz each side of the carrier

13.32 The diagram shows a simplified diagram of an AM sound transmitter. Block A is
- ❑ *a* An RF amplifier
- ❑ *b* An AF amplifier
- ❑ *c* An RF oscillator
- ❑ *d* A modulator

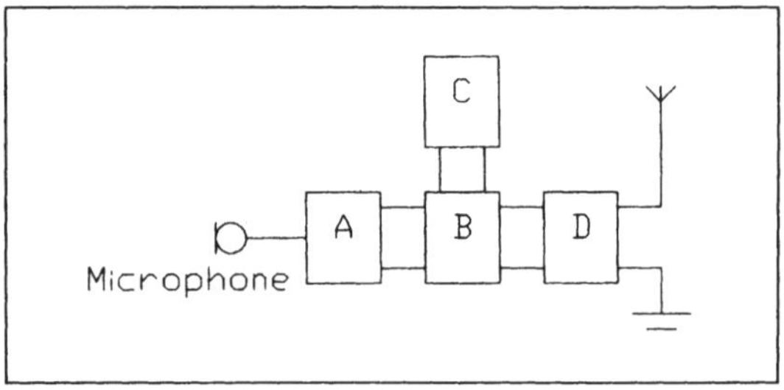

13.33 In Question 13.32 block B is
- ❑ *a* An AF amplifier
- ❑ *b* An RF oscillator
- ❑ *c* An RF amplifier
- ❑ *d* A modulator

13.34 In Question 13.32 block C is
- ❑ *a* An AF amplifier
- ❑ *b* An RF oscillator
- ❑ *c* A modulator
- ❑ *d* An RF amplifier

13.35 In Question 13.32 block D is
- ❑ *a* An RF amplifier
- ❑ *b* An RF oscillator
- ❑ *c* A modulator
- ❑ *d* An AF amplifier

13.36 The diagram shows the first four stages of a superhet receiver for AM signals. Block A is
- ❑ *a* An IF amplifier
- ❑ *b* An RF tuned amplifier
- ❑ *c* An AF voltage amplifier
- ❑ *d* A mixer

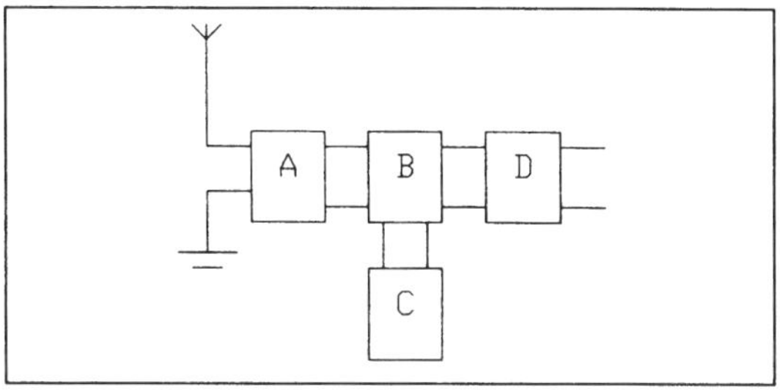

13.37 In Question 13.36 block B is
- ❑ *a* An AM modulator
- ❑ *b* An AF voltage amplifier
- ❑ *c* An AF power amplifier
- ❑ *d* A mixer

13.38 In Question 13.36 block C is
- ❑ *a* An RF amplifier
- ❑ *b* An oscillator
- ❑ *c* A mixer
- ❑ *d* An AF voltage amplifier

13.39 In Question 13.36 block D is
- ❑ *a* An IF amplifier
- ❑ *b* An RF amplifier
- ❑ *c* An AM demodulator
- ❑ *d* An AF voltage amplifier

13.40 The diagram shows the last three stages of a superhet receiver for AM signals driving the loudspeaker. Block E is
- ❑ *a* An AM demodulator
- ❑ *b* An oscillator
- ❑ *c* An AF voltage amplifier
- ❑ *d* An AF power amplifier

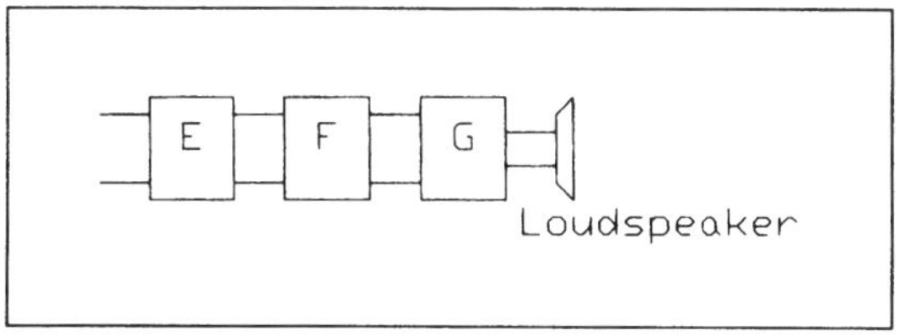

13.41 In Question 13.40 block F is
- ❑ *a* A mixer
- ❑ *b* An AM demodulator
- ❑ *c* An AF voltage amplifier
- ❑ *d* An AF power amplifier

13.42 In Question 13.40 block G is
- ❑ *a* An RF filter
- ❑ *b* An AM demodulator
- ❑ *c* A mixer
- ❑ *d* An AF power amplifier

13.43 In a superhet receiver for AM signals an automatic gain control voltage is fed
- ❑ *a* From the demodulator to the mixer
- ❑ *b* From the demodulator to the IF amplifier
- ❑ *c* From the power amplifier to the oscillator
- ❑ *d* From the voltage amplifier to the mixer

13.44 In a superhet receiver for AM signals the value of the intermediate frequency (IF) is
- ❑ *a* 170 kHz
- ❑ *b* 270 kHz
- ❑ *c* 470 kHz
- ❑ *d* 600 kHz

13.45 If for the superhet of Question 13.44 the receiver is tuned to a 1.2-MHz carrier, the oscillator frequency is
- ❑ *a* 1.37 MHz
- ❑ *b* 1.47 MHz
- ❑ *c* 1.57 MHz
- ❑ *d* 1.67 MHz

13.46 In frequency modulation the maximum variation of carrier frequency required for BBC radio is
- ❑ *a* ±25 000 Hz
- ❑ *b* ±50 000 Hz
- ❑ *c* ±75 000 Hz
- ❑ *d* ±90 000 Hz

13.47 If the maximum modulating frequency transmitted is 12 kHz for a BBC radio station, the modulation index is

❑ *a* 5.0
❑ *b* 6.25
❑ *c* 7.5
❑ *d* 8

13.48 Given that the modulation index is 5 and the maximum variation in carrier is 75 000 Hz, the maximum modulating frequency is

❑ *a* 10 000 Hz
❑ *b* 11 000 Hz
❑ *c* 12 000 Hz
❑ *d* 15 000 Hz

13.49 If the modulation index is 6 and the maximum modulating frequency is 10 000 Hz, the maximum variation in the carrier frequency is

❑ *a* 30 000 Hz
❑ *b* 40 000 Hz
❑ *c* 60 000 Hz
❑ *d* 75 000 Hz

13.50 In frequency modulation, if a carrier is modulated by a single frequency the number of sidebands produced is

❑ *a* 1
❑ *b* 2
❑ *c* 4
❑ *d* Greater than 4

13.51 In a frequency modulated system the upper and lower sidebands are spaced apart by an amount equal to

❑ *a* The modulating frequency
❑ *b* The carrier frequency
❑ *c* Twice the modulating frequency
❑ *d* Twice the carrier frequency

13.52 In Question 13.51, if the number of sidebands is 30 and the modulating frequency is 10 000 Hz the total bandwidth is

❑ *a* 100 000 Hz
❑ *b* 150 000 Hz
❑ *c* 300 000 Hz
❑ *d* 600 000 Hz

13.53 If in Question 13.51 the total bandwidth is 210 000 Hz and the number of sidebands is 20, the modulating frequency is

❑ *a* 10 000 Hz
❑ *b* 10 500 Hz
❑ *c* 11 000 Hz
❑ *d* 15 500 Hz

13.54 The total bandwidth required in an AM system for Question 13.52 is

❑ *a* 10 000 Hz
❑ *b* 20 000 Hz
❑ *c* 30 000 Hz
❑ *d* 40 000 Hz

13.55 The diagram shows a block diagram of a simple FM transmitter. Block A is

❑ *a* An IF amplifier
❑ *b* An RF amplifier
❑ *c* An AF amplifier
❑ *d* An RF oscillator

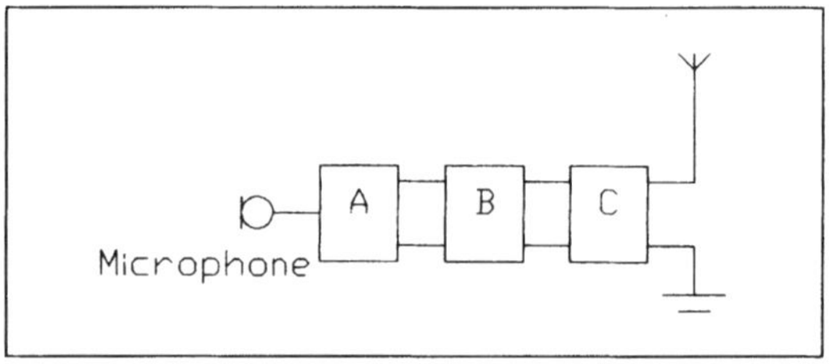

13.56 In the diagram in Question 13.55 block B is

❑ *a* An RF oscillator
❑ *b* An AF amplifier
❑ *c* An RF amplifier
❑ *d* An IF amplifier

13.57 In the diagram in Question 13.55 block C is

❑ *a* An RF oscillator
❑ *b* An RF amplifier
❑ *c* An AF amplifier
❑ *d* An IF amplifier

13.58 The block diagram shows the first four stages of a superhet FM receiver. Block A is
- ❑ *a* A mixer
- ❑ *b* An RF amplifier
- ❑ *c* An FM demodulator
- ❑ *d* An oscillator

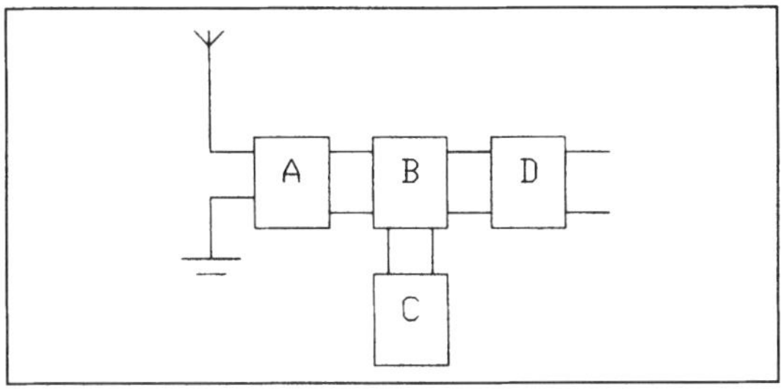

13.59 In Question 13.58 block B is
- ❑ *a* An IF amplifier
- ❑ *b* An FM demodulator
- ❑ *c* An AF power amplifier
- ❑ *d* A mixer

13.60 In Question 13.58 block C is
- ❑ *a* An oscillator
- ❑ *b* An IF amplifier
- ❑ *c* An RF amplifier
- ❑ *d* An AF voltage amplifier

13.61 In Question 13.58 block D is
- ❑ *a* An RF amplifier
- ❑ *b* An IF amplifier
- ❑ *c* An oscillator
- ❑ *d* An AF power amplifier

13.62 The diagram shows the last three stages of a superhet FM receiver. Block E is
- ❑ *a* An AF voltage amplifier
- ❑ *b* An RF amplifier
- ❑ *c* An FM demodulator
- ❑ *d* An AF power amplifier

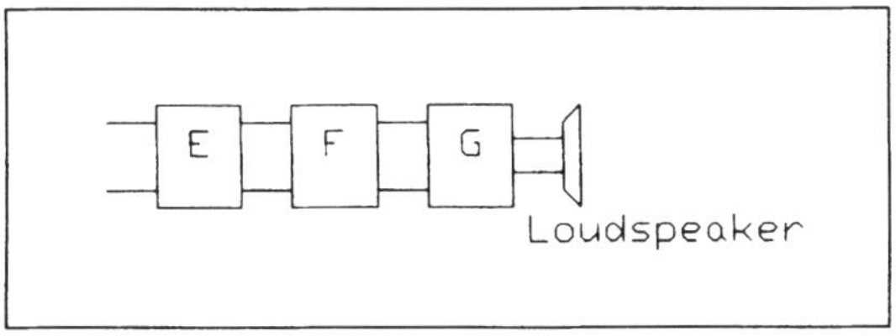

13.63 In Question 13.62 block F is
- ❑ *a* An FM demodulator
- ❑ *b* An AF voltage amplifier
- ❑ *c* An oscillator
- ❑ *d* An AF power amplifier

13.64 In Question 13.62 block G is
- ❑ *a* An AF power amplifier
- ❑ *b* An AF voltage amplifier
- ❑ *c* An FM demodulator
- ❑ *d* An RF amplifier

13.65 In frequency modulation, noise components which affect the amplitude can be eliminated using
- ❑ *a* An oscillator
- ❑ *b* An integrator
- ❑ *c* An FM limiter circuit
- ❑ *d* A differentiator

13.66 In the superhet FM receiver a ratio detector can be used in the
- ❑ *a* Oscillator
- ❑ *b* Demodulator
- ❑ *c* Mixer
- ❑ *d* RF amplifier

13.67 In an FM receiver an automatic frequency control (AFC) circuit is incorporated between
- ❑ *a* IF amplifier and oscillator
- ❑ *b* IF amplifier and mixer
- ❑ *c* Demodulator and oscillator
- ❑ *d* Demodulator and mixer

13.68 In a typical FM receiver the IF amplifier is tuned to a frequency of
- ❑ *a* 470 kHz + sidebands
- ❑ *b* 4.7 MHz + sidebands
- ❑ *c* 8.9 MHz + sidebands
- ❑ *d* 10.7 MHz + sidebands

13.69 In an FM receiver the RF amplifier is tuned to a frequency of 101 MHz. The oscillator frequency should be
- ❑ *a* 101.7 MHz
- ❑ *b* 102.7 MHz
- ❑ *c* 110.7 MHz
- ❑ *d* 111.7 MHz

13.70 If the output frequency from the mixer in an FM receiver is 10.7 MHz and the oscillator is running at 105.7 MHz, the receiver is tuned to a frequency of

- ❑ *a* 95 MHz
- ❑ *b* 96 MHz
- ❑ *c* 97 MHz
- ❑ *d* 98 MHz

13.71 All mains-operated receivers need a power supply unit which

- ❑ *a* Produces oscillations
- ❑ *b* Produces amplification
- ❑ *c* Changes the AC supply to DC voltages
- ❑ *d* Changes a DC supply into AC

13.72 All superhet receivers require an oscillator. An oscillator is a device which

- ❑ *a* Amplifies a sine wave
- ❑ *b* Differentiates a square wave
- ❑ *c* Rectifies an AC voltage
- ❑ *d* Produces AC from DC

13.73 The main difference between the circuits of the AM and FM receivers lies in the

- ❑ *a* Mixer circuit
- ❑ *b* Demodulator circuit
- ❑ *c* Oscillator circuit
- ❑ *d* Power amplifier circuit

13.74 In order to avoid distortion, the maximum percentage modulation should not exceed

- ❑ *a* 89%
- ❑ *b* 90%
- ❑ *c* 100%
- ❑ *d* 110%

13.75 The sensitivity of a radio receiver can be increased by

- ❑ *a* Employing more than one IF amplifier
- ❑ *b* Having more than one oscillator
- ❑ *c* Having more than one demodulator
- ❑ *d* Rectification

14 Television and tape recorders

14.1 The television system used in the UK is called
- ❑ *a* PCM
- ❑ *b* PAL
- ❑ *c* PPM
- ❑ *d* PWM

14.2 The number of lines in the colour television system is
- ❑ *a* 600
- ❑ *b* 605
- ❑ *c* 615
- ❑ *d* 625

14.3 PAL stands for
- ❑ *a* Phase amplitude line
- ❑ *b* Pulse alternate line
- ❑ *c* Phase alternate line
- ❑ *d* Pulse amplitude line

14.4 In North and South America, colour television has
- ❑ *a* 500 lines
- ❑ *b* 525 lines
- ❑ *c* 605 lines
- ❑ *d* 625 lines

14.5 The diagram shows a simplified block diagram of the electromagnetic spectrum. Television is radiated in block
- ❑ *a* 1
- ❑ *b* 2
- ❑ *c* 3
- ❑ *d* 4

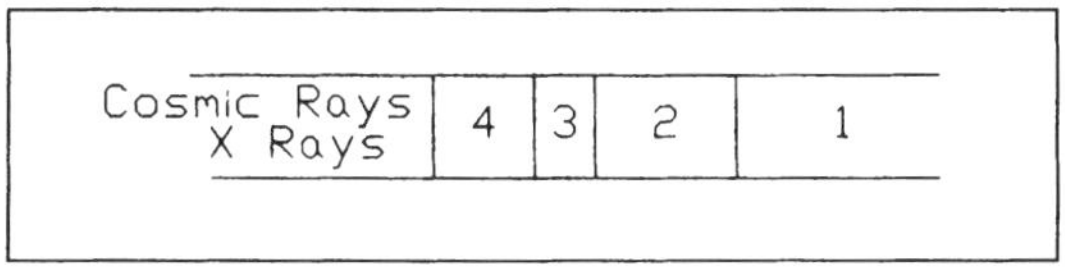

14.6 In Question 14.5 visible light is radiated in block
- ❑ *a* 1
- ❑ *b* 2
- ❑ *c* 3
- ❑ *d* 4

14.7 In British television the operating frequencies in Band 1 are
- ❑ *a* 20–30 MHz
- ❑ *b* 30–35 MHz
- ❑ *c* 41–68 MHz
- ❑ *d* 75–100 MHz

14.8 In British television the operating frequencies in Band 3 are
- ❑ *a* 174–216 MHz
- ❑ *b* 300–380 MHz
- ❑ *c* 390–470 MHz
- ❑ *d* 600–690 MHz

14.9 In British television the operating frequencies in Band 4 are
- ❑ *a* 200–300 MHz
- ❑ *b* 310–360 MHz
- ❑ *c* 400–450 MHz
- ❑ *d* 470–582 MHz

14.10 In British television the operating frequencies in Band 5 are
- ❑ *a* 614–854 MHz
- ❑ *b* 860–900 MHz
- ❑ *c* 900–940 MHz
- ❑ *d* 940–995 MHz

14.11 The Band 1 frequencies are in the
- ❑ *a* VLF band
- ❑ *b* UHF band
- ❑ *c* VHF band
- ❑ *d* SHF band

14.12 The Band 5 frequencies are in the

- ❑ *a* SHF band
- ❑ *b* UHF band
- ❑ *c* HF band
- ❑ *d* VHF band

14.13 The diagram shows the visible wave band. The colour in block 2 is

- ❑ *a* Orange
- ❑ *b* Yellow
- ❑ *c* Blue
- ❑ *d* Green

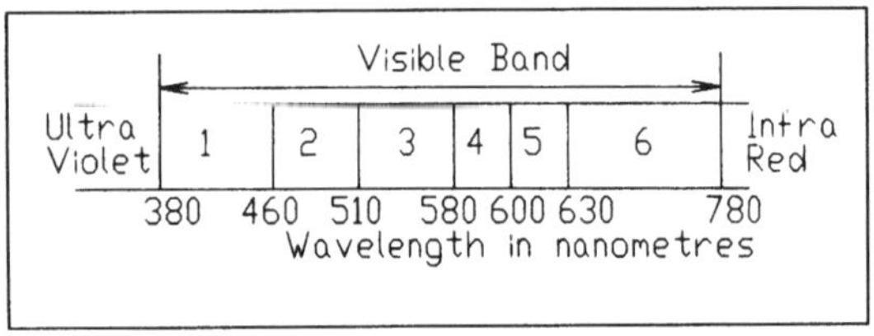

14.14 In Question 14.13 the colour in block 3 is

- ❑ *a* Green
- ❑ *b* Blue
- ❑ *c* Orange
- ❑ *d* Yellow

14.15 In Question 14.13 the colour in block 4 is

- ❑ *a* Green
- ❑ *b* Blue
- ❑ *c* Orange
- ❑ *d* Yellow

14.16 In Question 14.13 the colour in block 5 is

- ❑ *a* Green
- ❑ *b* Blue
- ❑ *c* Orange
- ❑ *d* Yellow

14.17 The wavelengths in nanometres (nm) of the green colour extend approximately between

- ❑ *a* 380 and 460
- ❑ *b* 460 and 510
- ❑ *c* 510 and 580
- ❑ *d* 630 and 780

14.18 The wavelengths of the red colour extend approximately between

- ❑ *a* 510 and 580 nm
- ❑ *b* 580 and 600 nm
- ❑ *c* 600 and 630 nm
- ❑ *d* 630 and 780 nm

14.19 The wavelengths of the orange colour extend approximately between

- ❑ *a* 510 and 580 nm
- ❑ *b* 580 and 600 nm
- ❑ *c* 600 and 630 nm
- ❑ *d* 630 and 780 nm

14.20 The wavelengths of the blue colour extend approximately between

- ❑ *a* 460 and 510 nm
- ❑ *b* 510 and 580 nm
- ❑ *c* 580 and 600 nm
- ❑ *d* 600 and 630 nm

14.21 The wavelengths of the yellow colour extend approximately between

- ❑ *a* 460 and 510 nm
- ❑ *b* 510 and 580 nm
- ❑ *c* 580 and 600 nm
- ❑ *d* 600 and 630 nm

14.22 The wavelengths of the violet colour extend approximately between

- ❑ *a* 460 and 510 nm
- ❑ *b* 380 and 460 nm
- ❑ *c* 510 and 580 nm
- ❑ *d* 630 and 780 nm

14.23 The frequency spectrum of visible light extends approximately from

- ❑ *a* 4.6×10^5 to 5.8×10^5 MHz
- ❑ *b* 3.8×10^6 to 7.8×10^6 MHz
- ❑ *c* 3.8×10^8 to 7.8×10^8 MHz
- ❑ *d* 4.6×10^8 to 5.8×10^8 MHz

14.24 A light which contains all the frequencies of the visible spectrum in the correct proportions is

- ❑ *a* Black
- ❑ *b* White
- ❑ *c* Cyan
- ❑ *d* Magenta

14.25 The three primary colours are produced by three camera tubes. These colours are
- ❑ *a* Blue, orange, yellow
- ❑ *b* Green, red, violet
- ❑ *c* Violet, blue, orange
- ❑ *d* Red, blue, green

14.26 Mixing green and blue light gives
- ❑ *a* Cyan
- ❑ *b* Yellow
- ❑ *c* Magenta
- ❑ *d* Violet

14.27 Mixing red and blue lights gives
- ❑ *a* Orange
- ❑ *b* Cyan
- ❑ *c* Green
- ❑ *d* Magenta

14.28 Mixing red and green lights gives
- ❑ *a* Cyan
- ❑ *b* White
- ❑ *c* Orange
- ❑ *d* Yellow

14.29 The three complementary or secondary colours are
- ❑ *a* White, yellow, cyan
- ❑ *b* Black, yellow, magenta
- ❑ *c* Yellow, magenta, cyan
- ❑ *d* White, magenta, cyan

14.30 When the colours cyan and magenta are mixed the resultant colour is
- ❑ *a* Yellow
- ❑ *b* White
- ❑ *c* Orange
- ❑ *d* Blue

14.31 When the colours yellow and magenta are mixed the resultant colour is
- ❑ *a* White
- ❑ *b* Orange
- ❑ *c* Blue
- ❑ *d* Red

14.32 When the colours cyan and yellow are mixed the resultant colour is
- ❑ *a* Red
- ❑ *b* White
- ❑ *c* Blue
- ❑ *d* Green

14.33 In a television camera the wavelength of the red signal is approximately
- ❑ *a* 640 nm
- ❑ *b* 650 nm
- ❑ *c* 700 nm
- ❑ *d* 770 nm

14.34 In a television camera the wavelength of the green signal is approximately
- ❑ *a* 546 nm
- ❑ *b* 566 nm
- ❑ *c* 570 nm
- ❑ *d* 575 nm

14.35 In a television camera the wavelength of the blue signal is approximately
- ❑ *a* 400 nm
- ❑ *b* 420 nm
- ❑ *c* 436 nm
- ❑ *d* 500 nm

14.36 A studio camera produces a chrominance signal. This signal contains information about
- ❑ *a* The musical content
- ❑ *b* The speech content
- ❑ *c* The brightness of the scene
- ❑ *d* The colour content of the scene

14.37 A studio camera also produces a luminance signal which gives information about
- ❑ *a* The colour content of the scene
- ❑ *b* The brightness content of the scene
- ❑ *c* The musical content
- ❑ *d* The speech content

14.38 In the British 625-line system the type of modulation used for the vision signal is
- ❑ *a* Frequency
- ❑ *b* Amplitude
- ❑ *c* Pulse duration
- ❑ *d* Pulse code

14.39 In the British 625-line system the type of modulation used for the sound signal is
- ❑ *a* Pulse position
- ❑ *b* Pulse code
- ❑ *c* Frequency
- ❑ *d* Amplitude

14.40 The diagram shows a simplified block diagram of the first six stages of a colour television receiver. Block A is
- ❑ *a* An IF amplifier
- ❑ *b* An AFC stage
- ❑ *c* A UHF tuner
- ❑ *d* A video detector

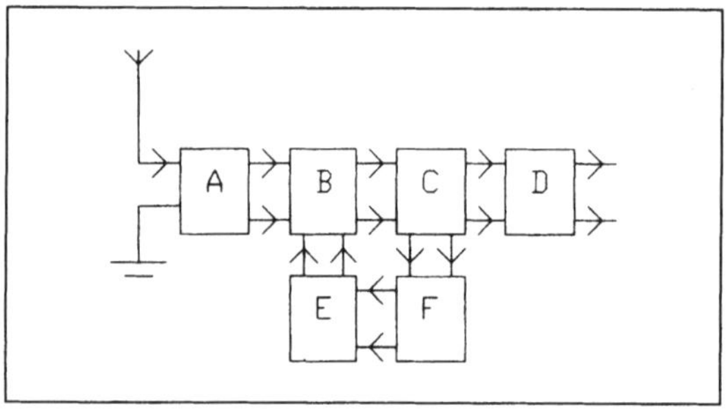

14.41 In Question 14.40 block B is
- ❑ *a* A UHF amplifier
- ❑ *b* An oscillator
- ❑ *c* An AFC stage
- ❑ *d* A mixer

14.42 In Question 14.40 block C is
- ❑ *a* An IF amplifier
- ❑ *b* A UHF amplifier
- ❑ *c* A mixer
- ❑ *d* An AFC stage

14.43 In Question 14.40 block D is
- ❑ *a* A mixer
- ❑ *b* A video detector
- ❑ *c* An oscillator
- ❑ *d* An AFC stage

14.44 In Question 14.40 block E is
- ❑ *a* A UHF amplifier
- ❑ *b* A video detector
- ❑ *c* An oscillator
- ❑ *d* A mixer

14.45 In Question 14.40 block F is
- ❑ *a* An AFC stage
- ❑ *b* A video detector
- ❑ *c* A UHF amplifier
- ❑ *d* An oscillator

14.46 The diagram shows the sound system in a television receiver. Block A is
- ❑ *a* A timebase
- ❑ *b* A sound amplifier
- ❑ *c* An FM demodulator
- ❑ *d* An oscillator

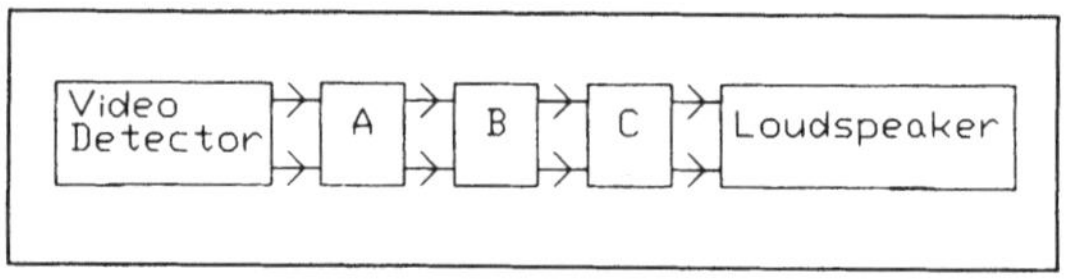

14.47 In Question 14.46 block B is
- ❑ *a* An AF amplifier
- ❑ *b* A sound amplifier
- ❑ *c* A timebase
- ❑ *d* An FM demodulator

14.48 In Question 14.46 block C is
- ❑ *a* An oscillator
- ❑ *b* An FM demodulator
- ❑ *c* An AF amplifier
- ❑ *d* A timebase

14.49 The diagram shows the inputs to the tricolour cathode ray tube's three-electron guns. Block A is
- ❑ *a* An FM modulator
- ❑ *b* A luminance amplifier
- ❑ *c* A chrominance amplifier
- ❑ *c* A colour decoder

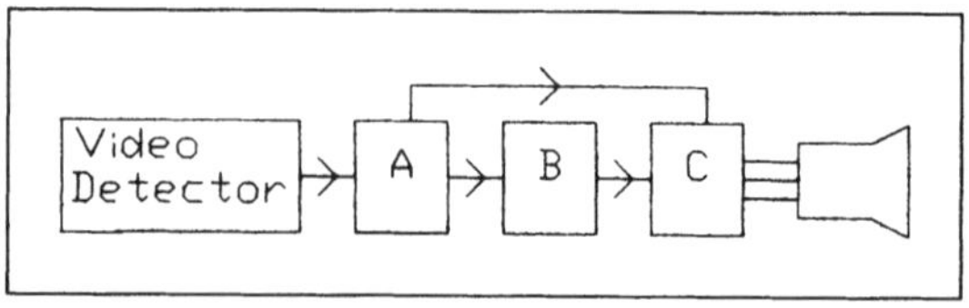

14.50 In Question 14.49 block B is
- ❑ *a* A colour decoder
- ❑ *b* A chrominance amplifier
- ❑ *c* A luminance amplifier
- ❑ *d* An IF amplifier

14.51 In Question 14.49 block C is
- ❑ *a* A colour decoder
- ❑ *b* A chrominance amplifier
- ❑ *c* An IF amplifier
- ❑ *d* An AF amplifier

14.52 The diagram shows the remaining inputs to the cathode ray tube. Block A is
- ❑ *a* A line timebase
- ❑ *b* A convergence circuit
- ❑ *c* A field timebase
- ❑ *d* A sync separator

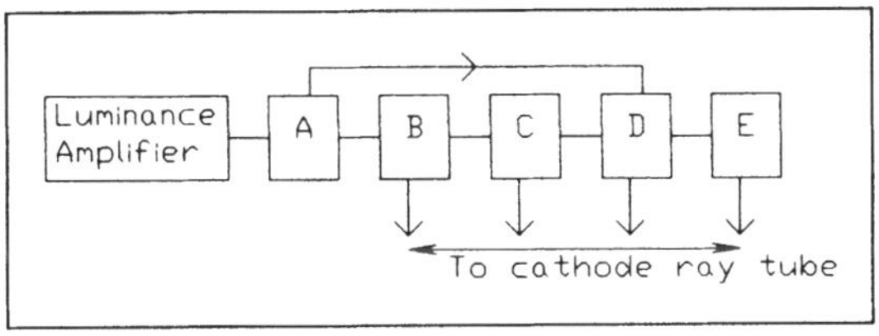

14.53 In Question 14.52 block B is
- ❑ *a* A line timebase
- ❑ *b* A field timebase
- ❑ *c* A sync separator
- ❑ *d* An EHT circuit

14.54 In Question 14.53 block C is
- ❑ *a* An EHT circuit
- ❑ *b* A sync separator
- ❑ *c* A convergence circuit
- ❑ *d* A line timebase

14.55 In Question 14.53 block D is
- ❑ *a* A line timebase
- ❑ *b* An EHT circuit
- ❑ *c* A sync separator
- ❑ *d* A field timebase

14.56 In Question 14.53 block E is
- ❑ *a* An EHT circuit
- ❑ *b* A sync separator
- ❑ *c* A field timebase
- ❑ *d* A line timebase

14.57 The line timebase scans 625 lines in
- ❑ *a* 0.01 s
- ❑ *b* 0.02 s
- ❑ *c* 0.03 s
- ❑ *d* 0.04 s

14.58 The frequency of the line timebase is
- ❑ *a* 10 625 Hz
- ❑ *b* 15 625 Hz
- ❑ *c* 18 425 Hz
- ❑ *d* 20 425 Hz

14.59 The frequency of the field timebase is
- ❑ *a* 25 Hz
- ❑ *b* 50 Hz
- ❑ *c* 100 Hz
- ❑ *d* 625 Hz

14.60 The cathode ray tube will need an EHT supply in the region of
- ❑ *a* 240–410 V
- ❑ *b* 600–800 V
- ❑ *c* 2–4 kV
- ❑ *d* 18–24 kV

14.61 If a television receiver is tuned to a station with a frequency of 700 MHz, the oscillator frequency will be approximately
- ❑ *a* 704.5 MHz
- ❑ *b* 739.5 MHz
- ❑ *c* 800.6 MHz
- ❑ *d* 850.5 MHz

14.62 The line timebase in a television receiver
- ❑ *a* Deflects the cathode ray tube spot horizontally
- ❑ *b* Deflects the cathode ray tube spot vertically
- ❑ *c* Changes the brightness of the spot
- ❑ *d* Focuses the cathode ray tube spot

14.63 The minimum spacing between channels in the British 625 line system is
- ❑ *a* 2 MHz
- ❑ *b* 4 MHz
- ❑ *c* 6 MHz
- ❑ *d* 8 MHz

14.64 The bandwidth of a video amplifier in a television receiver is

- ❑ *a* 2.4 MHz
- ❑ *b* 4.2 MHz
- ❑ *c* 5.5 MHz
- ❑ *d* 10.5 MHz

14.65 Plastic tapes used in tape recorders use

- ❑ *a* Particles of zinc
- ❑ *b* Particles of lead
- ❑ *c* Particles of iron oxide
- ❑ *d* Particles of manganin

14.66 The front gap width in a recording head must be very narrow in order to give a satisfactory

- ❑ *a* Low frequency response
- ❑ *b* High frequency response
- ❑ *c* Noise level
- ❑ *d* Amplification figure

14.67 With a tape speed of 7.5 inches per second, one cycle of a 20-kHz signal will occupy a length on the tape of

- ❑ *a* 1.5×10^{-2} inches
- ❑ *b* 1.85×10^{-3} inches
- ❑ *c* 2.15×10^{-3} inches
- ❑ *d* 3.75×10^{-4} inches

14.68 The frequency of the bias signal in a tape recorder lies in the range

- ❑ *a* 5–10 Hz
- ❑ *b* 5–15 kHz
- ❑ *c* 40–100 kHz
- ❑ *d* 2–4 MHz

14.69 The bias signal in a tape recorder is used to

- ❑ *a* Eliminate distortion
- ❑ *b* Eliminate noise
- ❑ *c* Eliminate wow
- ❑ *d* Eliminate flutter

14.70 The diagram shows a simplified block diagram of a tape recorder. Block 1 is

- ❑ *a* A playback head
- ❑ *b* An erase head
- ❑ *c* An audio amplifier
- ❑ *d* A playback amplifier

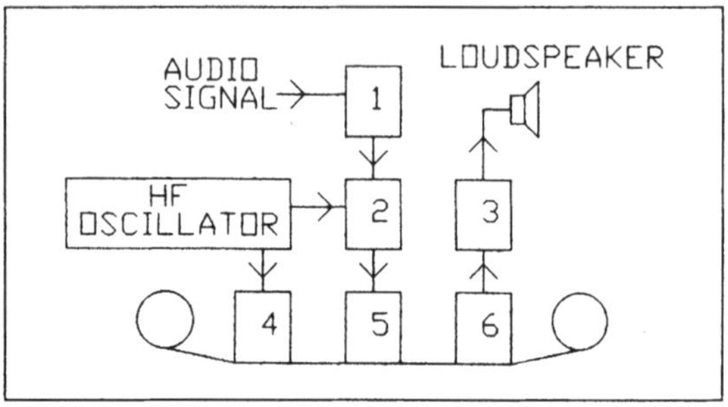

14.71 In Question 14.70 block 2 is

- ❑ *a* A playback amplifier
- ❑ *b* A mixer
- ❑ *c* A record head
- ❑ *d* An erase head

14.72 In Question 14.70 block 3 is

- ❑ *a* An erase head
- ❑ *b* A record head
- ❑ *c* A mixer
- ❑ *d* A playback amplifier

14.73 In Question 14.70 block 4 is

- ❑ *a* An erase head
- ❑ *b* A record head
- ❑ *c* A mixer
- ❑ *d* A playback head

14.74 In Question 14.70 block 5 is

- ❑ *a* A mixer
- ❑ *b* An erase head
- ❑ *c* A record head
- ❑ *d* A playback head

14.75 In Question 14.70 block 6 is

- ❑ *a* An erase head
- ❑ *b* A playback head
- ❑ *c* A mixer
- ❑ *d* A record head

15 Voltage supplies and health and safety

15.1 The diagram shows a two-phase three-wire voltage system where V_1 and V_2 are displaced by 90 °. If $V_1 = V_2 = 2000$ V, the voltage V_3 is

❑ *a* 2222 V
❑ *b* 2428 V
❑ *c* 2828 V
❑ *d* 5656 V

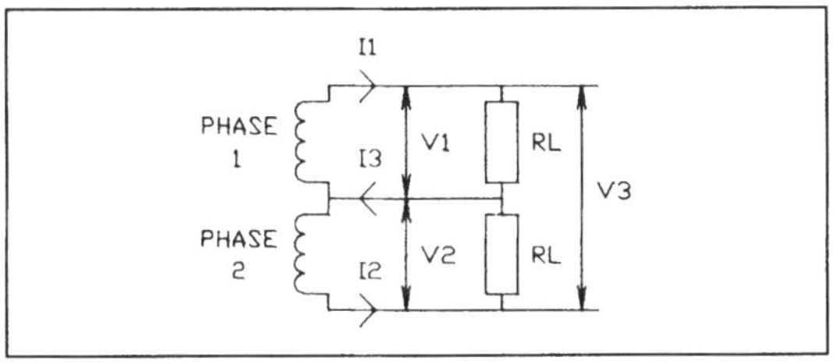

15.2 If in Question 15.1 the currents I_1 and I_2 are displaced by 90 ° and $I_1 = I_2 = 240$ A, the current I_3 is

❑ *a* 248.42 A
❑ *b* 339.36 A
❑ *c* 440.62 A
❑ *d* 480.24 A

15.3 If in Question 15.1 the phase voltages V_1 and V_2 are both 10 kV, the voltage between the outer conductors V_3 is

❑ *a* 14.14 kV
❑ *b* 18.18 kV
❑ *c* 20.14 kV
❑ *d* 28.28 kV

15.4 If in Question 15.1 the phase currents I_1 and I_2 are both 60 A, the current I_3 in the neutral wire is

❑ *a* 72.44 A
❑ *b* 76.32 A
❑ *c* 80.24 A
❑ *d* 84.84 A

15.5 A voltage supply system identical to the one in Question 15.1 has 5656 V between the outer conductors. The phase voltages are

❑ *a* 2 kV
❑ *b* 3 kV
❑ *c* 4 kV
❑ *d* 5 kV

15.6 The phase currents in a two-phase three-wire system are 120 A and 160 A. If the currents are in phase with their respective voltages, the current in the neutral wire I_3 is

❑ *a* 175 A
❑ *b* 180 A
❑ *c* 200 A
❑ *d* 210 A

15.7 If in Question 15.6 the currents are 15 A and 20 A, the current in the neutral is

❑ *a* 25 A
❑ *b* 30 A
❑ *c* 35 A
❑ *d* 40 A

15.8 If in Question 15.6 the currents are 60 A and 80 A, the current in the neutral is

❑ *a* 95 A
❑ *b* 100 A
❑ *c* 120 A
❑ *d* 125 A

15.9 If in a two-phase three-wire system the current in the neutral is 12.5 A and the current in phase 1 is 10 A, the current in phase 2 is

❑ *a* 6.4 A
❑ *b* 7.5 A
❑ *c* 8.5 A
❑ *d* 8.8 A

15.10 If in Question 15.9 the current in the neutral wire is 150 A and the current in phase 1 is 90 A, the current in phase 2 is

- ❑ *a* 100 A
- ❑ *b* 105 A
- ❑ *c* 110 A
- ❑ *d* 120 A

15.11 If in Question 15.9 the current in the neutral wire is 75 A and the current in phase 2 is 60 A, the current in phase 1 is

- ❑ *a* 40 A
- ❑ *b* 42 A
- ❑ *c* 45 A
- ❑ *d* 50 A

15.12 The diagram shows a star-connected secondary winding of a transformer. If the phase voltages are 231 V, the line voltage V_L is

- ❑ *a* 330 V
- ❑ *b* 360 V
- ❑ *c* 400 V
- ❑ *d* 440 V

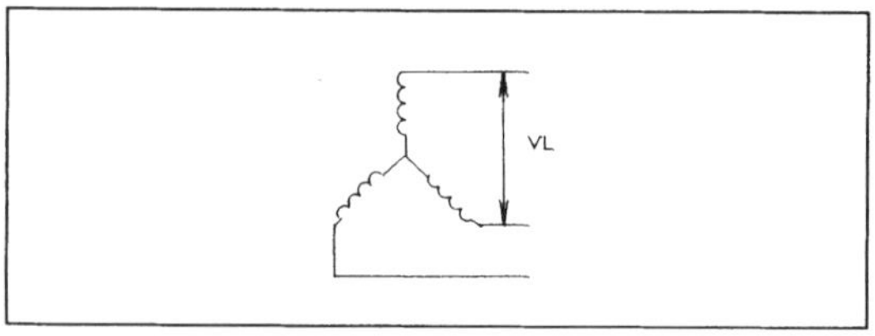

15.13 In Question 15.12, if the phase currents are 200 A, the line currents are

- ❑ *a* 200 A
- ❑ *b* 282 A
- ❑ *c* 346 A
- ❑ *d* 400 A

15.14 If in Question 15.12 the phase voltages are 500 V, the line voltages are

- ❑ *a* 470 V
- ❑ *b* 700 V
- ❑ *c* 760 V
- ❑ *d* 865 V

15.15 If in Question 15.12 the line voltages are 1038 V, the phase voltages are

- ❑ *a* 400 V
- ❑ *b* 600 V
- ❑ *c* 800 V
- ❑ *d* 900 V

15.16 A three-phase 240-V delta-connected induction motor takes a line current of 50 A at a power factor of 0.9. The input power which is the active power is

- ❑ *a* 16.6 kW
- ❑ *b* 18.7 kW
- ❑ *c* 19.2 kW
- ❑ *d* 19.8 kW

15.17 In Question 15.16 the apparent power is

- ❑ *a* 12.42 kVA
- ❑ *b* 14.81 kVA
- ❑ *c* 16.33 kVA
- ❑ *d* 20.76 kVA

15.18 In Question 15.16 the phase currents are

- ❑ *a* 26.6 A
- ❑ *b* 28.9 A
- ❑ *c* 33.2 A
- ❑ *d* 40.8 A

15.19 A three-phase 410-V delta-connected induction motor takes a line current of 45 A at a power factor of 0.9. The active power is

- ❑ *a* 24.308 kW
- ❑ *b* 26.612 kW
- ❑ *c* 28.727 kW
- ❑ *d* 31.604 kW

15.20 In Question 15.19 the apparent power is

- ❑ *a* 18.402 kVA
- ❑ *b* 22.612 kVA
- ❑ *c* 24.813 kVA
- ❑ *d* 31.919 kVA

15.21 In Question 15.19 the phase currents are

- ❑ *a* 26 A
- ❑ *b* 32 A
- ❑ *c* 34 A
- ❑ *d* 38 A

15.22 A three-phase 650-V delta-connected induction motor takes a line current of 30 A at a power factor of 0.8. The active power is

- ❑ *a* 24.667 kW
- ❑ *b* 26.988 kW
- ❑ *c* 30.667 kW
- ❑ *d* 42.343 kW

15.23 In Question 15.22 the apparent power is

- ❑ *a* 25.168 kVA
- ❑ *b* 27.882 kVA
- ❑ *c* 28.621 kVA
- ❑ *d* 33.735 kVA

15.24 In Question 15.22 the phase currents are

- ❑ *a* 17.34 A
- ❑ *b* 18.62 A
- ❑ *c* 20.47 A
- ❑ *d* 24.56 A

15.25 The diagram shows an alternator connected in star. If the phase voltage V_P is 200 V and the phase current on full load is 20 A, the line voltage V_L is

- ❑ *a* 300 V
- ❑ *b* 346 V
- ❑ *c* 400 V
- ❑ *d* 410 V

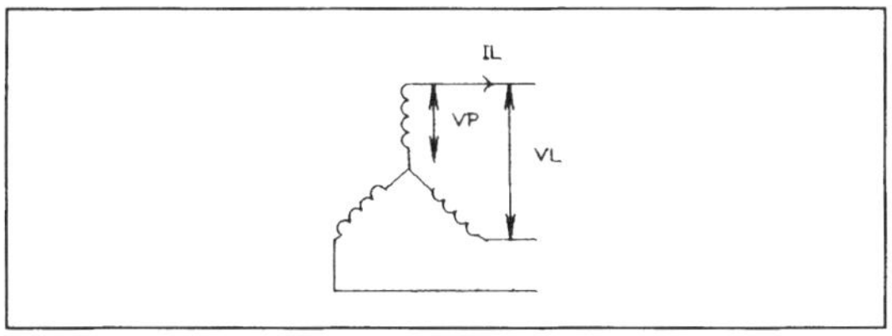

15.26 In Question 15.25 the line current I_L is

- ❑ *a* 20 A
- ❑ *b* 28 A
- ❑ *c* 34 A
- ❑ *d* 36 A

15.27 In Question 15.25 the apparent output power is

- ❑ *a* 10.12 kVA
- ❑ *b* 10.98 kVA
- ❑ *c* 11.97 kVA
- ❑ *d* 12.16 kVA

15.28 The diagram shows the alternator of Question 15.25 connected in delta and delivering the same current. The line voltage V_L is

- ❑ *a* 200 V
- ❑ *b* 310 V
- ❑ *c* 410 V
- ❑ *d* 450 V

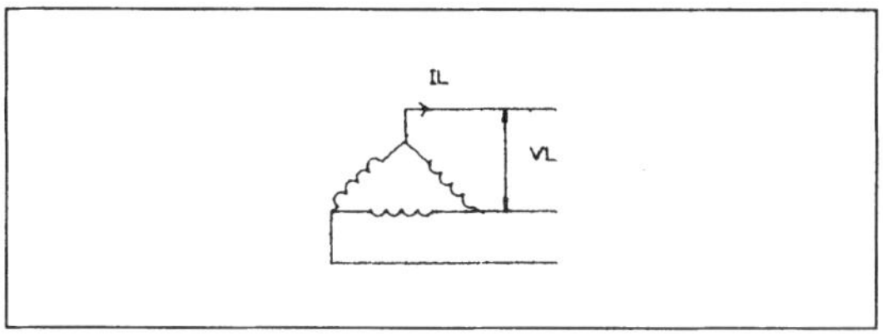

15.29 In Question 15.28 the line current I_L is

- ❑ *a* 26.4 A
- ❑ *b* 28.1 A
- ❑ *c* 30.3 A
- ❑ *d* 34.6 A

15.30 In Question 15.28 the apparent output power is

- ❑ *a* 10.67 kVA
- ❑ *b* 11.97 kVA
- ❑ *c* 12.62 kVA
- ❑ *d* 13.77 kVA

15.31 An alternator is connected in star. When fully loaded the phase current is 60 A. If the phase voltage is 110 V, the line voltage is

- ❑ *a* 150.4 V
- ❑ *b* 160.6 V
- ❑ *c* 175.6 V
- ❑ *d* 190.3 V

15.32 In Question 15.31 the line current is

- ❑ *a* 35 A
- ❑ *b* 45 A
- ❑ *c* 60 A
- ❑ *d* 65 A

15.33 In Question 15.31 the apparent output power is

- ❑ *a* 18.612 kVA
- ❑ *b* 19.753 kVA
- ❑ *c* 20.667 kVA
- ❑ *d* 22.334 kVA

15.34 If the alternator in Question 15.31 is connected in mesh supplying the same load, the line voltage is
- ❑ *a* 110 V
- ❑ *b* 190 V
- ❑ *c* 220 V
- ❑ *d* 240 V

15.35 In Question 15.34 the line current is
- ❑ *a* 88.8 A
- ❑ *b* 90.6 A
- ❑ *c* 103.8 A
- ❑ *d* 110.6 A

15.36 In Question 15.34 the apparent output power is
- ❑ *a* 16.824 kVA
- ❑ *b* 19.753 kVA
- ❑ *c* 20.381 kVA
- ❑ *d* 40.671 kVA

15.37 Soda acid and carbon dioxide are types of portable
- ❑ *a* Foam extinguishers
- ❑ *b* Water extinguishers
- ❑ *c* Vaporizing liquid extinguishers
- ❑ *d* Dry powder extinguishers

15.38 Foam fire extinguishers are coloured
- ❑ *a* Black
- ❑ *b* Red
- ❑ *c* Green
- ❑ *d* White or cream

15.39 Water fire extinguishers are coloured
- ❑ *a* Red
- ❑ *b* Green
- ❑ *c* Blue
- ❑ *d* Black

15.40 Dry powder fire extinguishers are coloured
- ❑ *a* Black
- ❑ *b* Red
- ❑ *c* Blue
- ❑ *d* Green

15.41 Vaporizing liquid fire extinguishers are coloured
- ❑ *a* Red
- ❑ *b* Blue
- ❑ *c* Black
- ❑ *d* Green

15.42 Carbon dioxide gas fire extinguishers are coloured
- ❑ *a* Blue
- ❑ *b* Black
- ❑ *c* Red
- ❑ *d* Green

15.43 Burning materials such as paper, wood and textiles constitute a
- ❑ *a* Class A fire
- ❑ *b* Class B fire
- ❑ *c* Class C fire
- ❑ *d* Class D fire

15.44 Burning petrol, paint, oils and fats constitute a
- ❑ *a* Class A fire
- ❑ *b* Class B fire
- ❑ *c* Class C fire
- ❑ *d* Class D fire

15.45 A fire involving live electrical equipment where there is a danger of shock is classed
- ❑ *a* A
- ❑ *b* B
- ❑ *c* C
- ❑ *d* D

15.46 The most effective agent for a fire involving wood and paper is
- ❑ *a* Carbon dioxide gas
- ❑ *b* Dry powder
- ❑ *c* Foam
- ❑ *d* Water

15.47 The most effective agent for a fire involving liquids, gases and electrical equipment is
- ❑ *a* Water
- ❑ *b* Dry powder
- ❑ *c* Vaporizing liquid
- ❑ *d* Dry powder

15.48 The most effective agent for a fire involving electrical and electronic equipment is

❑ *a* Foam
❑ *b* Dry powder
❑ *c* Carbon dioxide gas
❑ *d* Water

15.49 The most effective agent for a fire involving petrol, oil, fat and paint is

❑ *a* Foam
❑ *b* Carbon dioxide gas
❑ *c* Water
❑ *d* Dry powder

15.50 The most effective agent for a fire involving motor vehicles and laboratory equipment is

❑ *a* Water
❑ *b* Foam
❑ *c* Vaporizing liquid
❑ *d* Dry powder

15.51 An individual can be protected against electrostatic build-up by using

❑ *a* Goggles
❑ *b* Safety boots
❑ *c* A helmet
❑ *d* A face screen

15.52 An individual can be protected against inhaling toxic gases and vapours by using

❑ *a* Gloves
❑ *b* A respirator
❑ *c* A face screen
❑ *d* Proper clothing

15.53 An electrical engineer can be protected against electric shock by

❑ *a* Wearing spectacles
❑ *b* Face screens
❑ *c* Using earplugs
❑ *d* Wearing gloves

15.54 Protection against high-intensity noise can be achieved using

❑ *a* Goggles
❑ *b* Helmets
❑ *c* Earplugs or muffs
❑ *d* Gloves

15.55 Protection against falling or flying objects is achieved using

❑ *a* Helmets
❑ *b* Earplugs
❑ *c* Masks
❑ *d* Safety boots

15.56 The eyes can be protected against chemical splash using

❑ *a* Clogs
❑ *b* Aprons
❑ *c* Spectacles
❑ *d* Boiler suits

15.57 Nitrogen oxides and carbon monoxide are examples of hazardous

❑ *a* Vapours
❑ *b* Fumes
❑ *c* Dusts
❑ *d* Gases

15.58 Zinc cadmium is an example of a hazardous

❑ *a* Fume
❑ *b* Dust
❑ *c* Gas
❑ *d* Vapour

15.59 When averaged over an 8-hour day the HSE Code of Practice recommends that no one should be exposed to a noise level of more than

❑ *a* 75 dB(A)
❑ *b* 80 dB(A)
❑ *c* 90 dB(A)
❑ *d* 100 dB(A)

15.60 Welding, carbon arc and vapour lamps emit spectral radiation in the

❑ *a* Near infra-red
❑ *b* Ultraviolet
❑ *c* Far infra-red
❑ *d* VHF

15.61 Sunburn and skin cancer are caused by spectral radiation in the

❑ *a* Ultraviolet
❑ *b* RF
❑ *c* Audible
❑ *d* VLF

15.62 Electrical tools used outdoors are best worked from
- ❑ *a* A three-pin plug
- ❑ *b* An isolating transformer
- ❑ *c* A two-pin plug
- ❑ *d* A low-loss cable

15.63 Electricity from overhead power lines can flash over to objects placed
- ❑ *a* 10–15 m away
- ❑ *b* 24–26 m away
- ❑ *c* 60–65 m away
- ❑ *d* 90–95 m away

15.64 Using a ladder in the workplace requires that the ladder should extend a minimum distance above the landing place of
- ❑ *a* 0.1 m
- ❑ *b* 0.5 m
- ❑ *c* 1.0 m
- ❑ *d* 2.5 m

15.65 The maximum distance between vertical supports in scaffolds is
- ❑ *a* 10–12 m
- ❑ *b* 8–9 m
- ❑ *c* 6–7.5 m
- ❑ *d* 2–2.5 m

15.66 Guardrails must be provided above the top platform of a designed scaffold. The height of the guardrail above the platform must not be less than
- ❑ *a* 1.0 m
- ❑ *b* 2.5 m
- ❑ *c* 4.2 m
- ❑ *d* 5.0 m

15.67 Under the current colour code for a three-core flexible cable the live lead is coloured
- ❑ *a* Yellow
- ❑ *b* Green/yellow
- ❑ *c* Brown
- ❑ *d* Blue

15.68 Under the current colour code for a three-core flexible cable the neutral lead is coloured
- ❑ *a* Blue
- ❑ *b* Brown
- ❑ *c* Yellow
- ❑ *d* Pink

15.69 Under the current colour code for a three-core flexible cable the earth lead is coloured
- ❑ *a* Brown
- ❑ *b* Green/yellow
- ❑ *c* Blue
- ❑ *d* Pink

15.70 A suitable fuse for a 1-kW heater is
- ❑ *a* 2 A
- ❑ *b* 3 A
- ❑ *c* 5 A
- ❑ *d* 13 A

15.71 A suitable fuse for a 3-kW heater is
- ❑ *a* 2 A
- ❑ *b* 3 A
- ❑ *c* 5 A
- ❑ *d* 13 A

15.72 A three-phase supply from a substation will have a voltage between the lines of
- ❑ *a* 110 V
- ❑ *b* 240 V
- ❑ *c* 330 V
- ❑ *d* 415 V

15.73 In Question 15.72 the voltage between any phase and the neutral conductor is
- ❑ *a* 110 V
- ❑ *b* 200 V
- ❑ *c* 240 V
- ❑ *d* 310 V

15.74 The recommended fuse for a domestic mains-driven radio set is

- ❑ *a* 3 A
- ❑ *b* 5 A
- ❑ *c* 7 A
- ❑ *d* 13 A

15.75 In television receivers and cathode ray oscilloscopes the voltages can be as high as

- ❑ *a* 500 V
- ❑ *b* 800 V
- ❑ *c* 3 kV
- ❑ *d* 20 kV

16 Filters and attenuators

16.1 If the output and input powers of an electronic network are 2 W and 0.01 W respectively, the gain in decibels (dB) is

- ❑ *a* 10
- ❑ *b* 12
- ❑ *c* 23
- ❑ *d* 33

16.2 If an electronic network has an input of 40 W and an output of 40 W, the gain is

- ❑ *a* 0 dB
- ❑ *b* 6 dB
- ❑ *c* 12 dB
- ❑ *d* 18 dB

16.3 A network has an input of 0.3 mW and an output of 6 mW. The gain is

- ❑ *a* 2 dB
- ❑ *b* 3 dB
- ❑ *c* 6 dB
- ❑ *d* 13 dB

16.4 A network has a power gain of 60 dB. If the output power is 1460 W, the input power is

- ❑ *a* 1.26 mW
- ❑ *b* 1.46 mW
- ❑ *c* 2.46 mW
- ❑ *d* 4.24 mW

16.5 A network has a power gain of 120 dB. If the output power is 4 MW, the input power is

- ❑ *a* 4 mW
- ❑ *b* 40 mW
- ❑ *c* 4 μW
- ❑ *d* 40 μW

16.6 A network has an input of 8 W and an output of 2 W. The gain in decibels is

- ❑ *a* 6
- ❑ *b* −12
- ❑ *c* 3
- ❑ *d* −6

16.7 If a network has an input of 25 W and an output of 0.5 W, the gain in decibels is

- ❑ *a* 13
- ❑ *b* −10
- ❑ *c* −17
- ❑ *d* 20

16.8 If a network has an output of 40 W and an input of 200 W, the gain in decibels is

- ❑ *a* −7
- ❑ *b* 8
- ❑ *c* −9
- ❑ *d* 9

16.9 A filter has a power gain of −40 dB. If the input power is 100 W, the output power is

- ❑ *a* 0.001 W
- ❑ *b* 0.01 W
- ❑ *c* 0.1 W
- ❑ *d* 1.0 W

16.10 A network has a power gain of −1.5 dB. If the input power is 66 W, the output power is

- ❑ *a* 25.48 W
- ❑ *b* 26.17 W
- ❑ *c* 30.38 W
- ❑ *d* 46.72 W

16.11 A network has a power gain of –3 dB. If the input power is 100 W, the output power is

- ❑ *a* 50 W
- ❑ *b* 55 W
- ❑ *c* 60 W
- ❑ *d* 62 W

16.12 A network has a voltage gain of –6 dB. If the input voltage is 10 V, the output voltage is

- ❑ *a* 2 V
- ❑ *b* 3 V
- ❑ *c* 5 V
- ❑ *d* 7 V

16.13 A filter has a voltage gain of –12 dB. If the input voltage is 48 V, the output voltage is

- ❑ *a* 8 V
- ❑ *b* 10 V
- ❑ *c* 12 V
- ❑ *d* 24 V

16.14 A filter has a voltage gain of –60 dB. If the input voltage is 1500 V, the output voltage is

- ❑ *a* 1.2 V
- ❑ *b* 1.5 V
- ❑ *c* 10 V
- ❑ *d* 15 V

16.15 Measurements on a filter network showed that the input current was 16.5 μA and that the output current was 8.25 μA. The loss in decibels is

- ❑ *a* 6
- ❑ *b* 12
- ❑ *c* 18
- ❑ *d* 24

16.16 The input current to a network is 190 μA and the output current is 1.3 μA. The loss in decibels is

- ❑ *a* 20.2
- ❑ *b* 21.5
- ❑ *c* 28.6
- ❑ *d* 43.3

16.17 The input current to a filter is 0.6 mA and the output is 5 μA. The loss in decibels is

- ❑ *a* 36.67
- ❑ *b* 41.58
- ❑ *c* 48.24
- ❑ *d* 49.36

16.18 If four networks connected in series have gains of –60 dB, 78 dB, 10 dB and –29 dB, the overall gain is

- ❑ *a* –1 dB
- ❑ *b* 1 dB
- ❑ *c* 39 dB
- ❑ *d* –88 dB

16.19 If four networks connected in series have gains of –0.5 dB, –0.3 dB, –2 dB and +6.8 dB, the overall gain is

- ❑ *a* 2 dB
- ❑ *b* –2 dB
- ❑ *c* 4 dB
- ❑ *d* –4 dB

16.20 The diagram shows a two-stage network. The gain of stage 1 is

- ❑ *a* 1
- ❑ *b* 5
- ❑ *c* 6
- ❑ *d* 8

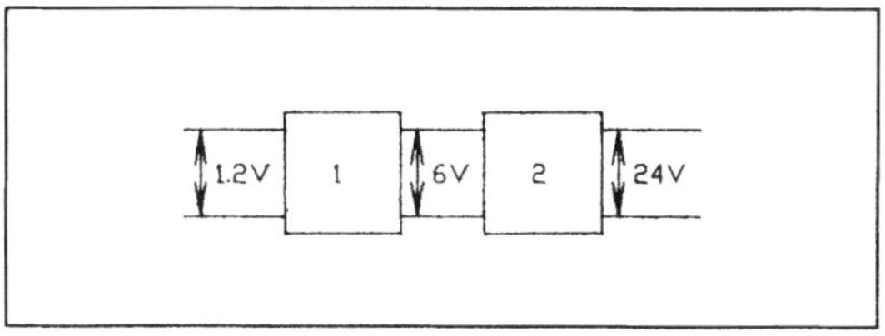

16.21 In Question 16.20 the gain of stage 2 is

- ❑ *a* 4
- ❑ *b* 8
- ❑ *c* 12
- ❑ *d* 20

16.22 In Question 16.20 the overall gain is

- ❑ *a* 12
- ❑ *b* 20
- ❑ *c* 2
- ❑ *d* 4

16.23 In Question 16.20 the gain of stage 1 in decibels is
- ❑ *a* 6.44
- ❑ *b* 10.22
- ❑ *c* 10.88
- ❑ *d* 13.98

16.24 In Question 16.20 the gain of stage 2 in decibels is
- ❑ *a* 12
- ❑ *b* 14
- ❑ *c* 18
- ❑ *d* 20

16.25 In Question 16.20 the overall gain in decibels is approximately
- ❑ *a* 18
- ❑ *b* 22
- ❑ *c* 26
- ❑ *d* 34

16.26 The diagram shows another two-stage network. The gain of stage 1 is
- ❑ *a* 10
- ❑ *b* 20
- ❑ *c* 100
- ❑ *d* 1000

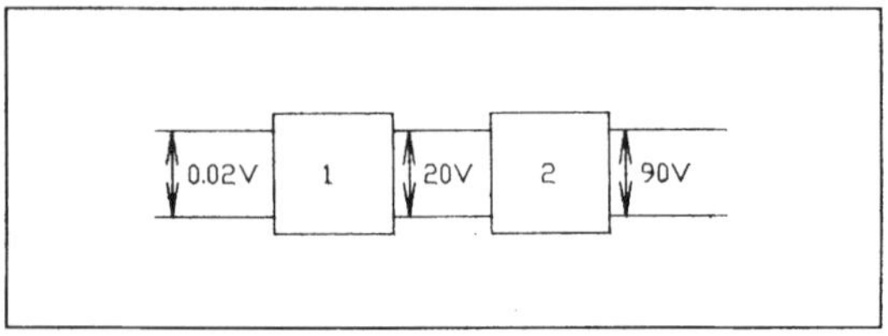

16.27 In Question 16.26 the gain of stage 2 is
- ❑ *a* 4
- ❑ *b* 4.5
- ❑ *c* 4.8
- ❑ *d* 5.0

16.28 In Question 16.26 the overall gain is
- ❑ *a* 400
- ❑ *b* 500
- ❑ *c* 4500
- ❑ *d* 5000

16.29 In Question 16.26 the gain of stage 1 in decibels is
- ❑ *a* 10
- ❑ *b* 20
- ❑ *c* 30
- ❑ *d* 60

16.30 In Question 16.26 the gain of stage 2 in decibels is
- ❑ *a* 13
- ❑ *b* 15
- ❑ *c* 21
- ❑ *d* 27

16.31 In Question 16.26 the overall gain in decibels is
- ❑ *a* 23
- ❑ *b* 45
- ❑ *c* 73
- ❑ *d* 87

16.32 The diagram shows another two-stage network. The overall gain is
- ❑ *a* 10
- ❑ *b* 12
- ❑ *c* 14
- ❑ *d* 20

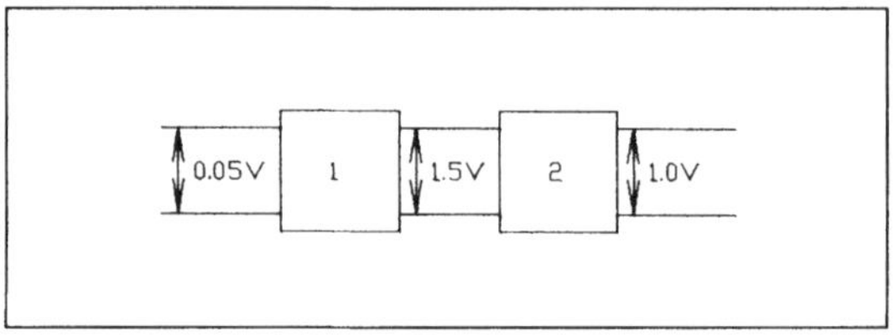

16.33 The gain of stage 2 in Question 16.32 is
- ❑ *a* 0.5
- ❑ *b* 0.67
- ❑ *c* 1.0
- ❑ *d* 1.5

16.34 The overall gain in Question 16.32 in decibels is
- ❑ *a* 18
- ❑ *b* 26
- ❑ *c* 30
- ❑ *d* 34

16.35 The diagram shows a three-stage network. The overall gain is

- ❑ *a* 75
- ❑ *b* 80
- ❑ *c* 100
- ❑ *d* 120

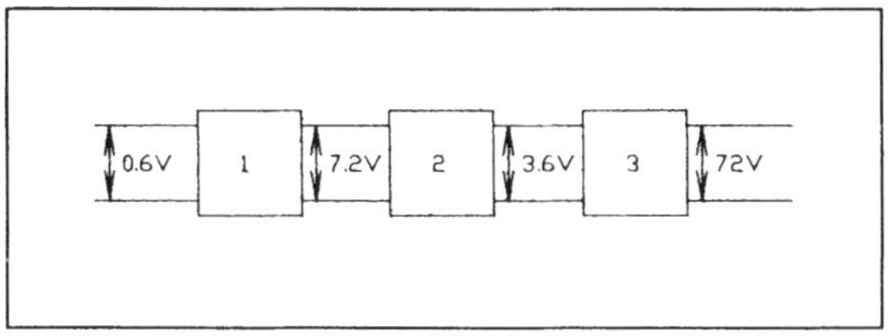

16.36 In Question 16.35 the gain of stage 2 in decibels is

- ❑ *a* –6
- ❑ *b* 6
- ❑ *c* –12
- ❑ *d* 12

16.37 In Question 16.35 the overall gain in decibels is

- ❑ *a* 25.74
- ❑ *b* 30.28
- ❑ *c* 36.64
- ❑ *d* 41.58

16.38 The diagram shows another three-stage network. The gain of stage 2 is

- ❑ *a* 0.1
- ❑ *b* 0.25
- ❑ *c* 1.25
- ❑ *d* 25

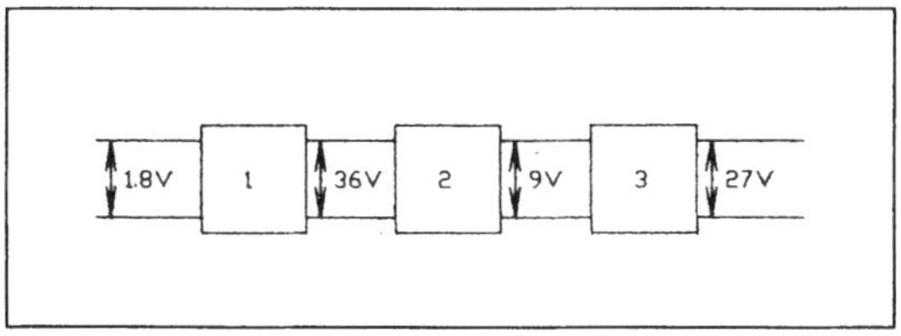

16.39 The overall gain in Question 16.38 is

- ❑ *a* 10
- ❑ *b* 12
- ❑ *c* 15
- ❑ *d* 20

16.40 The gain of stage 2 in Question 16.38 in decibels is

- ❑ *a* –6
- ❑ *b* –12
- ❑ *c* 6
- ❑ *d* 12

16.41 The overall gain of the network in Question 16.38 in decibels is

- ❑ *a* 17.16
- ❑ *b* 19.58
- ❑ *c* 21.83
- ❑ *d* 23.52

16.42 The diagram shows a single stage. The output power W_2 is

- ❑ *a* 14.4 W
- ❑ *b* 16.7 W
- ❑ *c* 19.9 W
- ❑ *d* 26.3 W

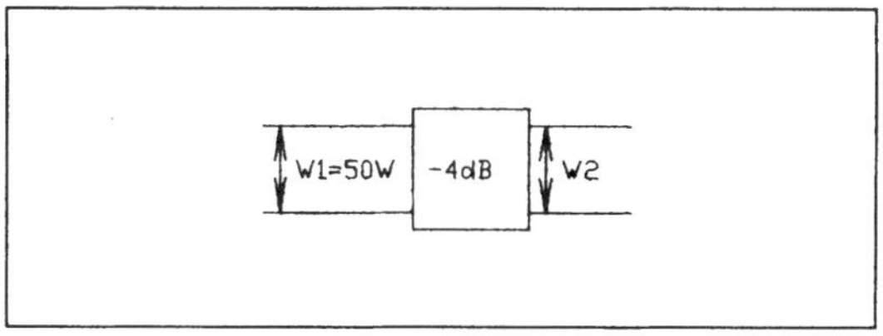

16.43 If in Question 16.42 the gain is –12 dB and the input power is 1.0 W, the output power is

- ❑ *a* 26 mW
- ❑ *b* 63 mW
- ❑ *c* 78 mW
- ❑ *d* 88 mW

16.44 If in Question 16.42 the gain is –30 dB and the output power is 1.0 W, the input power is

- ❑ *a* 10
- ❑ *b* 100
- ❑ *c* 500
- ❑ *d* 1000

16.45 The diagram shows a simple low-pass filter. The output voltage V_c at the –3-dB point is

❑ *a* 16.97 V
❑ *b* 18.42 V
❑ *c* 19.04 V
❑ *d* 20.68 V

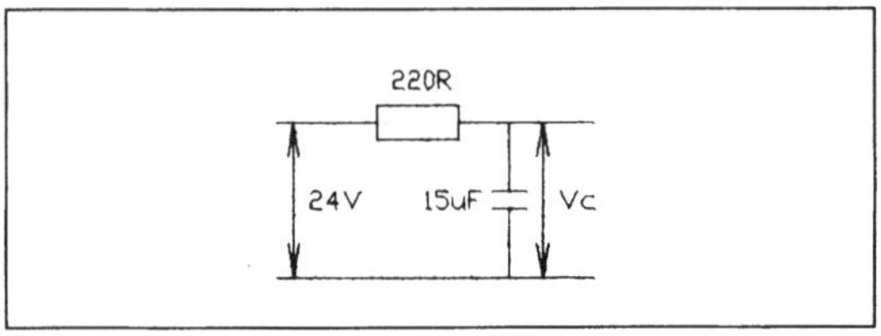

16.46 The voltage across the resistor in Question 16.45 at the –3-dB point is

❑ *a* 11.38 V
❑ *b* 12.67 V
❑ *c* 16.97 V
❑ *d* 19.62 V

16.47 In Question 16.45 the cut-off frequency at the –3-dB point is

❑ *a* 16.62 Hz
❑ *b* 48.23 Hz
❑ *c* 60.18 Hz
❑ *d* 65.33 Hz

16.48 The diagram shows another simple low-pass filter. The output voltage V_c at the –3-dB point is

❑ *a* 21.21 V
❑ *b* 42.42 V
❑ *c* 46.21 V
❑ *d* 48.42 V

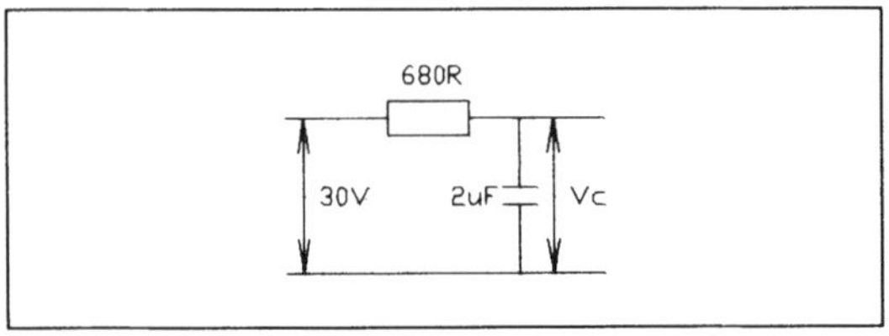

16.49 In Question 16.48 the voltage across the resistor at the –3-dB point is

❑ *a* 10.61 V
❑ *b* 16.82 V
❑ *c* 21.21 V
❑ *d* 24.42 V

16.50 In Question 16.48 the cut-off frequency at the –3-dB point is

❑ *a* 87 Hz
❑ *b* 101 Hz
❑ *c* 103 Hz
❑ *d* 117 Hz

16.51 The diagram shows a T low-pass filter, where the inductances have values of $L/2$. The filter is required to have a design impedance of 300 Ω and a cut-off frequency of 100 Hz. The value of L is

❑ *a* 863 mH
❑ *b* 900 mH
❑ *c* 955 mH
❑ *d* 996 mH

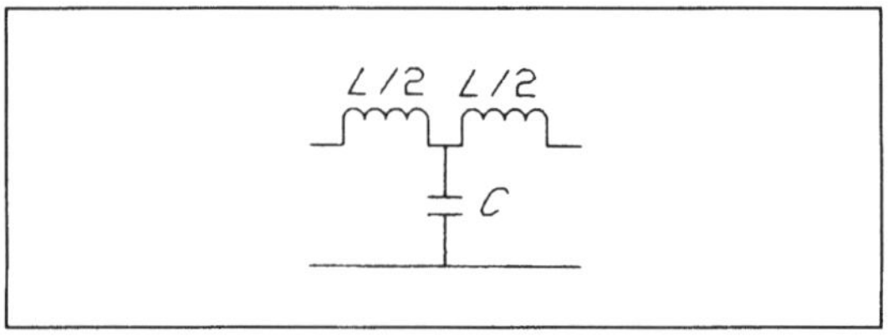

16.52 The value of C in Question 16.51 is

❑ *a* 10.05 μF
❑ *b* 10.61 μF
❑ *c* 14.34 μF
❑ *d* 17.82 μF

16.53 The diagram shows a low-pass π filter where the capacitors have values of $C/2$. It is a requirement for the design impedance to be 600 Ω and the cut-off frequency to be 50 Hz. The value of C is

❑ *a* 10.61 μF
❑ *b* 12.32 μF
❑ *c* 14.41 μF
❑ *d* 16.57 μF

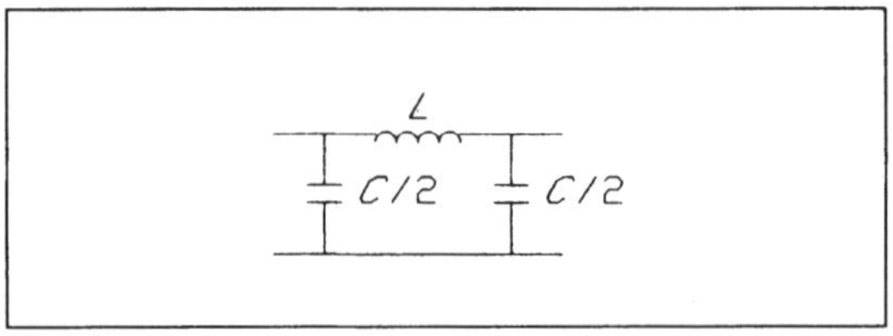

16.54 The value of L in Question 16.53 is
- ❑ *a* 2.84 H
- ❑ *b* 3.06 H
- ❑ *c* 3.82 H
- ❑ *d* 4.86 H

16.55 The diagram shows a high-pass T filter, where the capacitors have the values $2C$. If the design impedance is 600 Ω and the cut-off frequency is 150 Hz, the value of C is approximately
- ❑ *a* 0.62 μF
- ❑ *b* 0.69 μF
- ❑ *c* 0.88 μF
- ❑ *d* 0.92 μF

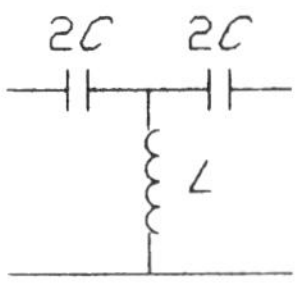

16.56 In Question 16.55 the value of L is approximately
- ❑ *a* 0.11 H
- ❑ *b* 0.16 H
- ❑ *c* 0.28 H
- ❑ *d* 0.32 H

16.57 The diagram shows a high-pass π filter, where the inductors have values $2L$. If the design impedance is 300 Ω and the cut-off frequency is 40 Hz, the value of C is approximately
- ❑ *a* 4.87 μF
- ❑ *b* 5.31 μF
- ❑ *c* 6.63 μF
- ❑ *d* 7.24 μF

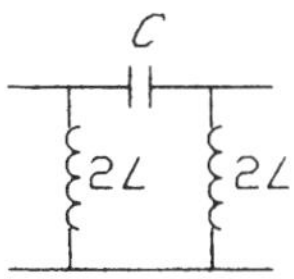

16.58 In Question 16.57 the value of L is approximately
- ❑ *a* 400 mH
- ❑ *b* 597 mH
- ❑ *c* 632 mH
- ❑ *d* 704 mH

16.59 The diagram shows a symmetrical T attenuator. The characteristic impedance R_o is
- ❑ *a* 10 Ω
- ❑ *b* 20 Ω
- ❑ *c* 40 Ω
- ❑ *d* 60 Ω

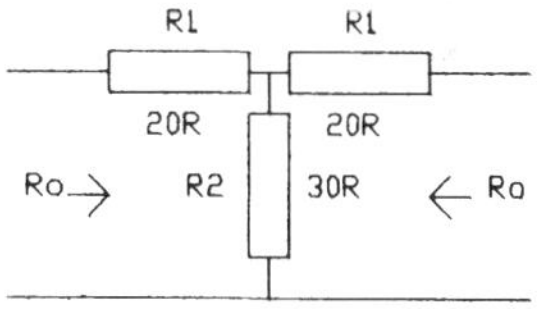

16.60 The attenuation of the network in Question 16.59 is
- ❑ *a* 3
- ❑ *b* 5
- ❑ *c* 7
- ❑ *d* 9

16.61 The voltage attenuation in Question 16.59 expressed in decibels is
- ❑ *a* 7.3
- ❑ *b* 9.5
- ❑ *c* 10.6
- ❑ *d* 12.7

16.62 If in Question 16.59 R_1 = 80 Ω and R_2 = 120 Ω, the characteristic impedance R_o is
- ❑ *a* 110 Ω
- ❑ *b* 120 Ω
- ❑ *c* 150 Ω
- ❑ *d* 160 Ω

16.63 The attenuation in Question 16.62 is
- ❑ *a* 2
- ❑ *b* 3
- ❑ *c* 6
- ❑ *d* 7

16.64 The voltage attenuation in Question 16.62 expressed in decibels is

❑ *a* 3.1
❑ *b* 4.8
❑ *c* 7.6
❑ *d* 9.5

16.65 The diagram shows a symmetrical T attenuator. It is a requirement for the attenuator to have an attenuation of 6 dB and a characteristic impedance of 600 Ω. The value of R_1 is approximately

❑ *a* 108 Ω
❑ *b* 199 Ω
❑ *c* 250 Ω
❑ *d* 360 Ω

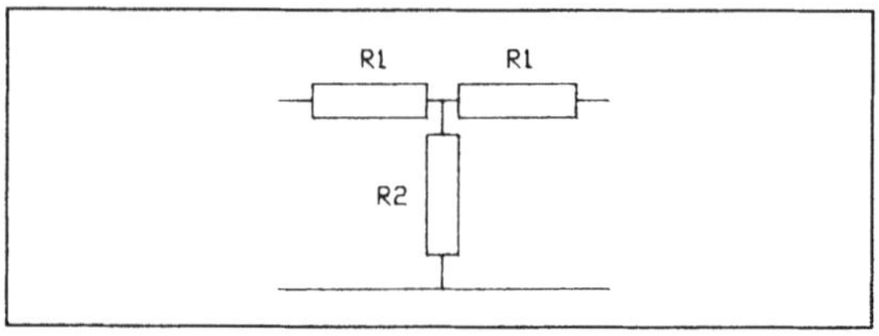

16.66 The value of R_2 in Question 16.65 is approximately

❑ *a* 608 Ω
❑ *b* 716 Ω
❑ *c* 803 Ω
❑ *d* 863 Ω

16.67 If it is a requirement in Question 16.65 for the attenuation to be 32 dB and the characteristic impedance to be 400 Ω, the value of R_1 is approximately

❑ *a* 264 Ω
❑ *b* 280 Ω
❑ *c* 320 Ω
❑ *d* 380 Ω

16.68 The value of R_2 in Question 16.67 is

❑ *a* 20 Ω
❑ *b* 42 Ω
❑ *c* 68 Ω
❑ *d* 97 Ω

16.69 The diagram shows an asymmetrical T attenuator. If the attenuation of the network is 10, the value of R_A is approximately

❑ *a* 163 Ω
❑ *b* 285 Ω
❑ *c* 367 Ω
❑ *d* 400 Ω

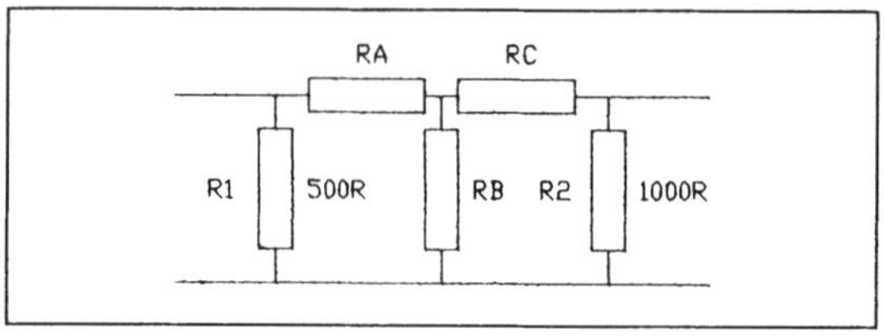

16.70 In Question 16.69 the value of R_B is approximately

❑ *a* 112 Ω
❑ *b* 143 Ω
❑ *c* 202 Ω
❑ *d* 287 Ω

16.71 In Question 16.69 the value of R_C is approximately

❑ *a* 408 Ω
❑ *b* 550 Ω
❑ *c* 648 Ω
❑ *d* 877 Ω

16.72 The diagram shows a symmetrical π attenuator. If $R_1 = 20\ \Omega$ and $R_2 = 30\ \Omega$, the characteristic impedance is

❑ *a* 10 Ω
❑ *b* 15 Ω
❑ *c* 20 Ω
❑ *d* 60 Ω

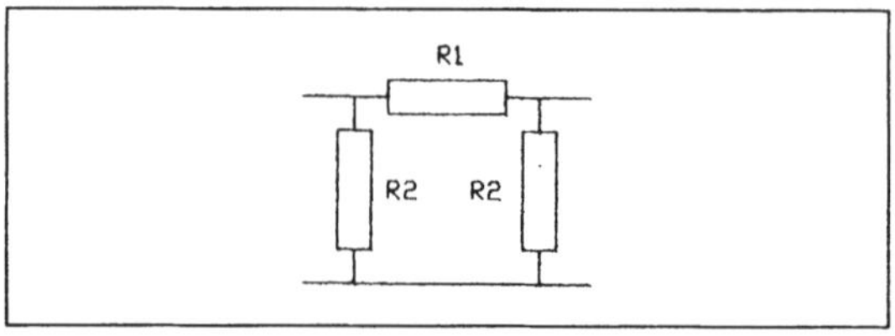

16.73 The diagram shows an asymmetrical π attenuator together with the source impedance R_1 and the load impedance R_2. If the attenuation of the network is 10, the value of R_A is approximately

- ❑ *a* 65 Ω
- ❑ *b* 82 Ω
- ❑ *c* 88 Ω
- ❑ *d* 97 Ω

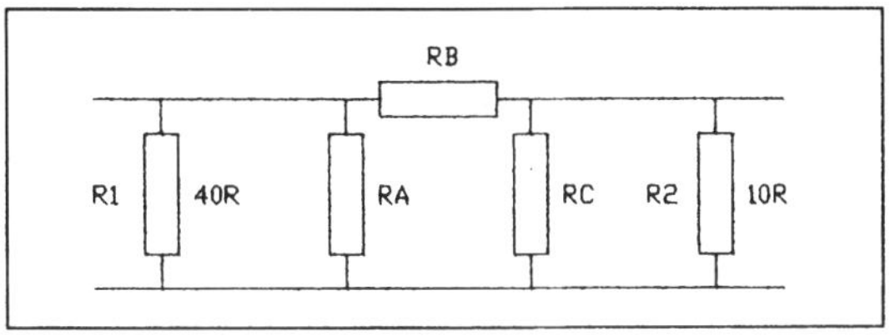

16.74 In Question 16.73 the value of R_B is approximately

- ❑ *a* 33 Ω
- ❑ *b* 60 Ω
- ❑ *c* 75 Ω
- ❑ *d* 99 Ω

16.75 In Question 16.73 the value of R_C is approximately

- ❑ *a* 11 Ω
- ❑ *b* 19 Ω
- ❑ *c* 29 Ω
- ❑ *d* 75 Ω

17 Denary, binary and logic gates

17.1 Denary 9 expressed in base 5 is
- ❑ *a* 12
- ❑ *b* 16
- ❑ *c* 20
- ❑ *d* 14

17.2 Denary 9 expressed in base 3 is
- ❑ *a* 16
- ❑ *b* 74
- ❑ *c* 100
- ❑ *d* 123

17.3 Denary 19 expressed in base 6 is
- ❑ *a* 31
- ❑ *b* 62
- ❑ *c* 72
- ❑ *d* 93

17.4 Denary 17 expressed in base 4 is
- ❑ *a* 101
- ❑ *b* 65
- ❑ *c* 142
- ❑ *d* 36

17.5 The number 111 in base 3 expressed in denary is
- ❑ *a* 12
- ❑ *b* 13
- ❑ *c* 24
- ❑ *d* 26

17.6 The number 33 in base 4 expressed in denary is
- ❑ *a* 4
- ❑ *b* 17
- ❑ *c* 9
- ❑ *d* 15

17.7 The number 30 in base 5 expressed in denary is
- ❑ *a* 5
- ❑ *b* 10
- ❑ *c* 15
- ❑ *d* 250

17.8 The number 251 in base 6 expressed in denary is
- ❑ *a* 64
- ❑ *b* 103
- ❑ *c* 127
- ❑ *d* 156

17.9 The denary number 7 when converted to binary is
- ❑ *a* 101
- ❑ *b* 111
- ❑ *c* 110
- ❑ *d* 11

17.10 Denary 25 converted to binary is
- ❑ *a* 11001
- ❑ *b* 1101
- ❑ *c* 11010
- ❑ *d* 10001

17.11 Denary 39 converted to binary is
- ❑ *a* 110000
- ❑ *b* 111101
- ❑ *c* 100010
- ❑ *d* 100111

17.12 Denary 61 converted to binary is
- ❑ *a* 1001
- ❑ *b* 11101
- ❑ *c* 111101
- ❑ *d* 111111

17.13 The binary number 101 when converted to denary is
- ❑ *a* 6
- ❑ *b* 12
- ❑ *c* 5
- ❑ *d* 7

17.14 The binary number 1110 converted to denary is
- ❑ *a* 14
- ❑ *b* 16
- ❑ *c* 12
- ❑ *d* 15

17.15 The binary number 11110 converted to denary is
- ❑ *a* 24
- ❑ *b* 26
- ❑ *c* 30
- ❑ *d* 42

17.16 The binary number 110011 converted to denary is
- ❑ *a* 51
- ❑ *b* 60
- ❑ *c* 28
- ❑ *d* 34

17.17 The addition of binary numbers 100 and 1100 gives
- ❑ *a* 1111
- ❑ *b* 10100
- ❑ *c* 11000
- ❑ *d* 10000

17.18 The addition of binary numbers 11 and 1111 gives
- ❑ *a* 11000
- ❑ *b* 10110
- ❑ *c* 10010
- ❑ *d* 10110

17.19 The addition of binary numbers 10001 and 10010 gives
- ❑ *a* 100011
- ❑ *b* 101011
- ❑ *c* 101010
- ❑ *d* 110001

17.20 The addition of binary numbers 1011 and 10100 gives
- ❑ *a* 11010
- ❑ *b* 11011
- ❑ *c* 11010
- ❑ *d* 11111

17.21 Binary 1010 subtracted from 1100 gives
- ❑ *a* 10
- ❑ *b* 11
- ❑ *c* 100
- ❑ *d* 101

17.22 Binary 11 subtracted from 1111 gives
- ❑ *a* 1001
- ❑ *b* 1100
- ❑ *c* 1000
- ❑ *d* 1010

17.23 Binary 10111 subtracted from 100000 gives
- ❑ *a* 1001
- ❑ *b* 1101
- ❑ *c* 1110
- ❑ *d* 1111

17.24 Binary 10111 subtracted from 101110 gives
- ❑ *a* 10001
- ❑ *b* 10010
- ❑ *c* 10111
- ❑ *d* 11011

17.25 The addition of the three binary numbers 100, 1111 and 11010 equals
- ❑ *a* 100001
- ❑ *b* 101101
- ❑ *c* 110001
- ❑ *d* 111011

17.26 The product of the binary numbers 11 and 101 is
- ❑ *a* 1000
- ❑ *b* 1100
- ❑ *c* 1010
- ❑ *d* 1111

17.27 The product of the binary numbers 110 and 111 is
- ❑ *a* 101010
- ❑ *b* 111010
- ❑ *c* 111101
- ❑ *d* 111110

17.28 The product of the binary numbers 110 and 1010 is
- ❑ *a* 100010
- ❑ *b* 101011
- ❑ *c* 111100
- ❑ *d* 111110

17.29 The product of the three binary numbers 11, 100 and 100 is
- ❑ *a* 10100
- ❑ *b* 110000
- ❑ *c* 1010
- ❑ *d* 1100

17.30 Binary 1100 divided by binary 100 is
- ❑ *a* 10
- ❑ *b* 11
- ❑ *c* 100
- ❑ *d* 101

17.31 Binary 10101 divided by binary 111 is
- ❑ *a* 11
- ❑ *b* 100
- ❑ *c* 101
- ❑ *d* 111

17.32 Binary 111100 divided by binary 11 is
- ❑ *a* 10010
- ❑ *b* 10100
- ❑ *c* 10101
- ❑ *d* 10111

17.33 The octal number 25 expressed in binary is
- ❑ *a* 010101
- ❑ *b* 001101
- ❑ *c* 010110
- ❑ *d* 110101

17.34 The octal number 40 expressed in binary is
- ❑ *a* 001100
- ❑ *b* 011101
- ❑ *c* 101111
- ❑ *d* 100000

17.35 The octal number 51.62 expressed in binary is
- ❑ *a* 100011.111001
- ❑ *b* 101000.110000
- ❑ *c* 101001.110010
- ❑ *d* 010000.110010

17.36 The binary number 1101 converted to octal is
- ❑ *a* 31
- ❑ *b* 15
- ❑ *c* 5
- ❑ *d* 25

17.37 The binary number 100.10111 converted to octal is
- ❑ *a* 4.53
- ❑ *b* 4.56
- ❑ *c* 4.57
- ❑ *d* 4.51

17.38 The binary number 11.1111 converted to octal is
- ❑ *a* 3.74
- ❑ *b* 2.71
- ❑ *c* 7.70
- ❑ *d* 7.74

17.39 The denary number 334 converted to hexadecimal is
- ❑ *a* 23A
- ❑ *b* 2AB
- ❑ *c* 13A
- ❑ *d* 14E

17.40 The denary number 29.75 converted to hexadecimal is
- ❑ *a* 1D.C
- ❑ *b* 1A.B
- ❑ *c* 1B.B
- ❑ *d* 1C.A

17.41 The denary number 252.5 converted to hexadecimal is
❑ *a* FC.8
❑ *b* B8.9
❑ *c* BF.F
❑ *d* AB.C

17.42 Hexadecimal 2AB converted to denary is
❑ *a* 324
❑ *b* 426
❑ *c* 683
❑ *d* 760

17.43 Hexadecimal D.A converted to denary is
❑ *a* 10.25
❑ *b* 12.5
❑ *c* 13.625
❑ *d* 14.75

17.44 Hexadecimal E1.4 converted to denary is
❑ *a* 120.4
❑ *b* 150.75
❑ *c* 175.5
❑ *d* 225.25

17.45 The hexadecimal number A2 converted to binary is
❑ *a* 10100001
❑ *b* 10100010
❑ *c* 10110011
❑ *d* 11010010

17.46 The hexadecimal number F.E converted to binary is
❑ *a* 1000.1111
❑ *b* 1111.1110
❑ *c* 1010.1100
❑ *d* 1001.1110

17.47 The hexadecimal number 1.AB converted to binary is
❑ *a* 0001.10101011
❑ *b* 0010.11101011
❑ *c* 0011.11110000
❑ *d* 0100.10011001

17.48 The binary number 10011011 converted to binary is
❑ *a* 6F
❑ *b* 7E
❑ *c* 8C
❑ *d* 9B

17.49 The binary number 11.11 converted to hexadecimal is
❑ *a* F.D
❑ *b* 7.C
❑ *c* 3.C
❑ *d* 7.7

17.50 The binary number 110.101 converted to hexadecimal is
❑ *a* 6.A
❑ *b* B.A
❑ *c* C.5
❑ *d* D.5

17.51 The truth table shown where *A* and *B* are the inputs and *Q* the output is for

	A	*B*	*Q*
❑ *a* An AND gate	0	0	1
❑ *b* A NAND gate	0	1	1
❑ *c* An OR gate	1	0	1
❑ *d* A NOR gate	1	1	0

17.52 In the diagram the gate is
❑ *a* A NAND
❑ *b* A NOR
❑ *c* An AND
❑ *d* A NOT

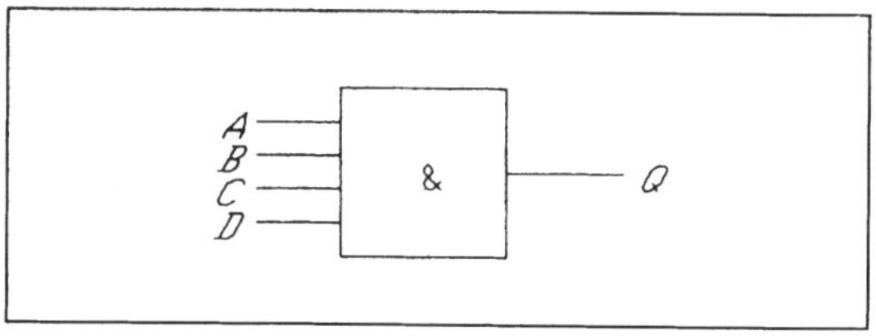

17.53 In the gate in Question 17.52 the number of possible input combinations is

- ❑ *a* 2
- ❑ *b* 4
- ❑ *c* 8
- ❑ *d* 16

17.54 In Question 17.52 the number of outputs Q at logic 1 will be

- ❑ *a* 1
- ❑ *b* 4
- ❑ *c* 8
- ❑ *d* 2

17.55 In the gate shown the output Q for all possible inputs listed as in Question 17.51 will be

- ❑ *a* 1000
- ❑ *b* 0001
- ❑ *c* 0111
- ❑ *d* 1110

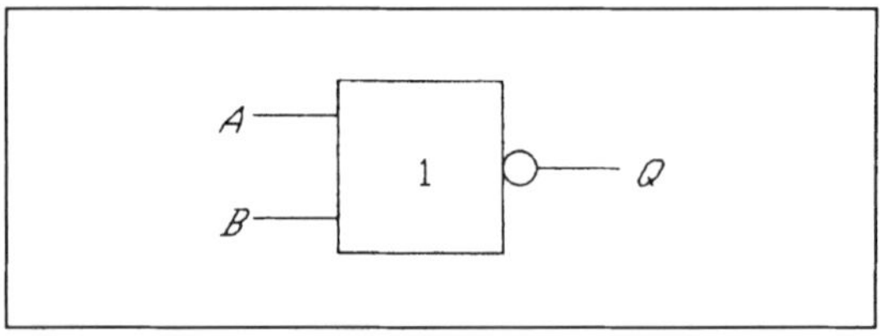

17.56 If the output of the gate in Question 17.55 is fed into a NOT gate, the output of the NOT gate will be

- ❑ *a* 1110
- ❑ *b* 1000
- ❑ *c* 0111
- ❑ *d* 0001

17.57 The truth table shown is for

	A	*B*	*Q*
❑ *a* An OR gate	0	0	0
❑ *b* An EXCLUSIVE–OR gate	0	1	1
❑ *c* A NOR gate	1	0	1
❑ *d* A NAND gate	1	1	0

17.58 Listing the inputs in ascending binary order for the gate combination shown, the output Q is

- ❑ *a* 0011
- ❑ *b* 1000
- ❑ *c* 1010
- ❑ *d* 1100

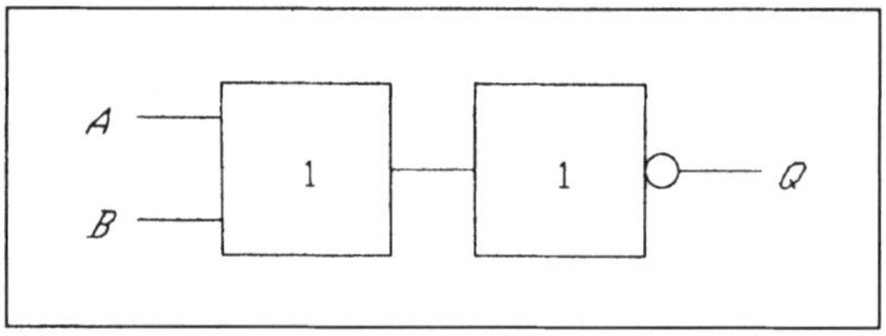

17.59 Which single gate would replace the combination in Question 17.58?

- ❑ *a* NAND
- ❑ *b* NOT
- ❑ *c* NOR
- ❑ *d* EXCLUSIVE–OR

17.60 In the diagram the input combination that gives a logic 1 output is

	A	*B*
❑ *a*	0	0
❑ *b*	0	1
❑ *c*	1	0
❑ *d*	1	1

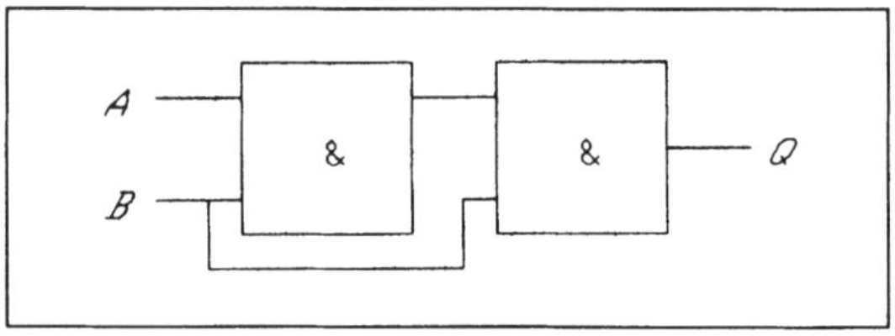

17.61 Listing the inputs in ascending binary order for the combination shown, the output Q is

❑ *a* 1110
❑ *b* 0001
❑ *c* 0011
❑ *d* 1100

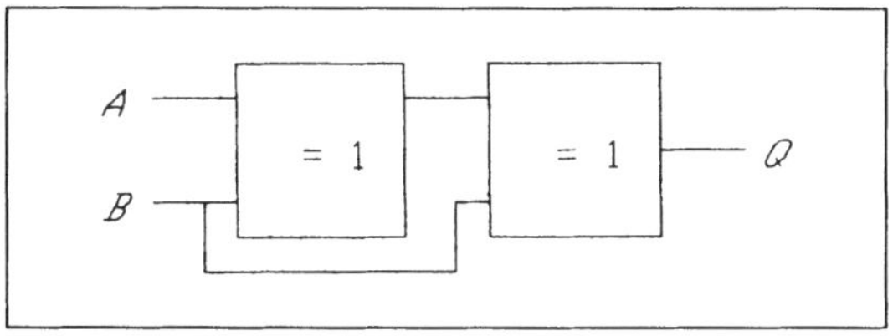

17.62 If Question 17.61 is repeated for the gate combination shown, the output Q is

❑ *a* 0101
❑ *b* 1001
❑ *c* 0110
❑ *d* 0001

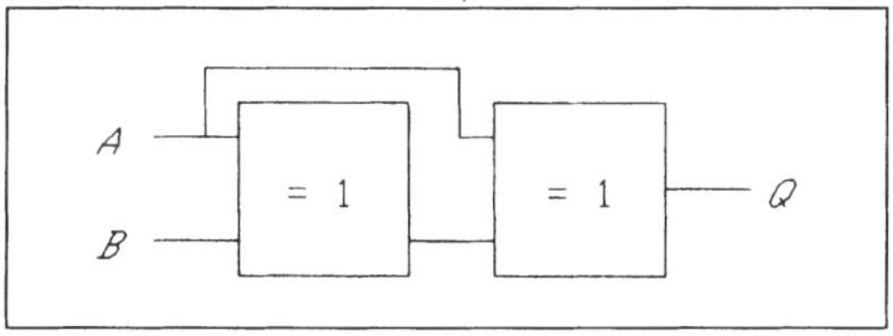

17.63 The diagram shows a combination of three gates. The input combination that gives a logic 0 output is

	A	B
❑ *a*	0	0
❑ *b*	0	1
❑ *c*	1	0
❑ *d*	1	1

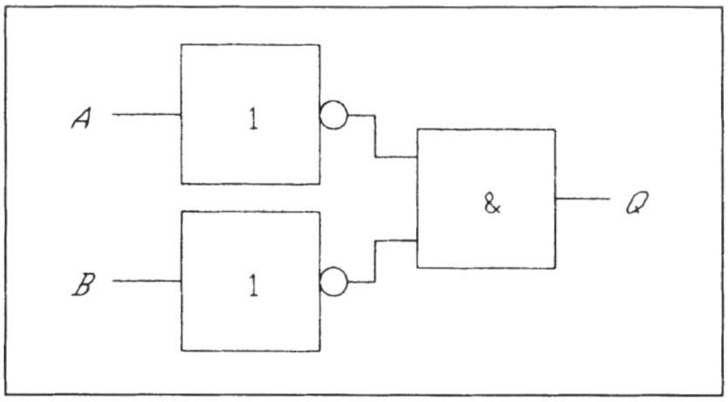

17.64 Which single gate could be used to replace the combination in Question 17.63?

❑ *a* A NOT
❑ *b* A NOR
❑ *c* An OR
❑ *d* An EXCLUSIVE–OR

17.65 The diagram shows a combination of two gates. The input combination that gives a logic 1 output is

	A	B	C
❑ *a*	0	0	0
❑ *b*	0	1	0
❑ *c*	1	1	1
❑ *d*	1	0	1

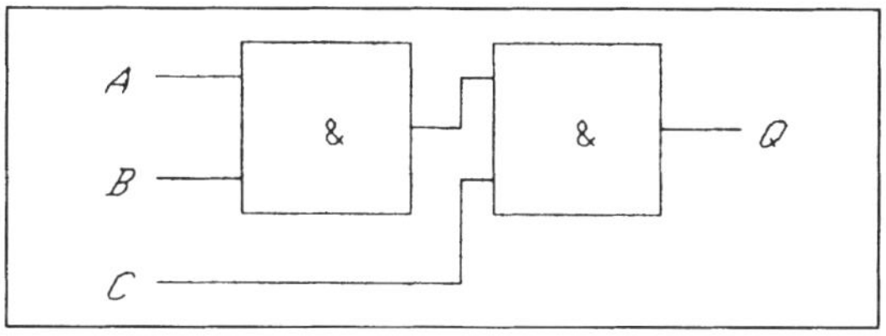

17.66 If the gate with inputs A and B in Question 17.65 is replaced by a NAND gate, the number of outputs Q at logic 1 will be

❑ *a* 1
❑ *b* 3
❑ *c* 5
❑ *d* 8

17.67 The diagram shows two NOR gates used in tandem. The input combination that will give a logic 1 output is

	A	B	C
❑ *a*	1	1	1
❑ *b*	0	0	0
❑ *c*	1	0	1
❑ *d*	0	1	0

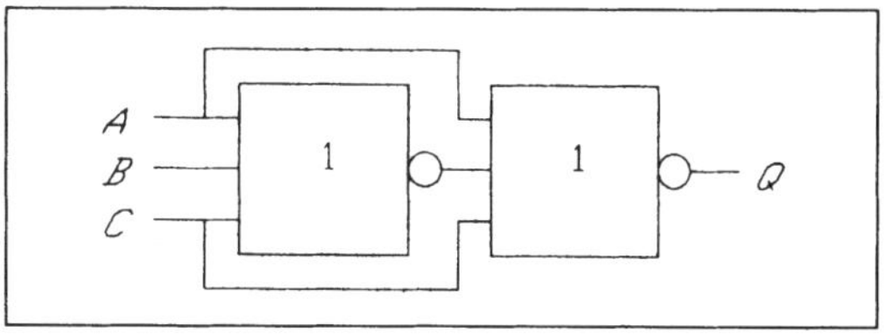

17.68 The diagram shows a two-gate combination. One of the input combinations that will give a logic 1 output is

	A	B	C
❑ *a*	0	0	0
❑ *b*	0	0	1
❑ *c*	0	1	0
❑ *d*	1	1	0

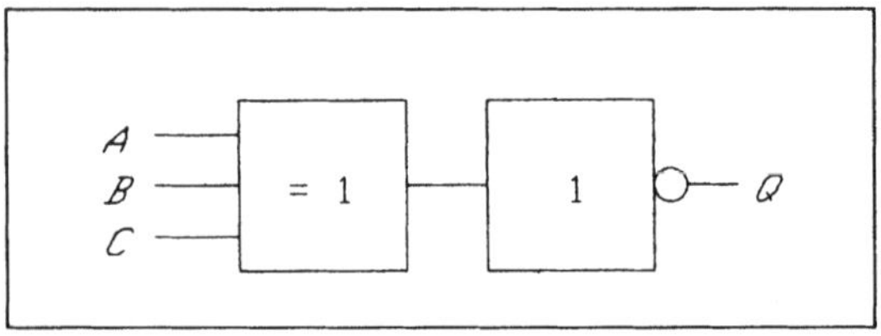

17.69 Denary 478 converted to hexadecimal is
❑ *a* 2AB
❑ *b* 1DE
❑ *c* ABC
❑ *d* A22

17.70 Denary 77.875 converted to hexadecimal is
❑ *a* 1A.B
❑ *b* 2D.E
❑ *c* 4D.E
❑ *d* 1A.A

17.71 Hexadecimal 3BC converted to denary is
❑ *a* 633
❑ *b* 728
❑ *c* 800
❑ *d* 956

17.72 Hexadecimal FE.D4 converted to denary is
❑ *a* 161.724442
❑ *b* 182.613413
❑ *c* 254.828125
❑ *d* 412.712712

17.73 Hexadecimal C4E converted to binary is
❑ *a* 100101001001
❑ *b* 000111001001
❑ *c* 010101000100
❑ *d* 110001001110

17.74 Binary 10001 multiplied by 111 is
❑ *a* 1010101
❑ *b* 1110111
❑ *c* 1000111
❑ *d* 1000000

17.75 Binary 1110 divided by 111 is
❑ *a* 10
❑ *b* 11
❑ *c* 100
❑ *d* 101

18 Background science

18.1 In the SI system the base unit for length is the
- ❑ *a* Centimetre
- ❑ *b* Millimetre
- ❑ *c* Metre
- ❑ *d* Yard

18.2 In the SI system the base unit for mass is the
- ❑ *a* Pound
- ❑ *b* Gram
- ❑ *c* Ounce
- ❑ *d* Kilogram

18.3 In the SI system the base unit for time is the
- ❑ *a* Second
- ❑ *b* Minute
- ❑ *c* Hour
- ❑ *d* Millisecond

18.4 In the SI system the base unit for current is the
- ❑ *a* Nanoampere
- ❑ *b* Microampere
- ❑ *c* Milliampere
- ❑ *d* Ampere

18.5 In the SI system the base unit for temperature is the
- ❑ *a* Degree Celsius
- ❑ *b* Degree centigrade
- ❑ *c* Degree Farenheit
- ❑ *d* Kelvin

18.6 Energy is measured in
- ❑ *a* Coulombs
- ❑ *b* Joules
- ❑ *c* Watts
- ❑ *d* Amperes

18.7 In the SI system density is measured in
- ❑ *a* $kg\,m^{-2}$
- ❑ *b* $kg\,m^{-3}$
- ❑ *c* $kg\,m^{2}$
- ❑ *d* $kg\,m^{3}$

18.8 A certain metal has a mass of 800 g and a volume of $40\,cm^3$. The density is
- ❑ *a* $2\ kg\,m^{-3}$
- ❑ *b* $20\ kg\,m^{-3}$
- ❑ *c* $0.02\ kg\,m^{-3}$
- ❑ *d* $0.2\ kg\,m^{-3}$

18.9 The SI unit for pressure is
- ❑ *a* $N\,m^{-2}$
- ❑ *b* $N\,m^{-3}$
- ❑ *c* $N\,m^{2}$
- ❑ *d* $N\,m^{4}$

18.10 The SI unit of force is the
- ❑ *a* Pound
- ❑ *b* Gram
- ❑ *c* Newton
- ❑ *d* Ton

18.11 If a force of 60 N acts over an area of $0.25\ m^2$, the pressure on this area is
- ❑ *a* $120\ N\,m^{-2}$
- ❑ *b* $240\ N\,m^{-2}$
- ❑ *c* $280\ N\,m^{-2}$
- ❑ *d* $300\ N\,m^{-2}$

18.12 If a force of 1 N moves an object a distance of 1 m in the direction of the force, the amount of work done is
- ❑ *a* 1 pound
- ❑ *b* 1 coulomb
- ❑ *c* 1 volt
- ❑ *d* 1 joule

18.13 Gravitational force in the SI units is equal to

- ❑ *a* 10 $N\,kg^2$
- ❑ *b* 10 $N\,kg^{-1}$
- ❑ *c* 20 $N\,kg^{-1}$
- ❑ *d* 25 $N\,kg^2$

18.14 The SI unit for power is the

- ❑ *a* Volt
- ❑ *b* Ampere
- ❑ *c* Watt
- ❑ *d* Ohm

18.15 The velocity of sound is approximately

- ❑ *a* 330 $m\,s^{-1}$
- ❑ *b* 400 $m\,s^{-1}$
- ❑ *c* 460 $m\,s^{-1}$
- ❑ *d* 700 $m\,s^{-1}$

18.16 The echo of a gun shot from a distant cliff is heard 4 s after the gun is fired. The distance to the cliff is

- ❑ *a* 330 m
- ❑ *b* 530 m
- ❑ *c* 660 m
- ❑ *d* 1320 m

18.17 The speed of sound in sea water is approximately

- ❑ *a* 1200 $m\,s^{-1}$
- ❑ *b* 1500 $m\,s^{-1}$
- ❑ *c* 2000 $m\,s^{-1}$
- ❑ *d* 2200 $m\,s^{-1}$

18.18 If a ship emits a short pulse of sound towards the sea bed and the echo is received back in 0.4 s, the sea depth under the ship is

- ❑ *a* 300 m
- ❑ *b* 400 m
- ❑ *c* 500 m
- ❑ *d* 600 m

18.19 Which one of the following is not an electromagnetic wave?

- ❑ *a* Infra-red
- ❑ *b* Radio
- ❑ *c* Light
- ❑ *d* Sound

18.20 In the visible spectrum the colour situated between orange and green is

- ❑ *a* Violet
- ❑ *b* Blue
- ❑ *c* Yellow
- ❑ *d* Red

18.21 The wavelength of emerald green is

- ❑ *a* 460 nm
- ❑ *b* 555 nm
- ❑ *c* 600 nm
- ❑ *d* 650 nm

18.22 In the electromagnetic spectrum, ultra-violet has a lower frequency than

- ❑ *a* X-rays
- ❑ *b* Visible light
- ❑ *c* Infra-red
- ❑ *d* Radio waves

18.23 Which of the following groups has the longest wavelengths?

- ❑ *a* X-rays
- ❑ *b* Infra-red radiation
- ❑ *c* Radio waves
- ❑ *d* Visible light

18.24 Which one of the following groups has the highest frequencies?

- ❑ *a* Radio waves
- ❑ *b* Infra-red radiation
- ❑ *c* Visible light
- ❑ *d* Ultraviolet radiation

18.25 If a light ray hits a plane mirror at an angle of 50°, the angle between the incident ray and the reflected ray is

- ❑ *a* 50°
- ❑ *b* 100°
- ❑ *c* 150°
- ❑ *d* 200°

18.26 A car rear-view mirror is

- ❑ *a* Convex
- ❑ *b* Concave
- ❑ *c* Plane
- ❑ *d* Parabolic

18.27 The simplest atom, having only one proton, is that of

- ❑ *a* Oxygen
- ❑ *b* Hydrogen
- ❑ *c* Copper
- ❑ *d* Silver

18.28 If a neutral atom has a nucleus containing 22 protons and 24 neutrons, the atomic number is

- ❑ *a* 11
- ❑ *b* 22
- ❑ *c* 24
- ❑ *d* 46

18.29 The mass number of the atom in Question 18.28 is

- ❑ *a* 20
- ❑ *b* 22
- ❑ *c* 24
- ❑ *d* 46

18.30 The number of electrons in Question 18.28 is

- ❑ *a* 22
- ❑ *b* 44
- ❑ *c* 46
- ❑ *d* 48

18.31 Two atoms of hydrogen and one atom of oxygen will combine to form a molecule of

- ❑ *a* Silver
- ❑ *b* Mercury
- ❑ *c* Water
- ❑ *d* Lead

18.32 Silicon has the atomic number

- ❑ *a* 2
- ❑ *b* 4
- ❑ *c* 14
- ❑ *d* 18

18.33 Germanium has the atomic number

- ❑ *a* 32
- ❑ *b* 36
- ❑ *c* 48
- ❑ *d* 64

18.34 15 °C converted to degrees Fahrenheit is

- ❑ *a* 20 °F
- ❑ *b* 27 °F
- ❑ *c* 32 °F
- ❑ *d* 59 °F

18.35 113 °F converted to degrees Celsius is

- ❑ *a* 38 °C
- ❑ *b* 40 °C
- ❑ *c* 45 °C
- ❑ *d* 81 °C

18.36 Absolute zero is

- ❑ *a* −100 °C
- ❑ *b* −132 °C
- ❑ *c* −273 °C
- ❑ *d* −300 °C

18.37 The freezing point of mercury is −37 °C. This temperature expressed in the Kelvin scale is

- ❑ *a* 140 K
- ❑ *b* 236 K
- ❑ *c* 246 K
- ❑ *d* 270 K

18.38 If a metal is cooled until it becomes a superconductor, the metal then

- ❑ *a* Has zero density
- ❑ *b* Has zero mass
- ❑ *c* Has zero electrical resistance
- ❑ *d* Has zero weight

18.39 The specific heat capacity of water measured in joules per kilogram per degree Celsius is

- ❑ *a* 1000
- ❑ *b* 1200
- ❑ *c* 3000
- ❑ *d* 4200

18.40 The amount of heat energy needed to heat 2 kg of water from 15 °C to 55 °C is

- ❑ *a* 168 kJ
- ❑ *b* 336 kJ
- ❑ *c* 400 kJ
- ❑ *d* 672 kJ

18.41 Which one of the following is not a good conductor of heat energy?
- ❑ *a* Copper
- ❑ *b* Iron
- ❑ *c* Glass
- ❑ *d* Silver

18.42 Heat is transferred to the tip of a soldering iron from the heating element by
- ❑ *a* Insulation
- ❑ *b* Radiation
- ❑ *c* Convection
- ❑ *d* Conduction

18.43 Heat transfer by the circulation of a gas or liquid is called
- ❑ *a* Radiation
- ❑ *b* Convection
- ❑ *c* Conduction
- ❑ *d* Insulation

18.44 The heat energy that can travel through a vacuum is called
- ❑ *a* Radiation
- ❑ *b* Conduction
- ❑ *c* Insulation
- ❑ *d* Convection

18.45 The time in minutes for the radiation from the sun to reach Earth is approximately
- ❑ *a* 4
- ❑ *b* 8
- ❑ *c* 16
- ❑ *d* 80

18.46 Which one of the following coloured surfaces is the best absorber of heat?
- ❑ *a* Blue
- ❑ *b* Green
- ❑ *c* White
- ❑ *d* Black

18.47 Which one of the following coloured surfaces is the best reflector of heat?
- ❑ *a* Black
- ❑ *b* White
- ❑ *c* Yellow
- ❑ *d* Green

18.48 Velocity has
- ❑ *a* Magnitude only
- ❑ *b* Direction only
- ❑ *c* Both magnitude and direction
- ❑ *d* Distance only

18.49 Speed has
- ❑ *a* Magnitude only
- ❑ *b* Distance only
- ❑ *c* Direction only
- ❑ *d* Magnitude and direction

18.50 A car travels at a constant speed a distance of 48 km in 37.5 m. The speed in kilometres per hour ($km\,h^{-1}$) is
- ❑ *a* 55.4
- ❑ *b* 59.2
- ❑ *c* 76.8
- ❑ *d* 80.4

18.51 The speed of the car in Question 18.50 expressed in metres per second ($m\,s^{-1}$) is
- ❑ *a* 21.33
- ❑ *b* 30.67
- ❑ *c* 42.67
- ❑ *d* 58.37

18.52 Acceleration is measured in
- ❑ *a* Metres per second
- ❑ *b* Metres per second per second
- ❑ *c* Metres per minute
- ❑ *d* Metres per hour

18.53 A vehicle starts from rest and has a speed of 30 $m\,s^{-1}$ after 10 s. The acceleration is
- ❑ *a* 0.3 $m\,s^{-2}$
- ❑ *b* 3.0 $m\,s^{-2}$
- ❑ *c* 10 $m\,s^{-2}$
- ❑ *d* 30 $m\,s^{-2}$

18.54 The definition of momentum is
- ❑ *a* Gravity × velocity
- ❑ *b* Distance × velocity
- ❑ *c* Mass × velocity
- ❑ *d* Mass × acceleration

18.55 The acceleration due to gravity is approximately
- ❑ *a* 10 m s^{-2}
- ❑ *b* 18 m s^{-2}
- ❑ *c* 20 m s^{-2}
- ❑ *d* 26 m s^{-2}

18.56 Electric current consists of
- ❑ *a* A flow of protons
- ❑ *b* A flow of atoms
- ❑ *c* A flow of molecules
- ❑ *d* A flow of electrons

18.57 An electron has
- ❑ *a* A positive and a negative charge
- ❑ *b* No charge
- ❑ *c* A negative charge
- ❑ *d* A positive charge

18.58 The voltage in an electrical circuit can be measured using
- ❑ *a* A voltmeter
- ❑ *b* An ammeter
- ❑ *c* A wattmeter
- ❑ *d* A signal generator

18.59 A rheostat is
- ❑ *a* A variable capacitor
- ❑ *b* A variable switch
- ❑ *c* A variable resistor
- ❑ *d* A fixed resistor

18.60 Tungsten is often used for the filament of a light bulb because
- ❑ *a* It has a low resistance
- ❑ *b* It can reach a high temperature without melting
- ❑ *c* It has a low capacitance
- ❑ *d* It has a low inductance

18.61 In a battery, electrical energy is produced from
- ❑ *a* Kinetic energy
- ❑ *b* Potential energy
- ❑ *c* Mechanical energy
- ❑ *d* Chemical energy

18.62 A primary battery cell which uses carbon and zinc electrodes produces
- ❑ *a* 0.5 V
- ❑ *b* 1.5 V
- ❑ *c* 6.5 V
- ❑ *d* 9.5 V

18.63 A lead–acid cell used in car batteries produces approximately
- ❑ *a* 0.6 V
- ❑ *b* 1. 0 V
- ❑ *c* 2.0 V
- ❑ *d* 3.0 V

18.64 When a car battery is being recharged, two gases are given out. These are
- ❑ *a* Oxygen and carbon dioxide
- ❑ *b* Hydrogen and carbon dioxide
- ❑ *c* Oxygen and hydrogen
- ❑ *d* Oxygen and carbon monoxide

18.65 An electrical appliance can be protected from excessive current and possible damage by using a
- ❑ *a* Fuse
- ❑ *b* Resistor
- ❑ *c* Transformer
- ❑ *d* Transistor

18.66 If a car headlamp bulb is rated as 12 V, 6 A, the wattage generated is
- ❑ *a* 18 W
- ❑ *b* 36 W
- ❑ *c* 54 W
- ❑ *d* 72 W

18.67 An electrical appliance consuming 2 kW from a mains supply would require a fuse with a rating of
- ❑ *a* 1 A
- ❑ *b* 3 A
- ❑ *c* 5 A
- ❑ *d* 10 A

18.68 Which one of the following is a magnetic material?
- ❑ *a* Copper
- ❑ *b* Iron
- ❑ *c* Lead
- ❑ *d* Glass

18.69 A plotting compass points to the
- ❑ *a* East
- ❑ *b* West
- ❑ *c* North
- ❑ *d* South

18.70 In an electric motor the electrical energy produces
- ❑ *a* Chemical energy
- ❑ *b* Potential energy
- ❑ *c* Kinetic energy
- ❑ *d* Solar energy

18.71 In an electrical generator electrical energy is produced from
- ❑ *a* Chemical energy
- ❑ *b* Kinetic energy
- ❑ *c* Solar energy
- ❑ *d* Potential energy

18.72 A lower alternating voltage can be produced from the mains supply by using a
- ❑ *a* Voltmeter
- ❑ *b* Capacitor
- ❑ *c* Rectifier
- ❑ *d* Transformer

18.73 The National Grid carries large amounts of power at high voltage in order to
- ❑ *a* Reduce the energy lost
- ❑ *b* Avoid capacitance effects
- ❑ *c* Avoid inductance effects
- ❑ *d* Reduce the number of cables required

18.74 The total resistance in a cable supplying 0.5 A is 1.5 Ω. The power lost in the cable is
- ❑ *a* 250 mW
- ❑ *b* 375 mW
- ❑ *c* 675 mW
- ❑ *d* 800 mW

18.75 If the current in Question 18.74 is halved, the power lost in the cable is
- ❑ *a* 22.25 mW
- ❑ *b* 45.65 mW
- ❑ *c* 50.25 mW
- ❑ *d* 93.75 mW

19 Calculations

19.1 $6^3 \times 6^2$ is
- ❑ *a* 36
- ❑ *b* 316
- ❑ *c* 1296
- ❑ *d* 7776

19.2 $6^3 \div 6^2$ is
- ❑ *a* 1
- ❑ *b* 6
- ❑ *c* 0
- ❑ *d* 36

19.3 20^0 is
- ❑ *a* 0
- ❑ *b* 1
- ❑ *c* 20
- ❑ *d* 40

19.4 $8^2 \times 8^1 \times 8^0$ is
- ❑ *a* 24
- ❑ *b* 32
- ❑ *c* 64
- ❑ *d* 512

19.5 $2^4 \times 2^{-2}$ is
- ❑ *a* 2
- ❑ *b* 4
- ❑ *c* 8
- ❑ *d* 16

19.6 $5^{-2} \div 5^{-5}$ is
- ❑ *a* 5
- ❑ *b* 25
- ❑ *c* 125
- ❑ *d* 625

19.7 $\sqrt{169}$ is
- ❑ *a* 11
- ❑ *b* 13
- ❑ *c* 42
- ❑ *d* 84.5

19.8 $6^{1.5}$ is
- ❑ *a* 10.3
- ❑ *b* 10.8
- ❑ *c* 11.2
- ❑ *d* 14.7

19.9 5^{-2} is
- ❑ *a* 0.04
- ❑ *b* 0.4
- ❑ *c* 10
- ❑ *d* 25

19.10 $\sqrt{24} \times \sqrt{24}$ is
- ❑ *a* 1
- ❑ *b* 12
- ❑ *c* 24
- ❑ *d* 48

19.11 The number 470 expressed in standard form is
- ❑ *a* 4.7×10^3
- ❑ *b* 47×10
- ❑ *c* 4.7×10^2
- ❑ *d* 47×10^2

19.12 The number 10 000 expressed in standard form is
- ❑ *a* 10×10^4
- ❑ *b* 10×10^3
- ❑ *c* 1×10^3
- ❑ *d* 1×10^4

19.13 The number 0.047 expressed in standard form is

- ❑ *a* 4.7×10^3
- ❑ *b* 4.7×10^2
- ❑ *c* 4.7×10^{-3}
- ❑ *d* 4.7×10^{-2}

19.14 $4x - 10x + 13x$ is

- ❑ *a* x
- ❑ *b* $-x$
- ❑ *c* $7x$
- ❑ *d* $-7x$

19.15 If $x = 10$ in Question 19.14, the sum is

- ❑ *a* 10
- ❑ *b* −10
- ❑ *c* 20
- ❑ *d* 70

19.16 $-10y^2 + 4y^2 - 4y^2 + 8y^2$ is

- ❑ *a* $-2y^2$
- ❑ *b* $2y^2$
- ❑ *c* $-6y^2$
- ❑ *d* $6y^2$

19.17 If $y = 10$ in Question 19.16, the sum is

- ❑ *a* 100
- ❑ *b* −100
- ❑ *c* −200
- ❑ *d* 200

19.18 $2x \times 10x \times x$ is

- ❑ *a* 20
- ❑ *b* $20x$
- ❑ *c* $20x^2$
- ❑ *d* $20x^3$

19.19 $-30y \times 4y^2$ is

- ❑ *a* $120y^2$
- ❑ *b* $-120y^2$
- ❑ *c* $-120y^3$
- ❑ *d* $120y^3$

19.20 $-12z^2 \times -5z^2$ is

- ❑ *a* $60z^4$
- ❑ *b* $-60z^4$
- ❑ *c* $60z^2$
- ❑ *d* $-60z^2$

19.21 $20xyz \div 4xy$ is

- ❑ *a* $20z$
- ❑ *b* $5z$
- ❑ *c* $4xy$
- ❑ *d* $5xy$

19.22 $-60x^2y^2 \div 10xy$ is

- ❑ *a* 6
- ❑ *b* $6xy$
- ❑ *c* $-6y$
- ❑ *d* $-6xy$

19.23 $-25xyz^2 \div -5xy$ is

- ❑ *a* $5xy$
- ❑ *b* $-5xy$
- ❑ *c* $5z^2$
- ❑ *d* $-5z^2$

19.24 If $3x + 4x = 28$, x is

- ❑ *a* 3
- ❑ *b* 4
- ❑ *c* 5
- ❑ *d* 6

19.25 If $4x - 2x - 6 = 0$, x is

- ❑ *a* 3
- ❑ *b* −3
- ❑ *c* 2
- ❑ *d* −2

19.26 If $-3y - 10y - 2y + 30 = 0$, y is

- ❑ *a* −2
- ❑ *b* 2
- ❑ *c* 5
- ❑ *d* −5

19.27 If $6a + 10b - 12c = 50$ and $a = 4$ and $b = 5$, the value of c is

- ❑ *a* 2
- ❑ *b* −4
- ❑ *c* 8
- ❑ *d* −9

19.28 Removing the bracket in $4(2 + x) + x$ gives

- ❑ *a* $6 + 2x$
- ❑ *b* $12 + x$
- ❑ *c* $8 + 2x$
- ❑ *d* $8 + 5x$

19.29 Removing the bracket in $-2(y + x)$ gives

- ❑ *a* $-2y + 2x$
- ❑ *b* $-2y - 2x$
- ❑ *c* $-2y - x$
- ❑ *d* $2y + 2x$

19.30 Removing the bracket in $-x(-a + b + c)$ gives

- ❑ *a* $ax + bx + cx$
- ❑ *b* $ax - bx - cx$
- ❑ *c* $-ax + bx + cx$
- ❑ *d* $-ax - bx - cx$

19.31 If $2x + 3y = 18$ and $4x + 2y = 20$, the values of x and y are

- ❑ *a* 1, 2
- ❑ *b* 2, 6
- ❑ *c* 3, 4
- ❑ *d* 2, 8

19.32 If $5x - 5y = 10$ and $2x + 2y = 36$, the values of x and y are

- ❑ *a* –2, 3
- ❑ *b* 4, –6
- ❑ *c* 5, 5
- ❑ *d* 10, 8

19.33 If $6x + 4y = -10$ and $4x - 8y = -60$, the values of x and y are

- ❑ *a* 4, 4
- ❑ *b* –5, 5
- ❑ *c* 6, –4
- ❑ *d* –6, 4

19.34 If $6x^2 = 486$, x is

- ❑ *a* 3
- ❑ *b* 9
- ❑ *c* 81
- ❑ *d* 90

19.35 If $x^3 + 20 = 84$, x is

- ❑ *a* 4
- ❑ *b* –4
- ❑ *c* 3
- ❑ *d* –3

19.36 If $2x^2 \times 5x^2 = 810$, x is

- ❑ *a* 2
- ❑ *b* 3
- ❑ *c* 4
- ❑ *d* 9

19.37 If $\sqrt{}(4x^2 - 19) = 9$, x is

- ❑ *a* 2
- ❑ *b* 4
- ❑ *c* 5
- ❑ *c* 9

19.38 If $\sqrt{}(x^2 + y^2) = 5$ and $y = 4$, the value of x is

- ❑ *a* 3
- ❑ *b* 4
- ❑ *c* 5
- ❑ *d* 2

19.39 If $\sqrt{}(a^2 + b^2) = 20$ and $a = 12$, the value of b is

- ❑ *a* 10
- ❑ *b* 12
- ❑ *c* 14
- ❑ *d* 16

19.40 If $4b \div xy = 4$ and $x = 2$ and $y = 5$, the value of b is

- ❑ *a* 6
- ❑ *b* 10
- ❑ *c* 12
- ❑ *d* 14

19.41 If $10z \div abc = 5$ and $a = 1$, $c = 4$ and $z = 6$, the value of b is

- ❑ *a* 25
- ❑ *b* 15
- ❑ *c* 9
- ❑ *d* 3

19.42 If $xy^2 \div ab^2 = 18$ and $x = 45$, $y = 4$ and $a = 10$, the value of b is

- ❑ *a* 2
- ❑ *b* 8
- ❑ *c* 10
- ❑ *d* 12

19.43 If $x \div (ab \times 10^{-6}) = 50$ and $x = 5$ and $a = 200$, the value of b is
- ❑ *a* 100
- ❑ *b* 250
- ❑ *c* 500
- ❑ *d* 1000

19.44 0.715 expressed as a percentage is
- ❑ *a* 0.715%
- ❑ *b* 7.15%
- ❑ *c* 71.5%
- ❑ *d* 715%

19.45 5% of 20 is
- ❑ *a* 1
- ❑ *b* 2
- ❑ *c* 5
- ❑ *d* 10

19.46 25% of a certain length is 150 mm. The complete length is
- ❑ *a* 200 mm
- ❑ *b* 250 mm
- ❑ *c* 400 mm
- ❑ *d* 600 mm

19.47 A 100-Ω resistor has a tolerance spread of ±2%. The spread is
- ❑ *a* ±1 Ω
- ❑ *b* ±2 Ω
- ❑ *c* ±10 Ω
- ❑ *d* ±20 Ω

19.48 A 10% increase in a certain resistance gives a value of 5500 Ω. The value of the resistance before the increase is
- ❑ *a* 3500 Ω
- ❑ *b* 4000 Ω
- ❑ *c* 4500 Ω
- ❑ *d* 5000 Ω

19.49 A current in a circuit is measured as 30 A. If the current is 10% below requirement, the required current is
- ❑ *a* 31.67 A
- ❑ *b* 32.33 A
- ❑ *c* 33.33 A
- ❑ *d* 34.67 A

19.50 If $1/R = 0.1 + 0.2 + 0.2$, R is
- ❑ *a* 0.5
- ❑ *b* 1.0
- ❑ *c* 1.5
- ❑ *d* 2.0

19.51 If $R = (R_1R_2) \div (R_1 + R_2)$ and $R_1 = 20\ \Omega$ and $R_2 = 30\ \Omega$, R is
- ❑ *a* 10 Ω
- ❑ *b* 12 Ω
- ❑ *c* 14 Ω
- ❑ *d* 15 Ω

19.52 If in Question 19.51 R_1 is 12 Ω and R is 3 Ω, the value of R_2 is
- ❑ *a* 4 Ω
- ❑ *b* 5 Ω
- ❑ *c* 6 Ω
- ❑ *d* 9 Ω

19.53 If in Question 19.51 R_2 is 48 Ω and R is 12 Ω, the value of R_1 is
- ❑ *a* 10 Ω
- ❑ *b* 14 Ω
- ❑ *c* 16 Ω
- ❑ *d* 18 Ω

19.54 45° expressed in radians is
- ❑ *a* π/4 rad
- ❑ *b* π rad
- ❑ *c* π/2 rad
- ❑ *d* 2π rad

19.55 210° expressed in radians is
- ❑ *a* π/6 rad
- ❑ *b* 7π/6 rad
- ❑ *c* 5π/6 rad
- ❑ *d* 3π/2 rad

19.56 5π/3 rad expressed in degrees is
- ❑ *a* 200°
- ❑ *b* 250°
- ❑ *c* 280°
- ❑ *d* 300°

19.57 2π/3 rad expressed in degrees is
- ❑ *a* 10°
- ❑ *b* 90°
- ❑ *c* 120°
- ❑ *d* 150°

19.58 Using logarithms to the base 10, 20 log 20 gives

❑ *a* 20
❑ *b* 26
❑ *c* 28
❑ *d* 30

19.59 Using logarithms to the base 10, if 30 log x = 30, x is

❑ *a* 1.5
❑ *b* 3.5
❑ *c* 7.5
❑ *d* 10

19.60 In Question 19.59, if 65 log x = 195, x is

❑ *a* 650
❑ *b* 1000
❑ *c* 1200
❑ *d* 1500

19.61 The diagram shows a triangle where the angle ABC is a right angle. If AB = 6 cm and BC = 8 cm, the length of AC is

❑ *a* 10 cm
❑ *b* 11 cm
❑ *c* 12 cm
❑ *d* 13 cm

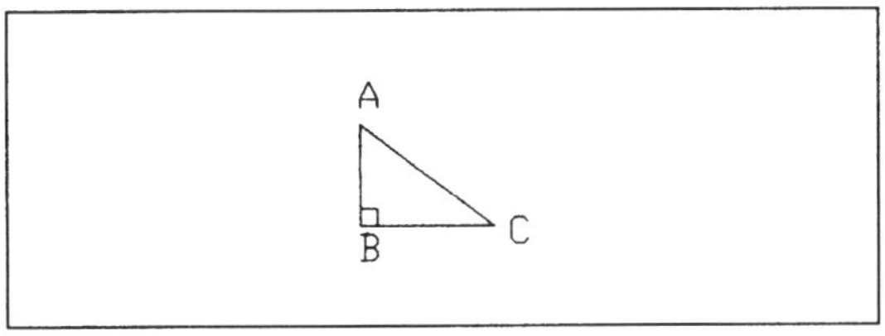

19.62 If in Question 19.61 AB = 12 cm and AC = 20 cm, the length BC is

❑ *a* 13 cm
❑ *b* 14 cm
❑ *c* 15 cm
❑ *d* 16 cm

19.63 If in Question 19.61 BC = 15 cm and AC = 25 cm, the length AB is

❑ *a* 17 cm
❑ *b* 18 cm
❑ *c* 20 cm
❑ *d* 21 cm

19.64 The diagram shows a triangle where angle ABC = 90° and angle ACB = 30°. If AB = 5 cm, the length AC is

❑ *a* 10 cm
❑ *b* 12 cm
❑ *c* 13 cm
❑ *d* 15 cm

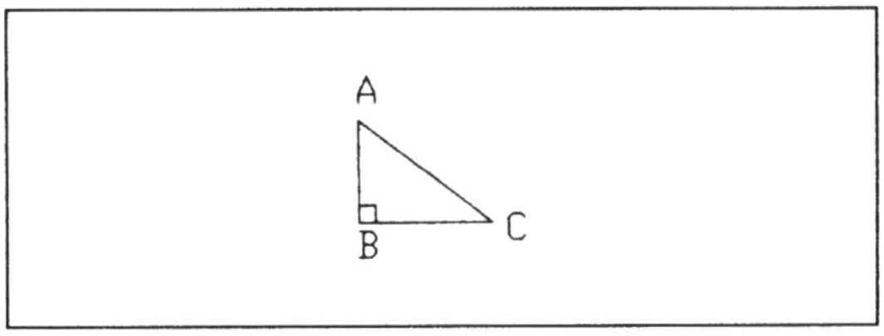

19.65 In Question 19.65 the angle CAB is equal to

❑ *a* 30°
❑ *b* 45°
❑ *c* 50°
❑ *d* 60°

19.66 In Question 19.65 the length of the side BC is

❑ *a* 7.16 cm
❑ *b* 7.42 cm
❑ *c* 8.66 cm
❑ *d* 9.33 cm

19.67 The diagram shows a triangle where angle ABC = 90° and angle ACB = 45°. If AB = 8 cm, the length BC is

❑ *a* 6 cm
❑ *b* 7 cm
❑ *c* 8 cm
❑ *d* 9 cm

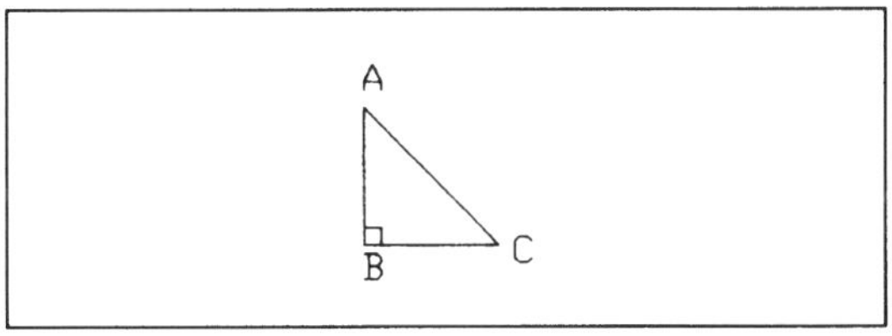

19.68 In Question 19.67 the length of side AC is approximately

❑ *a* 11.31 cm
❑ *b* 14.62 cm
❑ *c* 15.11 cm
❑ *d* 15.84 cm

19.69 The diagram shows a triangle where angle ABC = 90 ° and angle BAC = 60 °. If side AB = 12 cm, the length of side AC is

❑ *a* 21 cm
❑ *b* 24 cm
❑ *c* 25 cm
❑ *d* 26 cm

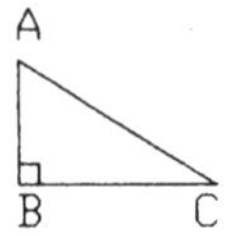

19.70 In Question 19.69 the length of side BC is approximately

❑ *a* 16.61 cm
❑ *b* 18.32 cm
❑ *c* 20.78 cm
❑ *d* 28.64 cm

19.71 The diagram shows a triangle where angle ABC = 90 °. If AB = 21 cm and BC = 28 cm, the length of side AC is

❑ *a* 31 cm
❑ *b* 35 cm
❑ *c* 37 cm
❑ *d* 40 cm

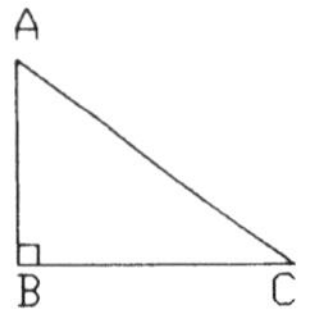

19.72 In Question 19.71 the angle BAC is

❑ *a* 14.62 °
❑ *b* 28.28 °
❑ *c* 46.14 °
❑ *d* 53.13 °

19.73 In Question 19.71 the angle ACB is

❑ *a* 36.87 °
❑ *b* 45.32 °
❑ *c* 51.16 °
❑ *d* 52.12 °

19.74 A circle has a diameter of 12 cm. The area is approximately

❑ *a* 102 cm^2
❑ *b* 113 cm^2
❑ *c* 160 cm^2
❑ *d* 170 cm^2

19.75 A circle has an area of 240 mm^2. The radius is approximately

❑ *a* 4.44 cm
❑ *b* 5.41 cm
❑ *c* 6.32 cm
❑ *d* 8.74 cm

20 Transposition of formulas

20.1 Given that $V = IR$ then I is
- ❑ *a* V/R
- ❑ *b* R/V
- ❑ *c* VR
- ❑ *d* $V + R$

20.2 In Question 20.1, R is
- ❑ *a* $V + I$
- ❑ *b* VI
- ❑ *c* I/V
- ❑ *d* V/I

20.3 Given that $P = I^2R$ then I is
- ❑ *a* $\sqrt{(R/P)}$
- ❑ *b* $\sqrt{(P/R)}$
- ❑ *c* P/R
- ❑ *d* R/P

20.4 In Question 20.3, R is
- ❑ *a* I^2/P
- ❑ *b* $\sqrt{(I^2/P)}$
- ❑ *c* P/I^2
- ❑ *d* $\sqrt{(P/I^2)}$

20.5 If $IL = \sqrt{(3)}IP$ then IP is
- ❑ *a* $IL/\sqrt{(3)}$
- ❑ *b* $\sqrt{(3)}/IL$
- ❑ *c* $IL/3$
- ❑ *d* $3/IL$

20.6 If $x = \sqrt{(y)}z$ then y is
- ❑ *a* $\sqrt{(x/z)}$
- ❑ *b* z^2/x^2
- ❑ *c* x^2/z^2
- ❑ *d* $\sqrt{(z/x)}$

20.7 In Question 20.6, if $x = 40$ and $z = 10$, y is
- ❑ *a* 2
- ❑ *b* 4
- ❑ *c* 8
- ❑ *d* 16

20.8 If $1/R = 1/R_1 + 1/R_2$, R is
- ❑ *a* $R_1 + R_2$
- ❑ *b* $R_1R_2/(R_1 + R_2)$
- ❑ *c* $(R_1 + R_2)/ R_1R_2$
- ❑ *d* R_1/R_1R_2

20.9 In Question 20.8, R_1 is
- ❑ *a* $RR_2/(R_2 + R)$
- ❑ *b* $RR_2/(R_2 - R)$
- ❑ *c* $(R_2 - R)/RR_2$
- ❑ *d* $(R_2 + R)/RR_2$

20.10 In Question 20.8, R_2 is
- ❑ *a* $RR_1/(R_1 - R)$
- ❑ *b* $(R_1 - R)/RR_1$
- ❑ *c* $RR_1/(R_1 + R)$
- ❑ *d* $(R_1 + R)/RR_1$

20.11 In Question 20.8, if $R = 3\ \Omega$ and $R_1 = 12\ \Omega$, R_2 is
- ❑ *a* 2 Ω
- ❑ *b* 3 Ω
- ❑ *c* 4 Ω
- ❑ *d* 8 Ω

20.12 If $R + R_m = V/I$, I is
- ❑ *a* $V(R + R_m)$
- ❑ *b* $(R + R_m)/V$
- ❑ *c* $V/(R + R_m)$
- ❑ *d* $V + R + R_m$

20.13 In Question 20.12, V is
- ❑ *a* $IR + R_m$
- ❑ *b* $(R + R_m)/I$
- ❑ *c* $I/(R + R_m)$
- ❑ *d* $I(R + R_m)$

20.14 In Question 20.12, R is
- ❑ *a* $(V/I) - R_m$
- ❑ *b* $(V/I) + R_m$
- ❑ *c* $(VI) - R_m$
- ❑ *d* $(VI) + R_m$

20.15 In Question 20.12, if $R = 25\ \Omega$, $R_m = 15\ \Omega$ and $V = 80$ V, I is
- ❑ *a* 1 A
- ❑ *b* 2 A
- ❑ *c* 8 A
- ❑ *d* 9 A

20.16 If $C = x[(N - 1)A]/d$, x is
- ❑ *a* $C[(N - 1)A]/d$
- ❑ *b* $Cd[(N - 1)A]$
- ❑ *c* $[(N - 1)A]/Cd$
- ❑ *d* $Cd/[(N - 1)A]$

20.17 In Question 20.16, N is
- ❑ *a* $(Cd/xA) + 1$
- ❑ *b* $(xA/Cd) + 1$
- ❑ *c* $(Cd/xA) - 1$
- ❑ *d* $(xA/Cd) - 1$

20.18 In Question 20.16, A is
- ❑ *a* $Cdx/(N - 1)$
- ❑ *b* $Cd[x(N - 1)]$
- ❑ *c* $Cd/[x(N - 1)]$
- ❑ *d* $[x(N - 1)]/Cd$

20.19 In Question 20.16, d is
- ❑ *a* $C[x(N - 1)]/A$
- ❑ *b* $C[x(N - 1)A]$
- ❑ *c* $[x(N - 1)A]/C$
- ❑ *d* $C/[x(N - 1)A]$

20.20 In Question 20.16, if $C = 6$, $x = 5$, $A = 10$ and $d = 25$, N is
- ❑ *a* 4
- ❑ *b* 7
- ❑ *c* 8
- ❑ *d* 9

20.21 If $e = 2LI/t$, t is
- ❑ *a* $LI/2e$
- ❑ *b* $e/2LI$
- ❑ *c* $2LI/e$
- ❑ *d* $2L/Ie$

20.22 In Question 20.21 L is
- ❑ *a* $2et/I$
- ❑ *b* $t/2eI$
- ❑ *c* $e/2It$
- ❑ *d* $et/2I$

20.23 In Question 20.21, I is
- ❑ *a* $et/2L$
- ❑ *b* $2L/et$
- ❑ *c* $L/2et$
- ❑ *d* $t/2eL$

20.24 In Question 20.21, if $L = 6$, $I = 8$ and $e = 3$, t is
- ❑ *a* 20
- ❑ *b* 28
- ❑ *c* 32
- ❑ *d* 40

20.25 If $X = 2\pi fL$, L is
- ❑ *a* $2\pi f/X$
- ❑ *b* $X/2\pi f$
- ❑ *c* $2\pi/fX$
- ❑ *d* $2f/\pi f$

20.26 In Question 20.25, f is
- ❑ *a* $2\pi/LX$
- ❑ *b* $2\pi L/X$
- ❑ *c* $2L/\pi X$
- ❑ *d* $X/2\pi L$

20.27 In Question 20.26, if $X = 3142$ and $f = 50$, L is
- ❑ *a* 10
- ❑ *b* 15
- ❑ *c* 25
- ❑ *d* 50

20.28 If $X = 1/2\pi fC$, C is
- ❑ *a* $2\pi fX$
- ❑ *b* $1/2\pi fX$
- ❑ *c* $2\pi f/X$
- ❑ *d* $X/2\pi f$

20.29 In Question 20.28, f is
- ❑ *a* $2\pi CX$
- ❑ *b* $2\pi/CX$
- ❑ *c* $CX/2\pi$
- ❑ *d* $1/2\pi CX$

20.30 In Question 20.28, if $f = 50$ and $X = 0.002$, C is
- ❑ *a* 1.06
- ❑ *b* 1.21
- ❑ *c* 1.59
- ❑ *d* 2.68

20.31 If $y = x^2z^2$, x is
- ❑ *a* $\sqrt{(z^2/y)}$
- ❑ *b* $\sqrt{(y/z^2)}$
- ❑ *c* y/z^2
- ❑ *d* z^2/y

20.32 In Question 20.31, z is
- ❑ *a* $\sqrt{(x^2/y)}$
- ❑ *b* x^2/y
- ❑ *c* $\sqrt{(y/x^2)}$
- ❑ *d* y/x^2

20.33 In Question 20.31, if $y = 400$ and $x = 4$, z is
- ❑ *a* 1
- ❑ *b* 2
- ❑ *c* 4
- ❑ *d* 5

20.34 If $A/BC = 1/N$, N is
- ❑ *a* BC/A
- ❑ *b* A/BC
- ❑ *c* BAC
- ❑ *d* B/CA

20.35 In Question 20.34, A is
- ❑ *a* BC/N
- ❑ *b* N/BC
- ❑ *c* BN/C
- ❑ *d* C/BN

20.36 In Question 20.34, B is
- ❑ *a* CAN
- ❑ *b* AN/C
- ❑ *c* C/AN
- ❑ *d* CN/A

20.37 In Question 20.34, C is
- ❑ *a* BA/N
- ❑ *b* B/AN
- ❑ *c* BAN
- ❑ *d* AN/B

20.38 In Question 20.34, if $A = 10$, $N = 7$ and $C = 5$, B is
- ❑ *a* 10
- ❑ *b* 12
- ❑ *c* 14
- ❑ *d* 24

20.39 If $v^2 = u^2 + 2as$, u is
- ❑ *a* $\sqrt{(v^2 + 2as)}$
- ❑ *b* $\sqrt{(v^2 - 2as)}$
- ❑ *c* $v^2 - 2as$
- ❑ *d* $2as + v^2$

20.40 In Question 20.39, a is
- ❑ *a* $(v^2 - u^2)/2s$
- ❑ *b* $2(v^2 - u^2)/s$
- ❑ *c* $2s/(v^2 - u^2)$
- ❑ *d* $2s(v^2 - u^2)$

20.41 In Question 20.39, s is
- ❑ *a* $2a/(v^2 - u^2)$
- ❑ *b* $2a(v^2 - u^2)$
- ❑ *c* $(v^2 - u^2)/2a$
- ❑ *d* $2(v^2 - u^2)/a$

20.42 In Question 20.39, if $v = 9$, $a = 8$ and $s = 2$, u is
- ❑ *a* 5
- ❑ *b* 7
- ❑ *c* 9
- ❑ *d* 13

20.43 If $A = (B - C)/(D - E)$, B is
- ❑ *a* $[(D - C)/A] + C$
- ❑ *b* $(AD + AE)/C$
- ❑ *c* $AD + AE + C$
- ❑ *d* $AD - AE + C$

20.44 In Question 20.43, C is
- ❑ *a* $B + AD + AE$
- ❑ *b* $B - AD - AE$
- ❑ *c* $B - AD + AE$
- ❑ *d* $B + AD - AE$

20.45 In Question 20.43, D is
- ❑ *a* $[(B - C)/A] + E$
- ❑ *b* $[(B - C)/A] - E$
- ❑ *c* $[A/(B - C)] + E$
- ❑ *d* $[A/(B - C)] - E$

20.46 In Question 20.43, E is
- ❑ *a* $[(C - B)/A] - D$
- ❑ *b* $[(B - C)/A] - D$
- ❑ *c* $[(B - C)/A] + D$
- ❑ *d* $[(C - B)/A] + D$

20.47 In Question 20.43, if $A = 5$, $C = 10$, $D = 11$ and $E = 3$, the value of B is
- ❑ *a* 10
- ❑ *b* 50
- ❑ *c* 70
- ❑ *d* 80

20.48 If $P_1 \times V_1/T_1 = (P_2 \times V_2)/T_2$, P_1 is
- ❑ *a* $P_2V_1V_2/T_1T_2$
- ❑ *b* $P_2T_2V_1/T_1V_2$
- ❑ *c* $P_2T_1T_2/V_1V_2$
- ❑ *d* $P_2V_2T_1/V_1T_2$

20.49 In Question 20.48, V_1 is
- ❑ *a* $P_1P_2V_2/T_1T_2$
- ❑ *b* $V_2T_1T_2/P_1P_2$
- ❑ *c* $P_2V_2T_1/P_1T_2$
- ❑ *d* $P_1V_2T_1/P_2T_2$

20.50 In Question 20.48, T_1 is
- ❑ *a* $P_1V_1T_2/P_2V_2$
- ❑ *b* $P_1P_2T_2/V_1V_2$
- ❑ *c* $P_1V_1V_2/P_2T_2$
- ❑ *d* $V_1V_2T_2/P_1P_2$

20.51 In Question 20.48, if $P_1 = 30$, $P_2 = 4$, $V_1 = 5$, $V_2 = 20$ and $T_1 = 15$, T_2 is
- ❑ *a* 6
- ❑ *b* 8
- ❑ *c* 18
- ❑ *d* 24

20.52 If $(C \times 9/5) + 32 = F$, C is
- ❑ *a* $9(F - 32)/5$
- ❑ *b* $5(F - 32)/9$
- ❑ *c* $9/5(F - 32)$
- ❑ *d* $5/9(F - 32)$

20.53 In Question 20.52, if $F = 194$, C is
- ❑ *a* 20
- ❑ *b* 60
- ❑ *c* 90
- ❑ *d* 98

20.54 In Question 20.52, if $F = 32$, C is
- ❑ *a* 0
- ❑ *b* 1
- ❑ *c* 32
- ❑ *d* 100

20.55 In Question 20.52, if $C = 20$, F is
- ❑ *a* 40
- ❑ *b* 48
- ❑ *c* 50
- ❑ *d* 68

20.56 In Question 20.52, if $C = 100$, F is
- ❑ *a* 142
- ❑ *b* 152
- ❑ *c* 200
- ❑ *d* 212

20.57 If $A = (v - u)/t$, v is
- ❑ *a* $At + u$
- ❑ *b* $A + ut$
- ❑ *c* $uA + t$
- ❑ *d* $At - u$

20.58 In Question 20.57, u is
- ❑ *a* $At - v$
- ❑ *b* $At + v$
- ❑ *c* $v - At$
- ❑ *d* $Av + t$

20.59 In Question 20.57, t is
- ❑ *a* $(v + u)/A$
- ❑ *b* $(v - u)/A$
- ❑ *c* $A(v + u)$
- ❑ *d* $A(v - u)$

20.60 In Question 20.57, if $A = 5$, $u = 9$ and $t = 8$, v is
- ❑ *a* 18
- ❑ *b* 29
- ❑ *c* 36
- ❑ *d* 49

20.61 If $E = \frac{1}{2}mv^2$, m is
- ❑ *a* $2v^2/E$
- ❑ *b* $2E/v^2$
- ❑ *c* $E/2v^2$
- ❑ *d* $v^2/2E$

20.62 In Question 20.61, v is
- ❑ *a* $\sqrt{(m/2E)}$
- ❑ *b* $\sqrt{(2E/m)}$
- ❑ *c* $2m/E$
- ❑ *d* $m/2E$

20.63 In Question 20.61, if $E = 192$ and $m = 6$, V is
- ❑ *a* 8
- ❑ *b* 10
- ❑ *c* 12
- ❑ *d* 24

20.64 If $A = (3N + 2)/(N + 1)$, N is
- ❑ *a* $(A - 2)/(A - 3)$
- ❑ *b* $(2 - A)/(A - 3)$
- ❑ *c* $(A - 2)/(3 - A)$
- ❑ *d* $(2 - A)/(3 - A)$

20.65 In Question 20.64, if $A = 2.667$, N is
- ❑ *a* 0.5
- ❑ *b* 1.2
- ❑ *c* 1.8
- ❑ *d* 2.0

20.66 In Question 20.64, if $A = 1.0$, N is
- ❑ *a* −0.5
- ❑ *b* −1.0
- ❑ *c* 0.5
- ❑ *d* 1.0

20.67 If $A = S(C - B)/C$, C is
- ❑ *a* $(S - B)/SA$
- ❑ *b* $SA/(S - B)$
- ❑ *c* $SB/(S - A)$
- ❑ *d* $(S - A)/SB$

20.68 In Question 20.67, S is
- ❑ *a* $(C - A)/CB$
- ❑ *b* $(C - B)/CA$
- ❑ *c* $(C + A)/CB$
- ❑ *d* $CA/(C - B)$

20.69 In Question 20.67, B is
- ❑ *a* $S/(SC - CA)$
- ❑ *b* $S/(SC + CA)$
- ❑ *c* $(SC - CA)/S$
- ❑ *d* $(SC + CA)/S$

20.70 In Question 20.67, if $A = 5$, $B = 1$ and $S = 10$, C is
- ❑ *a* 2
- ❑ *b* 4
- ❑ *c* 8
- ❑ *d* 10

20.71 If $D/A = (F + P)/(F - P)$, D is
- ❑ *a* $A(F - P)/(F + P)$
- ❑ *b* $A(F + P)/(F - P)$
- ❑ *c* $(F + P)/A(F - P)$
- ❑ *d* $(F + P)/(P + F)$

20.72 In Question 20.71, A is
- ❑ *a* $D(F - P)/(F + P)$
- ❑ *b* $D(F + P)/(F - P)$
- ❑ *c* $(F - P)/(F + P)$
- ❑ *d* $(F + P)/(F - P)$

20.73 In Question 20.71, F is
- ❑ *a* $(D - A)/(AP + DP)$
- ❑ *b* $(A + D)/(AP + DP)$
- ❑ *c* $(A + D)/(AP - DP)$
- ❑ *d* $(AP + DP)/(D - A)$

20.74 In Question 20.71, P is
- ❑ *a* $(A - D)/(DF + AF)$
- ❑ *b* $(A + D)/(DF - AF)$
- ❑ *c* $(DF + AF)/(A - D)$
- ❑ *d* $(DF - AF)/(A + D)$

20.75 In Question 20.71, if $D = 40$, $F = 35$ and $P = 5$, A is
- ❑ *a* 15
- ❑ *b* 30
- ❑ *c* 40
- ❑ *d* 45

Appendix A Sample test paper

A.1 In the diagram R is equal to

❑ *a* 30 Ω
❑ *b* 50 Ω
❑ *c* 70 Ω
❑ *d* 90 Ω

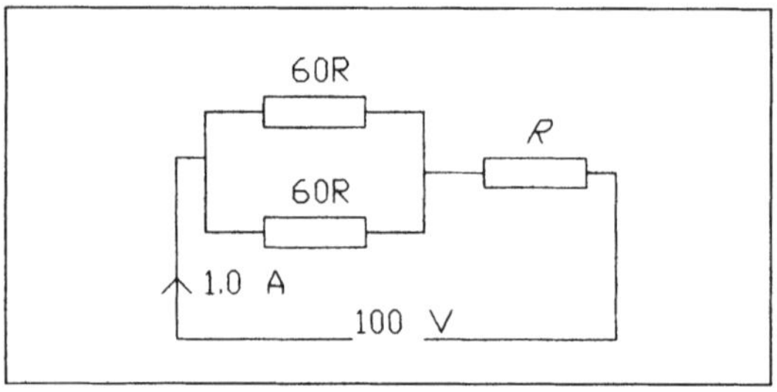

A.2 A radio station is found on 362 metres. This is equivalent to a frequency of

❑ *a* 650 kHz
❑ *b* 720 kHz
❑ *c* 788 kHZ
❑ *d* 828 kHz

A.3 An r.m.s. voltage of 1000 V will have a peak value of

❑ *a* 1200 V
❑ *b* 1414 V
❑ *c* 1528 V
❑ *d* 1614 V

A.4 A 3 : 1 step-up 240-V r.m.s. transformer will have a peak to peak voltage across the secondary of

❑ *a* 113 V
❑ *b* 720 V
❑ *c* 678 V
❑ *d* 2036 V

A.5 The ripple voltage produced by a 50-Hz half-wave power supply unit has a frequency of

❑ *a* 25 Hz
❑ *b* 50 Hz
❑ *c* 75 Hz
❑ *d* 100 Hz

A.6 A bridge rectifier uses

❑ *a* One diode
❑ *b* Two diodes
❑ *c* Three diodes
❑ *d* Four diodes

A.7 One cycle of a 500-Hz sine wave has a duration of

❑ *a* 2 μs
❑ *b* 200 ms
❑ *c* 2 ms
❑ *d* 5 ms

A.8 A zener diode is used in a power supply unit to

❑ *a* Reduce noise
❑ *b* Reduce the voltage ripple
❑ *c* Regulate the output voltage
❑ *d* Double the output voltage

A.9 In the diagram R is

❑ *a* 3R
❑ *b* 30R
❑ *c* 12R
❑ *d* 300R

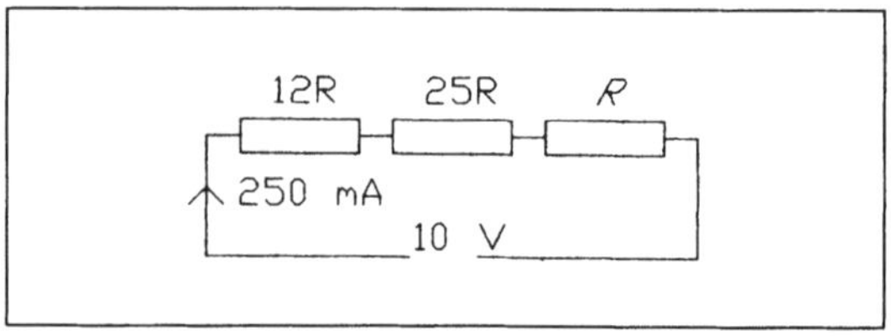

A.10 A 10:1 step-down mains transformer will have a peak voltage on the secondary of
- ❑ *a* 17 V
- ❑ *b* 34 V
- ❑ *c* 51 V
- ❑ *d* 68 V

A.11 A 2200-Ω resistor will have a colour code of
- ❑ *a* Red black red
- ❑ *b* Brown red red
- ❑ *c* Red red red
- ❑ *d* Black red orange

A.12 A 1-kW fire has a fused mains plug. The correct fuse rating is
- ❑ *a* 3 A
- ❑ *b* 5 A
- ❑ *c* 7 A
- ❑ *d* 13 A

A.13 In a half-wave rectifier the average value of the voltage is
- ❑ *a* $V_p/2\pi$
- ❑ *b* V_p/π
- ❑ *c* $2V_p/\pi$
- ❑ *d* $V_p/2$

A.14 All appliance flexes bear the modern colour coding where the live wire is coloured
- ❑ *a* Red
- ❑ *b* Blue
- ❑ *c* Brown
- ❑ *d* Green

A.15 A resistor has the following colour code: first colour blue, second colour grey, third colour red. The value is
- ❑ *a* 560R
- ❑ *b* 4700R
- ❑ *c* 6800R
- ❑ *d* 8200R

A.16 The time constant *t* (in seconds) for a capacitor and a resistor is equal to *C* (farads) × *R* (ohms). What will the time constant of 4 μF and 3 MΩ be?
- ❑ *a* 1 s
- ❑ *b* 10 s
- ❑ *c* 12 s
- ❑ *d* 24 s

A.17 Three resistors of 14R, 28R and 56R are connected in parallel. The combined resistance is
- ❑ *a* 2R
- ❑ *b* 6R
- ❑ *c* 8R
- ❑ *d* 12R

A.18 A filter unit in a power supply is
- ❑ *a* Low pass
- ❑ *b* High pass
- ❑ *c* Band pass
- ❑ *d* Band stop

A.19 An IF amplifier in an AM receiver handles frequencies around
- ❑ *a* 20 kHz
- ❑ *b* 470 kHz
- ❑ *c* 10.7 MHz
- ❑ *d* 37.5 MHz

A.20 Radio waves travel at a speed of
- ❑ *a* 1 000 000 $m\,s^{-1}$
- ❑ *b* 3 000 000 $m\,s^{-1}$
- ❑ *c* 100 000 000 $m\,s^{-1}$
- ❑ *d* 300 000 000 $m\,s^{-1}$

A.21 The primary colours used in television are
- ❑ *a* Red green yellow
- ❑ *b* Green blue yellow
- ❑ *c* Red green blue
- ❑ *d* Red blue white

A.22 A radio wave has a frequency of 100 MHz. The wavelength is
- ❑ *a* 3 m
- ❑ *b* 30 m
- ❑ *c* 100 m
- ❑ *d* 33 m

A.23 A device which will convert DC power to AC power is

- ❑ *a* A frequency changer
- ❑ *b* A demodulator
- ❑ *c* A rectifier
- ❑ *d* An oscillator

A.24 The peak to peak value of a 240 V AC voltage is

- ❑ *a* 120 V
- ❑ *b* 340 V
- ❑ *c* 480 V
- ❑ *d* 680 V

A.25 The current in the ammeter in the diagram is

- ❑ *a* 0.1 A
- ❑ *b* 0.2 A
- ❑ *c* 0.3 A
- ❑ *d* 1.0 A

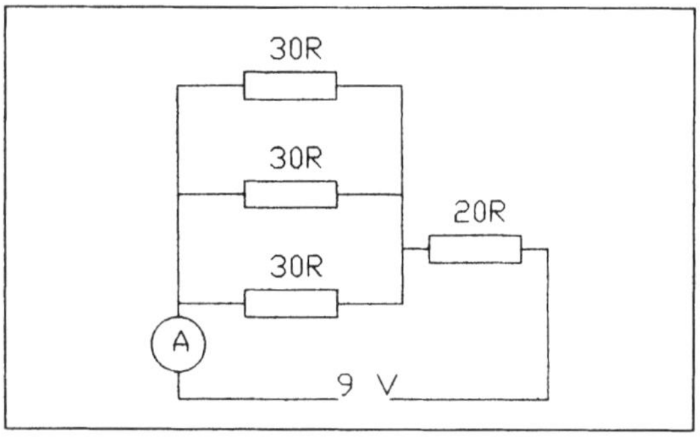

A.26 For mains operation a 3-kW radiator should be fitted with a

- ❑ *a* 3-A fuse
- ❑ *b* 5-A fuse
- ❑ *c* 7-A fuse
- ❑ *d* 13-A fuse

A.27 Toxic fumes are

- ❑ *a* Always visible
- ❑ *b* Non-poisonous
- ❑ *c* Healthy
- ❑ *d* Poisonous

A.28 Given a non-sinusoidal waveform at 750 Hz, the fourth harmonic is

- ❑ *a* 1500 Hz
- ❑ *b* 2150 Hz
- ❑ *c* 3000 Hz
- ❑ *d* 3750 Hz

A.29 The medium AM waveband extends from

- ❑ *a* 20 kHz to 50 kHz
- ❑ *b* 100 kHz to 250 kHz
- ❑ *c* 500 kHz to 1.5 MHz
- ❑ *d* 1.5 MHz to 2.5 MHz

A.30 When a carrier of frequency f_c is modulated by a sinusoidal voltage of frequency f_m, the following are produced:

- ❑ *a* One frequency (f_c)
- ❑ *b* Two frequencies (f_c and ($f_c + f_m$))
- ❑ *c* Three frequencies (f_c, ($f_c + f_m$) and ($f_c - f_m$))
- ❑ *d* Four frequencies (f_c, $2f_c$, ($f_c + f_m$) and ($f_c - f_m$))

A.31 If a 1-MHz carrier is amplitude modulated with a modulating voltage of frequency 10 kHz, the lower sideband will be at a frequency of

- ❑ *a* 999 000 Hz
- ❑ *b* 990 000 Hz
- ❑ *c* 998 000 Hz
- ❑ *d* 980 000 Hz

A.32 In frequency modulation the maximum variation of the carrier frequency is 75 000 Hz. If a modulating frequency of 15 000 Hz is considered, the modulation index is

- ❑ *a* 5
- ❑ *b* 10
- ❑ *c* 15
- ❑ *d* 25

A.33 In the circuit shown, V is

- ❑ *a* 60 V
- ❑ *b* 80 V
- ❑ *c* 100 V
- ❑ *d* 240 V

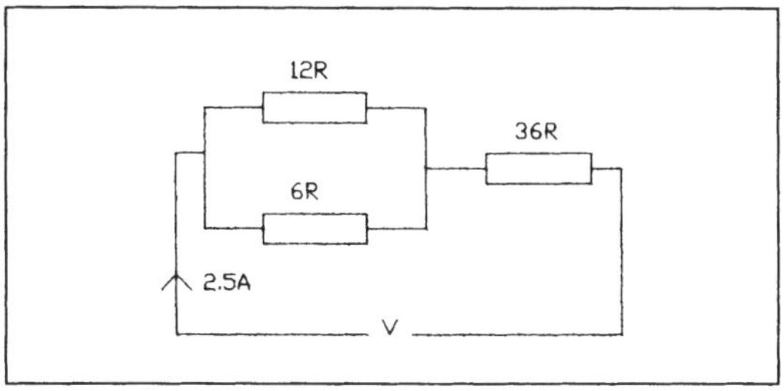

A.34 In BBC television the system of modulation for the vision signal is

- ❑ *a* Frequency
- ❑ *b* Amplitude
- ❑ *c* Pulse duration
- ❑ *d* Phase

A.35 Given that an amplifier has a voltage gain of 30 and a power gain of 450, the current gain is

- ❑ *a* 10
- ❑ *b* 15
- ❑ *c* 30
- ❑ *d* 45

A.36 Three resistors of 25 Ω, 50 Ω and 250 Ω are placed in parallel. The combined resistance is

- ❑ *a* 10.6
- ❑ *b* 12.8
- ❑ *c* 15.6
- ❑ *d* 18.4

A.37 Magenta is a mixture of

- ❑ *a* Red and green
- ❑ *b* Red and blue
- ❑ *c* Blue and green
- ❑ *d* Blue and white

A.38 In the circuit shown, R_1 is

- ❑ *a* 10 Ω
- ❑ *b* 25 Ω
- ❑ *c* 50 Ω
- ❑ *d* 100 Ω

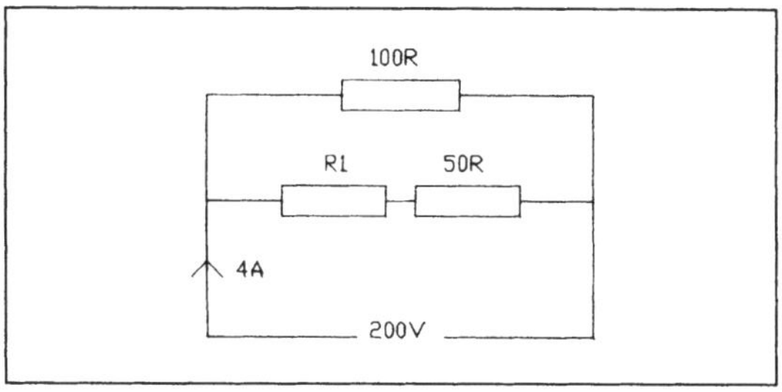

A.39 In a demodulator in a typical superhet broadcast receiver (medium and long wave) the frequency of the input will be approximately

- ❑ *a* 10 MHz
- ❑ *b* 10 kHz
- ❑ *c* 500 kHz
- ❑ *d* 1000 Hz

A.40 An output is obtained from a four-input AND gate when

- ❑ *a* A signal is applied to all four inputs
- ❑ *b* A signal is applied to three inputs
- ❑ *c* A signal is applied to one input
- ❑ *d* There is no signal applied to any input

A.41 The binary number 110110 represents in decimal

- ❑ *a* 36
- ❑ *b* 54
- ❑ *c* 56
- ❑ *d* 62

A.42 The frequency of the line timebase in a 625-line television receiver is

- ❑ *a* 50 Hz
- ❑ *b* 625 Hz
- ❑ *c* 15 625 Hz
- ❑ *d* 5.5 MHz

A.43 In the diagram the voltage drop across the 10 Ω is

- ❑ *a* 1.5 V
- ❑ *b* 2 V
- ❑ *c* 2.5 V
- ❑ *d* 5 V

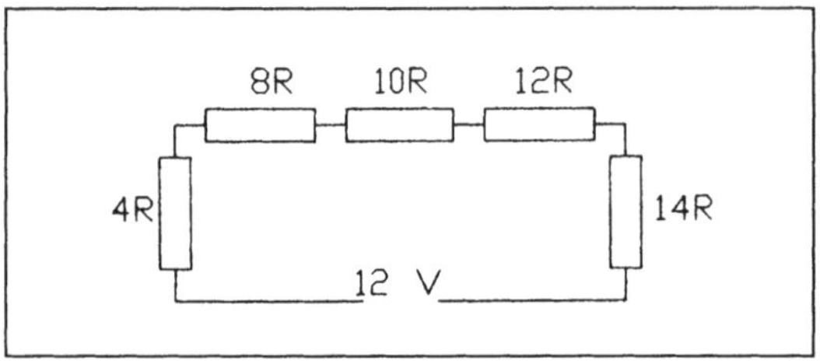

A.44 A wavelength of 3 m is equivalent to a frequency of

- ❑ *a* 100 kHz
- ❑ *b* 300 kHz
- ❑ *c* 300 MHz
- ❑ *d* 100 MHz

A.45 A complete cycle of a waveform takes 20 μs. The frequency is

- ❑ *a* 50 kHz
- ❑ *b* 500 Hz
- ❑ *c* 20 MHz
- ❑ *d* 20 kHz

A.46 In the diagram the power dissipated in R_3 is

- ❑ *a* 1.5 W
- ❑ *b* 5 W
- ❑ *c* 4.5 W
- ❑ *d* 12 W

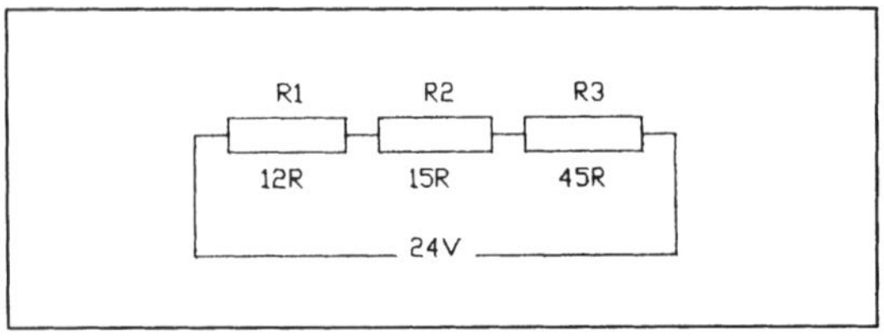

A.47 Which of the following is not suitable in dealing with a flammable liquid fire?

- ❑ *a* Carbon dioxide
- ❑ *b* Alcohol
- ❑ *c* Foam
- ❑ *d* Dry powder

A.48 The typical frequency of a bias oscillator in a magnetic tape recorder is

- ❑ *a* 200 Hz
- ❑ *b* 4500 Hz
- ❑ *c* 20 000 Hz
- ❑ *d* 45 000 Hz

A.49 In principle all amplifiers are essentially

- ❑ *a* Current amplifiers
- ❑ *b* Inverting amplifiers
- ❑ *c* Power amplifiers
- ❑ *d* Non-inverting amplifiers

A.50 Decimal 15 plus decimal 16 in binary is

- ❑ *a* 10001
- ❑ *b* 11000
- ❑ *c* 11011
- ❑ *d* 11111

Appendix B Sample test paper

B.1 Given a waveform of frequency 750 kHz, the time for one cycle is

- ❑ *a* 1.33 μs
- ❑ *b* 2.66 ms
- ❑ *c* 75.0 μs
- ❑ *d* 75.0 ms

B.2 12 + 18 + 20 in binary is

- ❑ *a* 110010
- ❑ *b* 101100
- ❑ *c* 111011
- ❑ *d* 110000

B.3 Given the truth table shown, where *A* and *B* are the inputs and *Q* the output, the gate is the

	A	*B*	*Q*
❑ *a* OR	0	0	1
❑ *b* NOR	0	1	1
❑ *c* AND	1	0	1
❑ *d* NAND	1	1	0

B.4 In the circuit shown, R_1 is

- ❑ *a* 4R
- ❑ *b* 6R
- ❑ *c* 12R
- ❑ *d* 24R

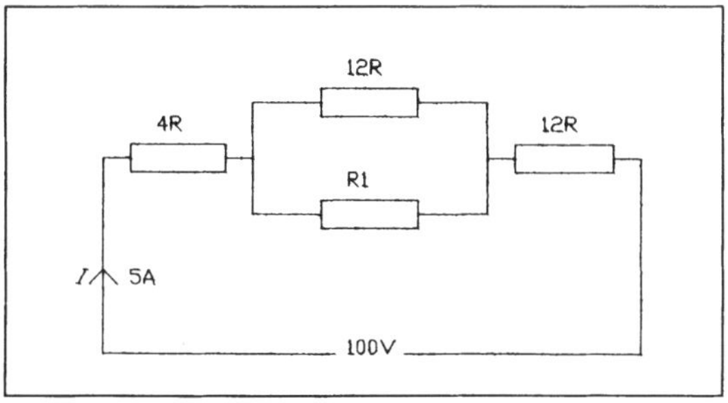

B.5 A transducer is

- ❑ *a* A device to convert a signal, e.g. mechanical to electrical
- ❑ *b* A special transformer
- ❑ *c* A transistor
- ❑ *d* A special radio receiver

B.6 The body of a portable carbon dioxide fire extinguisher is normally painted

- ❑ *a* Pink
- ❑ *b* White
- ❑ *c* Blue
- ❑ *d* Black

B.7 In frequency modulation, given that the modulation index is 5, the number of significant sidebands is

- ❑ *a* 5
- ❑ *b* 10
- ❑ *c* 16
- ❑ *d* 28

B.8 The input power to an amplifier is 20 mW. Given that the gain is 50, the output power is

- ❑ *a* 1000 mW
- ❑ *b* 4 mW
- ❑ *c* 1 W
- ❑ *d* 1000 W

B.9 In the circuit shown, the input current is

- ❑ *a* 8.2 A
- ❑ *b* 10.8 A
- ❑ *c* 12.6 A
- ❑ *d* 14.2 A

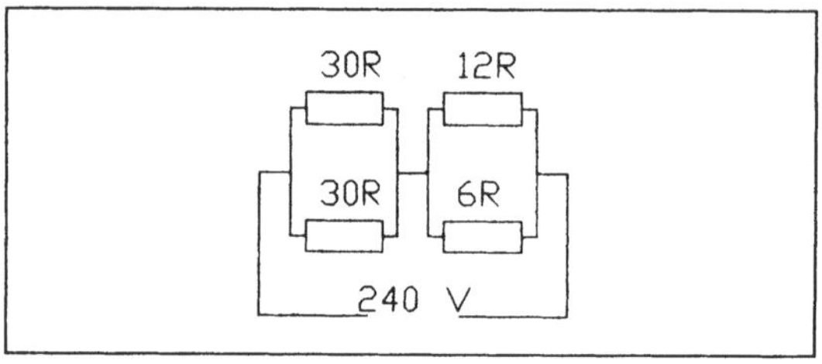

B.10 The diagram shown performs the function of

- ❑ *a* An AND gate
- ❑ *b* A NAND gate
- ❑ *c* An OR gate
- ❑ *d* A NOR gate

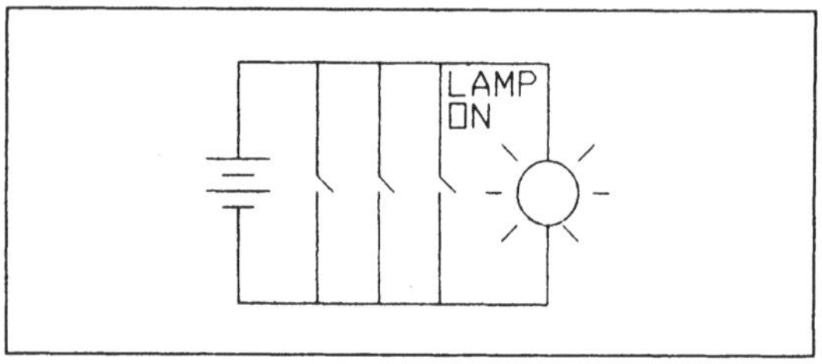

B.11 Radio 4 is on a wavelength of 1500 m. This is equivalent to a frequency in kilohertz of

- ❑ *a* 200
- ❑ *b* 750
- ❑ *c* 100
- ❑ *d* 150

B.12 In the electromagnetic spectrum the infra-red lies between

- ❑ *a* Ultraviolet and visible light
- ❑ *b* Visible light and heat
- ❑ *c* Heat and radio waves
- ❑ *d* X-rays and ultraviolet

B.13 In the diagram the power dissipated in the 24R resistor is

- ❑ *a* 24 W
- ❑ *b* 80 W
- ❑ *c* 216 W
- ❑ *d* 440 W

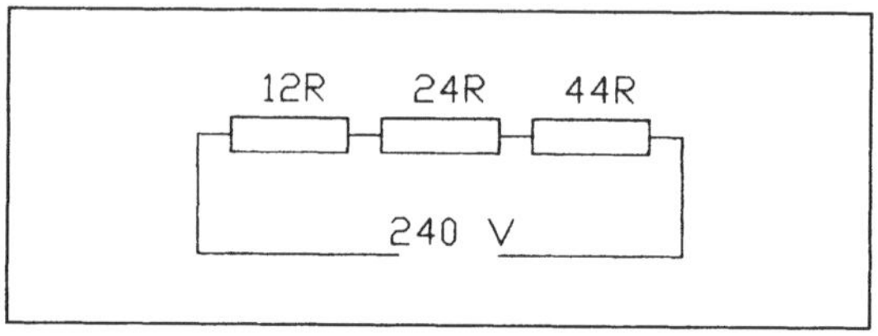

B.14 Binary 10010011 in decimal is

- ❑ *a* 123
- ❑ *b* 135
- ❑ *c* 141
- ❑ *d* 147

B.15 A video amplifier used in a TV receiver can deal with frequencies in the range of

- ❑ *a* 0 Hz to 30 MHz
- ❑ *b* 0 Hz to 500 MHz
- ❑ *c* 0 Hz to 5.5 MHz
- ❑ *d* 0 Hz to 897 MHz

B.16 A buffer amplifier is used to

- ❑ *a* Provide a mismatch
- ❑ *b* Isolate one circuit from another
- ❑ *c* Compare two voltages
- ❑ *d* Smooth ripple in a power supply

B.17 A differential amplifier is normally used for

- ❑ *a* Comparing the amplitude of one signal with another
- ❑ *b* Blocking a particular frequency
- ❑ *c* Producing sawtooth waveforms
- ❑ *d* Generating square waves

B.18 A bistable oscillator is basically capable of

- ❑ *a* Multiplying by 2 only
- ❑ *b* Multiplying by even numbers
- ❑ *c* Multiplying by odd numbers
- ❑ *d* Dividing by a factor of 2

B.19 The time constant of a 10-pF capacitor in series with a 10-MΩ resistor is

❑ *a* 100 s
❑ *b* 100 ms
❑ *c* 100 μs
❑ *d* 1 s

B.20 The voltage drop across the 50-Ω resistor in the diagram is

❑ *a* 50 V
❑ *b* 100 V
❑ *c* 150 V
❑ *d* 250 V

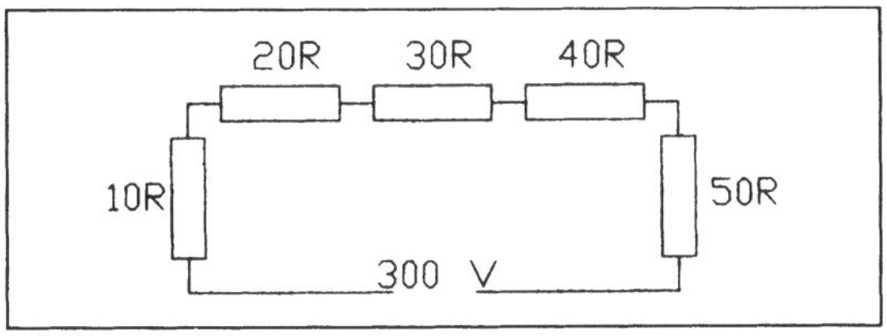

B.21 Given that the reactance of a capacitor is $1/2\pi fC$ Ω where *C* is in farads and *f* is in hertz, the reactance of a 1000-μF at 50 Hz in ohms is

❑ *a* 3.18
❑ *b* 62.8
❑ *c* 100
❑ *d* 311

B.22 In the circuit shown R_{in} equals

❑ *a* 1R
❑ *b* 1.5R
❑ *c* 2.0R
❑ *d* 2.5R

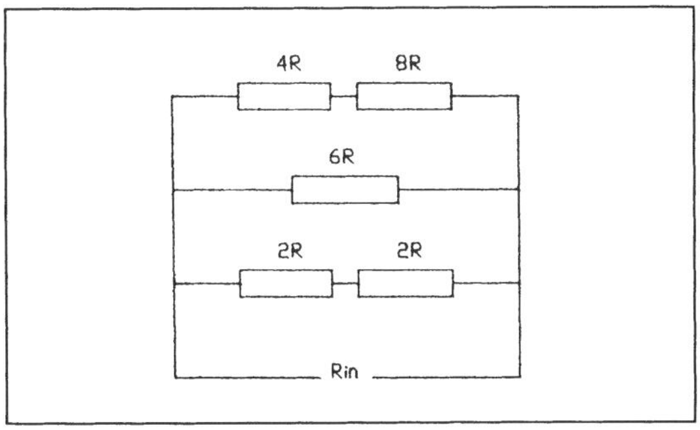

B.23 Given that the reactance of an inductor is given by $2\pi fL$ Ω where *f* is in hertz and *L* is in henrys, the reactance of a 250-mH choke at 250 Hz will be

❑ *a* 25.5 Ω
❑ *b* 392 Ω
❑ *c* 82 Ω
❑ *d* 100 Ω

B.24 If the voltage drop across a resistor is 0.25 V and the current through it 10 mA, the power dissipated in it will be

❑ *a* 2.5 W
❑ *b* 25 mW
❑ *c* 250 mW
❑ *d* 2.5 mW

B.25 The value of the resistor in Question B.24 is

❑ *a* 25 Ω
❑ *b* 250 Ω
❑ *c* 2.5 Ω
❑ *d* 0.25 Ω

B.26 Denary 24 in binary is

❑ *a* 11000
❑ *b* 11010
❑ *c* 11011
❑ *d* 11111

B.27 If the fourth ring on a resistor is coloured red, the tolerance is

❑ *a* ±1%
❑ *b* ±5%
❑ *c* ±2%
❑ *d* ±8%

B.28 In the diagram the value of R_2 is

❑ *a* 20R
❑ *b* 22R
❑ *c* 25R
❑ *d* 30R

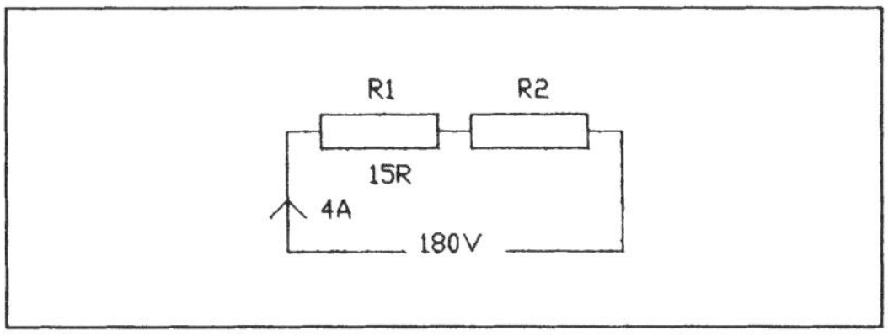

B.29 If a sine wave has a peak value of 30 V, the r.m.s. value is approximately

❑ *a* 18.3 V
❑ *b* 20.4 V
❑ *c* 21.2 V
❑ *d* 25.3 V

B.30 In the diagram the frequency of the waveform is

❑ *a* 2 Hz
❑ *b* 5 Hz
❑ *c* 10 Hz
❑ *d* 20 Hz

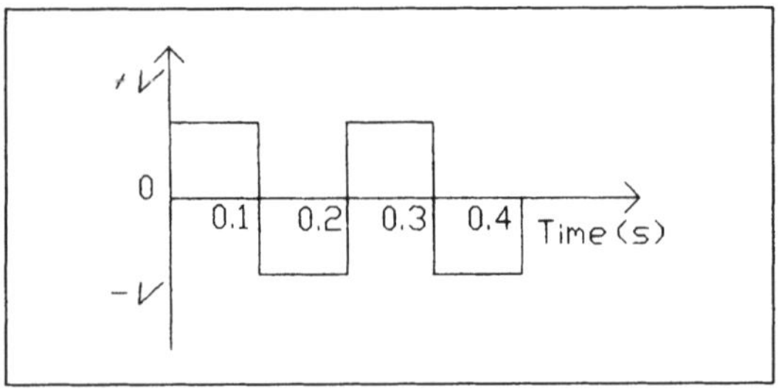

B.31 If $x^2 + 4 = 85$, the value of x is

❑ *a* 6
❑ *b* 7
❑ *c* 8
❑ *d* 9

B.32 In the diagram the value of R is

❑ *a* 20R
❑ *b* 24R
❑ *c* 30R
❑ *d* 32R

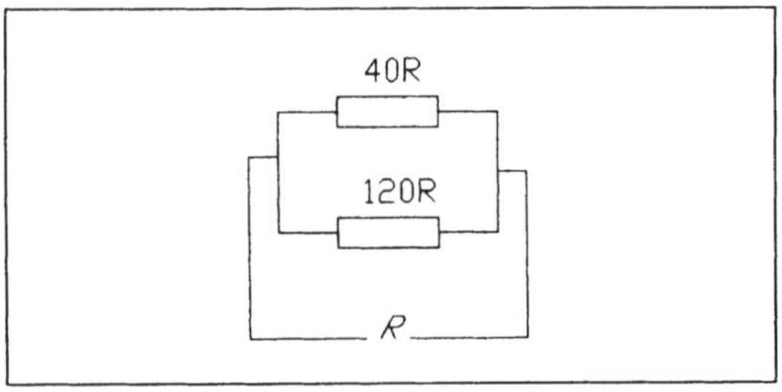

B.33 A radio station has a frequency of 300 kHz. The wavelength is

❑ *a* 1000 m
❑ *b* 1500 m
❑ *c* 3000 m
❑ *d* 4500 m

B.34 An electronic device is rated at 240 V, 120 W. The current consumed is

❑ *a* 0.1 A
❑ *b* 0.3 A
❑ *c* 0.4 A
❑ *d* 0.5 A

B.35 The charge on a capacitor is measured in

❑ *a* Watts
❑ *b* Joules
❑ *c* Coulombs
❑ *d* Ohms

B.36 Denary 33 expressed in binary is

❑ *a* 100000
❑ *b* 100001
❑ *c* 101010
❑ *d* 101110

B.37 A ±5% resistor will have a tolerance band coloured

❑ *a* Brown
❑ *b* Red
❑ *c* Gold
❑ *d* Silver

B.38 In the diagram the current I is

❑ *a* 2 mA
❑ *b* 20 mA
❑ *c* 2 A
❑ *d* 20 A

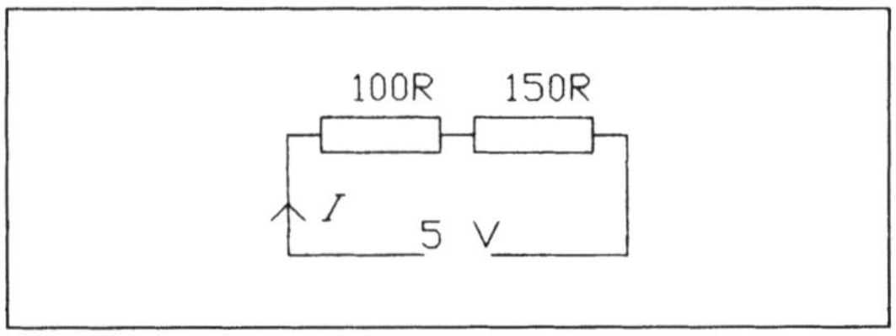

B.39 If a sine wave has a peak to peak value of 100 V, the r.m.s. value is

❑ *a* 35.36 V
❑ *b* 50.24 V
❑ *c* 70.72 V
❑ *d* 80.36 V

B.40 In the diagram the wavelength is
- ❑ *a* 150 m
- ❑ *b* 300 m
- ❑ *c* 400 m
- ❑ *d* 800 m

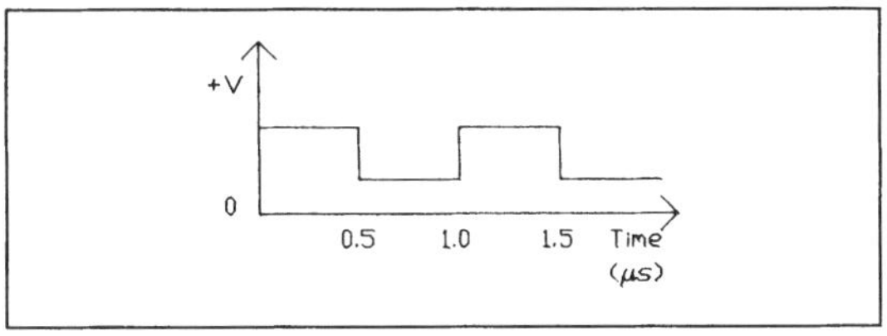

B.41 If $y^2 + 9 = 106 + 3$, y is
- ❑ *a* 20
- ❑ *b* 40
- ❑ *c* 30
- ❑ *d* 10

B.42 In the diagram the value of R_1 is
- ❑ *a* 40R
- ❑ *b* 50R
- ❑ *c* 70R
- ❑ *d* 80R

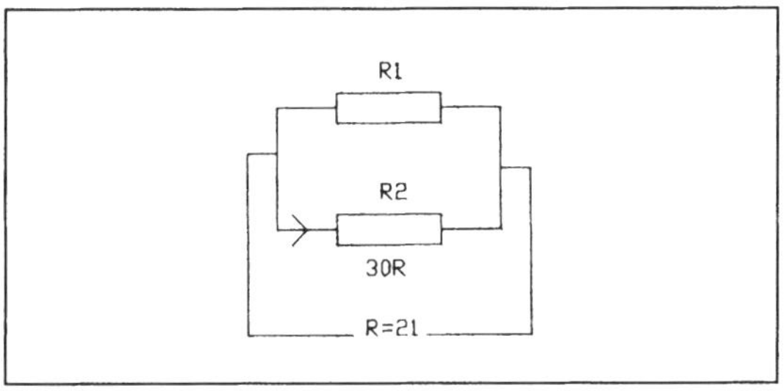

B.43 A wavelength of 3000 m is equivalent to a frequency of
- ❑ *a* 100 000 kHz
- ❑ *b* 150 000 kHz
- ❑ *c* 250 000 kHz
- ❑ *d* 300 000 kHz

B.44 A 150-Ω resistor consumes a current of 50 mA. The applied voltage is
- ❑ *a* 1.5 V
- ❑ *b* 2.0 V
- ❑ *c* 5.0 V
- ❑ *d* 7.5 V

B.45 The energy stored in a capacitor is measured in
- ❑ *a* Farads
- ❑ *b* Joules
- ❑ *c* Coulombs
- ❑ *d* Volts

B.46 Denary 24.5 expressed in binary is
- ❑ *a* 10110.0
- ❑ *b* 10101.1
- ❑ *c* 11000.1
- ❑ *d* 11101.0

B.47 A four-band resistor has the colour bands red, red, orange, silver. The value is
- ❑ *a* 220 Ω ± 5%
- ❑ *b* 22 kΩ ± 2%
- ❑ *c* 22 kΩ ± 10%
- ❑ *d* 220 kΩ ± 2%

B.48 In the diagram the value of R_2 is
- ❑ *a* 50R
- ❑ *b* 100R
- ❑ *c* 150R
- ❑ *d* 200R

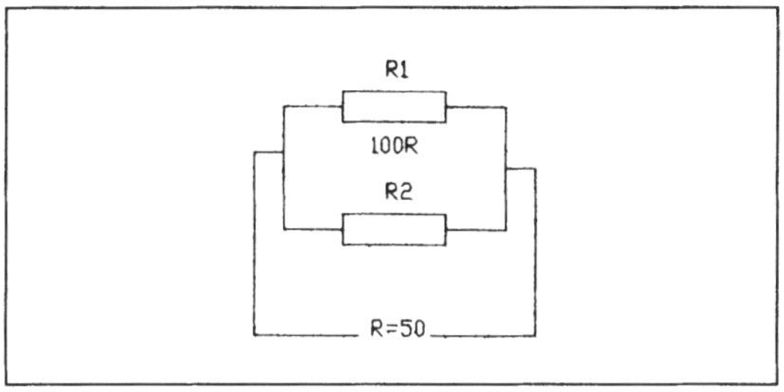

B.49 A waveform has an r.m.s. value of 28.28 V. The peak value is approximately
- ❑ *a* 36 V
- ❑ *b* 40 V
- ❑ *c* 50 V
- ❑ *d* 70 V

B.50 If 10 cycles of a sinusoidal voltage occur in 0.1 s, the frequency is
- ❑ *a* 100 Hz
- ❑ *b* 110 Hz
- ❑ *c* 200 Hz
- ❑ *d* 500 Hz

Appendix C Sample test paper

C.1 If $x^3 = 27$, x is

- ❑ *a* 0
- ❑ *b* 1
- ❑ *c* 2
- ❑ *d* 3

C.2 In the diagram the value of I_1 is

- ❑ *a* 0.1 A
- ❑ *b* 0.6 A
- ❑ *c* 1.0 A
- ❑ *d* 10 A

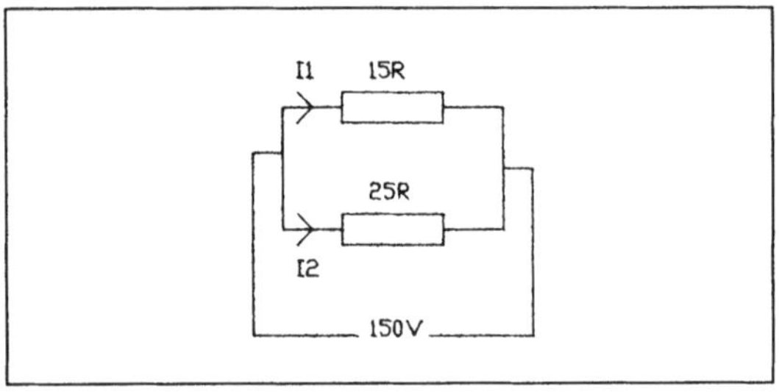

C.3 Wavelength is measured in

- ❑ *a* Newtons
- ❑ *b* Ohms
- ❑ *c* Metres
- ❑ *d* Hertz

C.4 A resistor consumes 60 W from a 240-V supply. The resistance value is

- ❑ *a* 240 R
- ❑ *b* 480 R
- ❑ *c* 800 R
- ❑ *d* 960 R

C.5 The SI unit of force is the

- ❑ *a* Kilogram
- ❑ *b* Newton
- ❑ *c* Joule
- ❑ *d* Weber

C.6 Denary 100 expressed in binary is

- ❑ *a* 1010110
- ❑ *b* 1100010
- ❑ *c* 1100100
- ❑ *d* 1110100

C.7 A 120-Ω resistor has a tolerance ring coloured silver. The upper limit of the resistance is

- ❑ *a* 122 Ω
- ❑ *b* 128 Ω
- ❑ *c* 132 Ω
- ❑ *d* 140 Ω

C.8 In the diagram the current *I* is

- ❑ *a* 3 A
- ❑ *b* 6 A
- ❑ *c* 8 A
- ❑ *d* 9 A

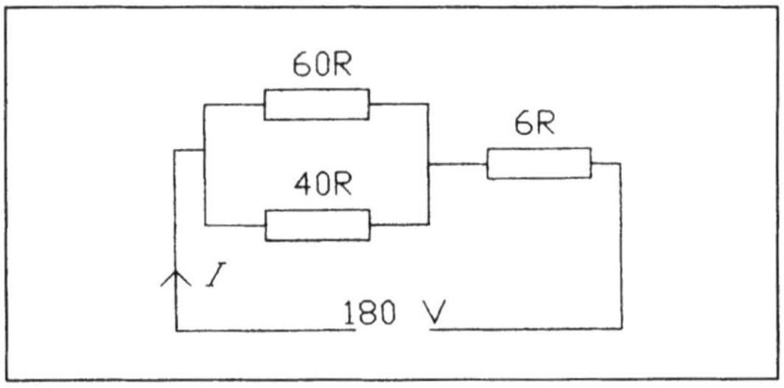

C.9 A sine wave has an r.m.s. value of 40 V. The peak to peak value is approximately

- ❑ *a* 108 V
- ❑ *b* 110 V
- ❑ *c* 113 V
- ❑ *d* 226 V

C.10 A frequency of 300 MHz has a wavelength of

- ❑ *a* 1 m
- ❑ *b* 10 m
- ❑ *c* 30 m
- ❑ *d* 100 m

C.11 If $x + 3x - x - 15 = 0$, x is

- ❑ *a* −3
- ❑ *b* 3
- ❑ *c* 5
- ❑ *d* −5

C.12 The diagram shows a series-parallel circuit. The current in R_2 is

- ❑ *a* 40 mA
- ❑ *b* 400 mA
- ❑ *c* 480 mA
- ❑ *d* 800 mA

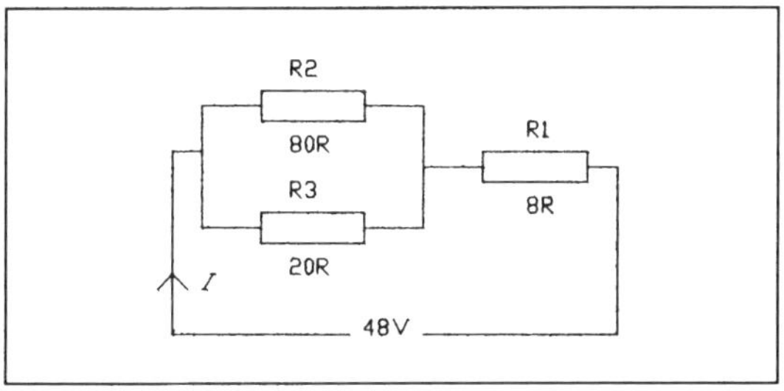

C.13 In Question C.12 the current in R_3 is

- ❑ *a* 1.1 A
- ❑ *b* 1.2 A
- ❑ *c* 1.4 A
- ❑ *d* 1.6 A

C.14 In Question C.12 the current I is

- ❑ *a* 1.0 A
- ❑ *b* 1.2 A
- ❑ *c* 2 A
- ❑ *d* 4 A

C.15 In Question C.12 the total power consumed by the circuit is

- ❑ *a* 24 W
- ❑ *b* 48 W
- ❑ *c* 60 W
- ❑ *d* 96 W

C.16 In Question C.12 the power consumed by R_2 is

- ❑ *a* 10.4 W
- ❑ *b* 12.8 W
- ❑ *c* 14.2 W
- ❑ *d* 24.4 W

C.17 In Question C.12 the power consumed by R_3 is

- ❑ *a* 51.2 W
- ❑ *b* 80.4 W
- ❑ *c* 88.8 W
- ❑ *d* 96.2 W

C.18 Binary 110 added to binary 111 is

- ❑ *a* 1001
- ❑ *b* 1010
- ❑ *c* 1100
- ❑ *d* 1101

C.19 In Question C.18 the answer in denary is

- ❑ *a* 12
- ❑ *b* 13
- ❑ *c* 24
- ❑ *d* 33

C.20 A resistor R that has no tolerance band has a value of

- ❑ *a* $R \pm 2\%$
- ❑ *b* $R \pm 5\%$
- ❑ *c* $R \pm 10\%$
- ❑ *d* $R \pm 20\%$

C.21 In the diagram with the switch S1 open, the input current is

- ❑ *a* 1.5 A
- ❑ *b* 2.0 A
- ❑ *c* 4.4 A
- ❑ *d* 8.0 A

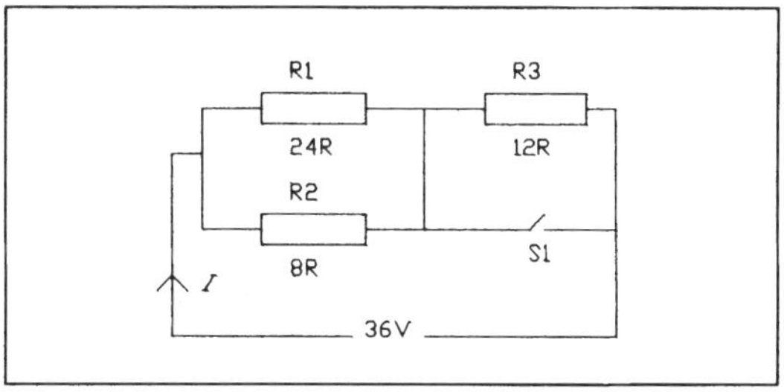

C.22 In Question C.21 with the switch open the input power is

- ❑ *a* 36 W
- ❑ *b* 48 W
- ❑ *c* 72 W
- ❑ *d* 88 W

C.23 In Question C.21 with the switch closed the current in R_1 is

- ❑ *a* 0.5 A
- ❑ *b* 1.5 A
- ❑ *c* 2.5 A
- ❑ *d* 4.5 A

C.24 In Question C.21 with the switch closed the current in R_3 is

- ❑ *a* 0 A
- ❑ *b* 1 A
- ❑ *c* 2 A
- ❑ *d* 3 A

C.25 A temperature of 75°C expressed in degrees Fahrenheit is

- ❑ *a* 100 °F
- ❑ *b* 110 °F
- ❑ *c* 135 °F
- ❑ *d* 167 °F

C.26 The maximum current which may be taken by a 4-Ω 100-W resistor is

- ❑ *a* 2 A
- ❑ *b* 5 A
- ❑ *c* 8 A
- ❑ *d* 9 A

C.27 The unit of inductance is the

- ❑ *a* Weber
- ❑ *b* Farad
- ❑ *c* Henry
- ❑ *d* Tesla

C.28 The diagram shows two capacitors connected in series. The value of *C* is

- ❑ *a* 1.2 μF
- ❑ *b* 2.4 μF
- ❑ *c* 10 μF
- ❑ *d* 24 μF

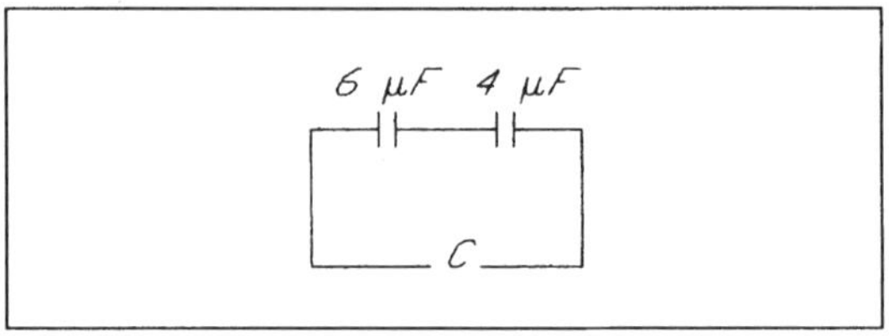

C.29 The SI unit for mass is the

- ❑ *a* Gram
- ❑ *b* Kelvin
- ❑ *c* Kilogram
- ❑ *d* Metre

C.30 If the capacitors in Question C.28 are connected in parallel, the combined capacitance is

- ❑ *a* 4 μF
- ❑ *b* 6 μF
- ❑ *c* 8 μF
- ❑ *d* 10 μF

C.31 The SI unit for length is the

- ❑ *a* Yard
- ❑ *b* Metre
- ❑ *c* Inch
- ❑ *d* Kilometre

C.32 A temperature of 122°F expressed in degrees centigrade is

- ❑ *a* 32 °C
- ❑ *b* 42 °C
- ❑ *c* 50 °C
- ❑ *d* 82 °C

C.33 The temperature in Question C.32 expressed in kelvins is

- ❑ *a* 273 K
- ❑ *b* 300 K
- ❑ *c* 305 K
- ❑ *d* 323 K

C.34 The unit of capacitance is the

- ❑ *a* Ohm
- ❑ *b* Farad
- ❑ *c* Henry
- ❑ *d* Ampere

C.35 The unit of impedance is the

- ❑ *a* Watt
- ❑ *b* Weber
- ❑ *c* Ohm
- ❑ *d* Volt

C.36 In the diagram *C* is equal to

- ❑ *a* 9 μF
- ❑ *b* 12 μF
- ❑ *c* 26 μF
- ❑ *d* 48 μF

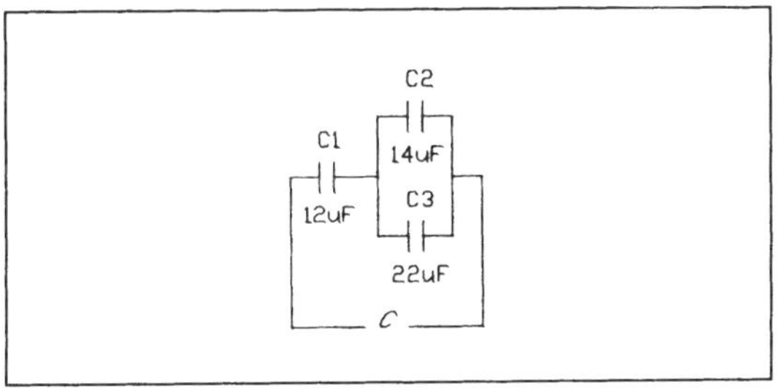

C.37 If in Question C.36 C_2 is increased to 38 μF, the value of *C* then becomes

- ❑ *a* 8 μF
- ❑ *b* 10 μF
- ❑ *c* 60 μF
- ❑ *d* 72 μF

C.38 The diagram shows three capacitors connected in series. The combined capacitance *C* is

- ❑ *a* 2 μF
- ❑ *b* 9 μF
- ❑ *c* 29 μF
- ❑ *d* 40 μF

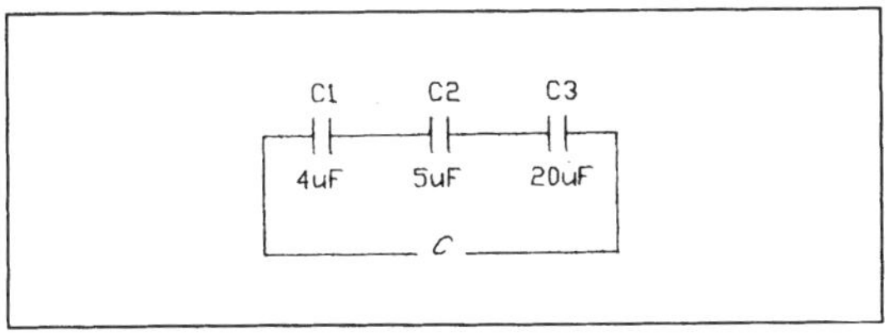

C.39 0.02 V expressed in microvolts is

- ❑ *a* 20 μV
- ❑ *b* 200 μV
- ❑ *c* 2000 μV
- ❑ *d* 20 000 μV

C.40 The resistance of a conductor is inversely proportional to the

- ❑ *a* Length
- ❑ *b* Type of material
- ❑ *c* Area
- ❑ *d* Age

C.41 Which one of the following is a good insulator?

- ❑ *a* Silver
- ❑ *b* Copper
- ❑ *c* Iron
- ❑ *d* Polythene

C.42 The diagram shows a circuit under test. Meter M1 is measuring the

- ❑ *a* Inductance
- ❑ *b* Wattage
- ❑ *c* Current
- ❑ *d* Voltage

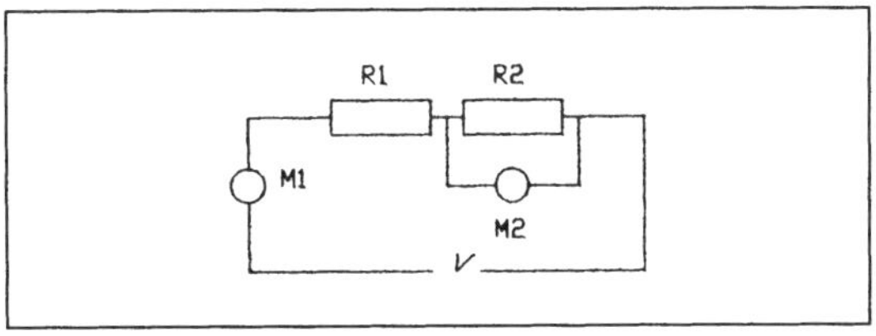

C.43 In Question C.42 meter M2 measures the

- ❑ *a* Voltage
- ❑ *b* Current
- ❑ *c* Wattage
- ❑ *d* Impedance

C.44 If in Question C.42 V = 120 V, R_1 = 400 Ω and R_2 = 600 Ω, meter M1 must be set to the

- ❑ *a* 1.0-mA range
- ❑ *b* 10-mA range
- ❑ *c* 100-mA range
- ❑ *d* 1.0-A range

C.45 In Question C.44 the reading in M2 would be

❑ *a* 48 V
❑ *b* 60 V
❑ *c* 72 V
❑ *d* 80 V

C.46 4.5 mV expressed in volts is

❑ *a* 0.45 V
❑ *b* 0.045 V
❑ *c* 0.0045 V
❑ *d* 0.00045 V

C.47 470 μV expressed in millivolts is

❑ *a* 0.047 mV
❑ *b* 0.47 mV
❑ *c* 4.7 mV
❑ *d* 47 mV

C.48 0.7 kV expressed in volts is

❑ *a* 7 V
❑ *b* 70 V
❑ *c* 700 V
❑ *d* 7000 V

C.49 Four 1000-Ω resistors are connected in parallel. The combined resistance is

❑ *a* 250 Ω
❑ *b* 1000 Ω
❑ *c* 2000 Ω
❑ *d* 4000 Ω

C.50 Three resistors having values of 40 Ω, 50 Ω and 200 Ω are connected in parallel. The combined resistance is

❑ *a* 10 Ω
❑ *b* 20 Ω
❑ *c* 25 Ω
❑ *d* 30 Ω

Appendix D Sample test paper

D.1 If $2x + 5x = 77$, x is

- ❑ *a* 10
- ❑ *b* 11
- ❑ *c* 12
- ❑ *d* 13

D.2 If $y + 3y - 40 = 0$, y is

- ❑ *a* 0
- ❑ *b* 10
- ❑ *c* −10
- ❑ *d* −40

D.3 In the diagram the total current supplied by the battery is

- ❑ *a* 2.5 A
- ❑ *b* 5.0 A
- ❑ *c* 7.5 A
- ❑ *d* 8.5 A

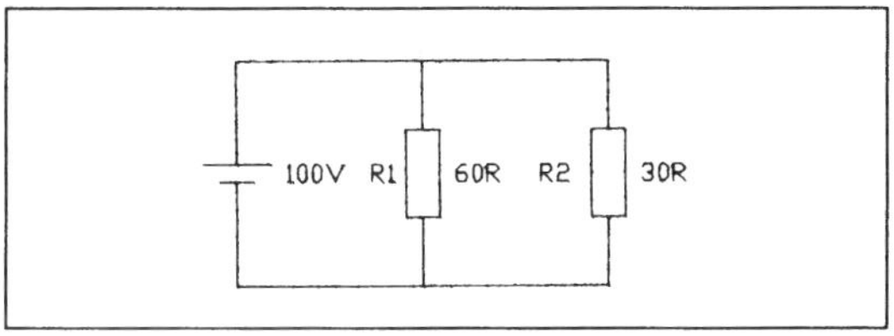

D.4 If in Question D.3 the resistor R_2 is removed from the circuit, the current supplied by the battery is then

- ❑ *a* 1.11 A
- ❑ *b* 1.24 A
- ❑ *c* 1.67 A
- ❑ *d* 2.67 A

D.5 If in the circuit shown in Question D.3 both resistors had values of 30 Ω, the total current supplied by the battery would be

- ❑ *a* 1.67 A
- ❑ *b* 2.67 A
- ❑ *c* 5.33 A
- ❑ *d* 6.67 A

D.6 The denary number 8 expressed in base 4 is

- ❑ *a* 12
- ❑ *b* 20
- ❑ *c* 22
- ❑ *d* 31

D.7 The denary number 8 expressed in base 3 is

- ❑ *a* 16
- ❑ *b* 20
- ❑ *c* 21
- ❑ *d* 22

D.8 The denary number 8 expressed in base 2 is

- ❑ *a* 1000
- ❑ *b* 1010
- ❑ *c* 1011
- ❑ *d* 1111

D.9 A 0.01-μF capacitance expressed in nanofarads is

- ❑ *a* 10 nF
- ❑ *b* 100 nF
- ❑ *c* 1000 nF
- ❑ *d* 10 000 nF

D.10 A 0.024-H inductance expressed in millihenrys is
- ❑ *a* 0.24 mH
- ❑ *b* 2.4 mH
- ❑ *c* 24 mH
- ❑ *d* 240 mH

D.11 A 0.15-mH inductance expressed in microhenrys is
- ❑ *a* 0.015 μH
- ❑ *b* 1.5 μH
- ❑ *c* 15 μH
- ❑ *d* 150 μH

D.12 The diagram shows a gate which is
- ❑ *a* An OR
- ❑ *b* A NAND
- ❑ *c* A NOR
- ❑ *d* A NOT

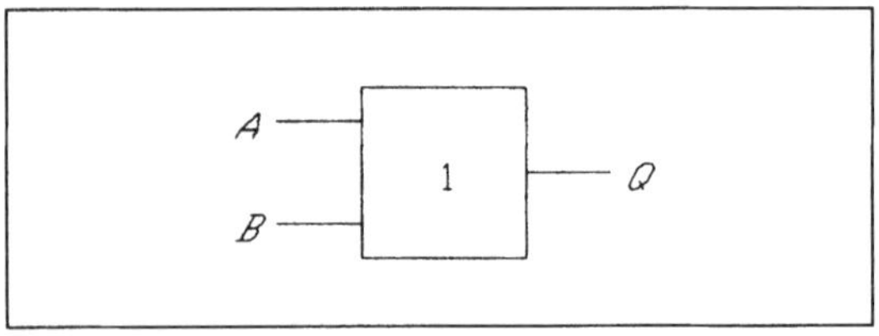

D.13 The denary number 20 expressed in base 8 is
- ❑ *a* 22
- ❑ *b* 23
- ❑ *c* 24
- ❑ *d* 32

D.14 The denary number 20 expressed in base 6 is
- ❑ *a* 32
- ❑ *b* 34
- ❑ *c* 42
- ❑ *d* 44

D.15 The denary number 20 expressed in base 4 is
- ❑ *a* 101
- ❑ *b* 102
- ❑ *c* 103
- ❑ *d* 110

D.16 The denary number 20 expressed as a binary number is
- ❑ *a* 10100
- ❑ *b* 10101
- ❑ *c* 10110
- ❑ *d* 10111

D.17 The diagram shows
- ❑ *a* A transformer
- ❑ *b* A diode
- ❑ *c* A transistor
- ❑ *d* An amplifier

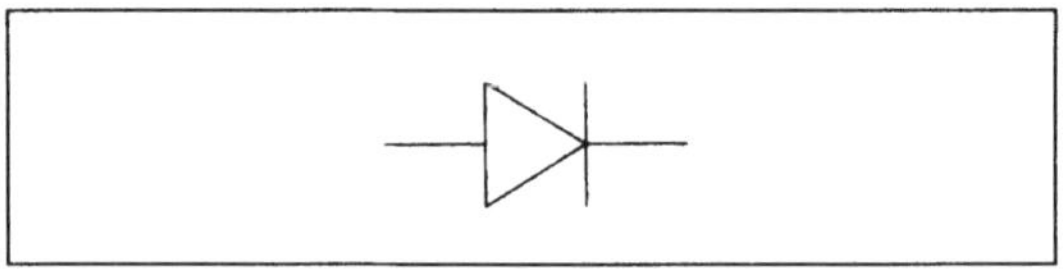

D.18 An angle of 45° expressed in radians is
- ❑ *a* 0.625 rad
- ❑ *b* 0.785 rad
- ❑ *c* 0.825 rad
- ❑ *d* 0.915 rad

D.19 5π/6 radians expressed in degrees is
- ❑ *a* 100°
- ❑ *b* 120°
- ❑ *c* 150°
- ❑ *d* 180°

D.20 The SI unit of time is the
- ❑ *a* Second
- ❑ *b* Minute
- ❑ *c* Hour
- ❑ *d* Day

D.21 In the diagram the value of *I* is
- ❑ *a* 2 μA
- ❑ *b* 20 μA
- ❑ *c* 2 mA
- ❑ *d* 20 mA

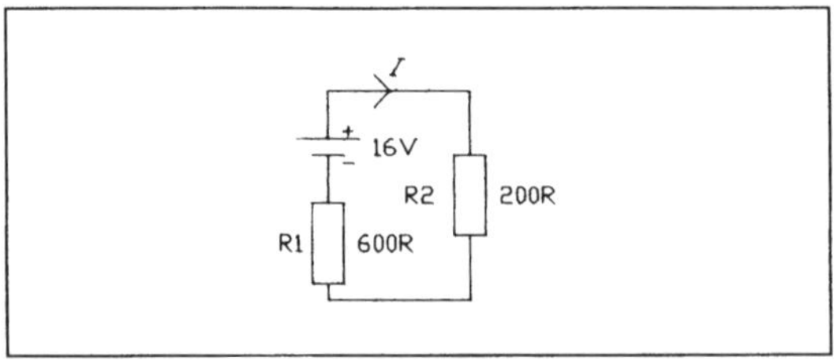

D.22 In Question D.21 the power dissipated in R_1 is

- ❑ *a* 200 mW
- ❑ *b* 240 mW
- ❑ *c* 200 μW
- ❑ *d* 240 μW

D.23 In Question D.21 the power dissipated in R_2 is

- ❑ *a* 80 μW
- ❑ *b* 100 μW
- ❑ *c* 80 mW
- ❑ *d* 200 mW

D.24 0.03 s expressed in milliseconds is

- ❑ *a* 0.3 ms
- ❑ *b* 3 ms
- ❑ *c* 30 ms
- ❑ *d* 300 ms

D.25 0.16 ms expressed in microseconds is

- ❑ *a* 0.016 μs
- ❑ *b* 1.6 μs
- ❑ *c* 16 μs
- ❑ *d* 160 μs

D.26 250 μs expressed in milliseconds is

- ❑ *a* 0.025
- ❑ *b* 0.25
- ❑ *c* 2.5
- ❑ *d* 25

D.27 In the diagram the voltage between points A and B is

- ❑ *a* 0 V
- ❑ *b* 8 V
- ❑ *c* 16 V
- ❑ *d* 24 V

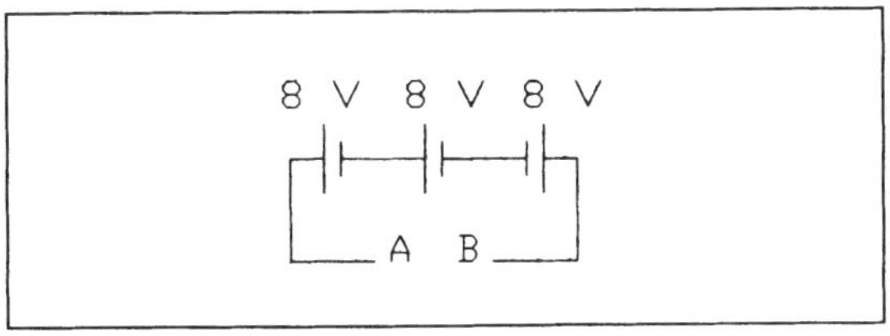

D.28 If in Question D.27 a 160-Ω resistor is connected between A and B, the current flowing is

- ❑ *a* 10 mA
- ❑ *b* 20 mA
- ❑ *c* 50 mA
- ❑ *d* 80 mA

D.29 In the diagram the voltage between points A and B is

- ❑ *a* 0 V
- ❑ *b* 12 V
- ❑ *c* 24 V
- ❑ *d* 36 V

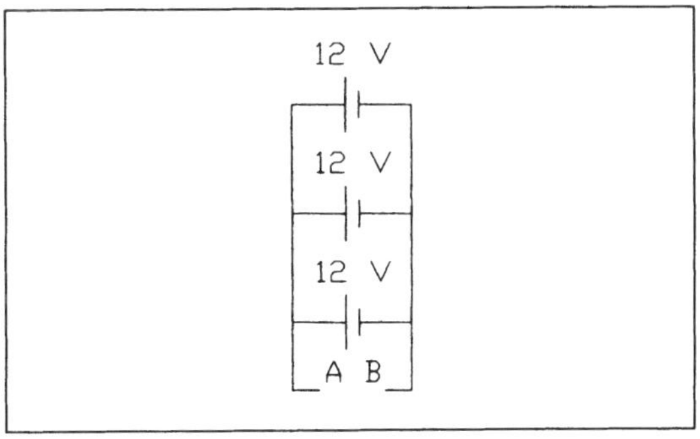

D.30 If in Question D.29 one battery is removed, the voltage between A and B is then

- ❑ *a* 0 V
- ❑ *b* 12 V
- ❑ *c* 18 V
- ❑ *d* 24 V

D.31 A 0.2-μF capacitor is connected to a 250-V supply. The charge on the plates is

- ❑ *a* 50 μC
- ❑ *b* 500 μC
- ❑ *c* 50 mC
- ❑ *d* 500 mC

D.32 In Question D.31 the energy stored in the capacitor is

- ❑ *a* 4.22 μJ
- ❑ *b* 5.25 μJ
- ❑ *c* 6.25 mJ
- ❑ *d* 8.36 mJ

D.33 $40^2 \div 40^3 =$
- ❑ *a* 0.025
- ❑ *b* 0.2
- ❑ *c* 20
- ❑ *d* 40

D.34 $x^4 \times x^{-4} =$
- ❑ *a* x
- ❑ *b* x^8
- ❑ *c* $-x$
- ❑ *d* 1

D.35 Four 0.08-μF capacitors are connected in series. The combined capacitance is
- ❑ *a* 0.01 μF
- ❑ *b* 0.02 μF
- ❑ *c* 0.32 μF
- ❑ *d* 0.82 μF

D.36 If in Question D.35 the capacitors are connected in parallel, the combined capacitance is
- ❑ *a* 0.01 μF
- ❑ *b* 0.02 μF
- ❑ *c* 0.32 μF
- ❑ *d* 0.82 μF

D.37 If $5x^2 = 80$, x is
- ❑ *a* 2
- ❑ *b* 3
- ❑ *c* 4
- ❑ *d* 5

D.38 A 680-pF capacitor is equal to
- ❑ *a* 680×10^{-3} F
- ❑ *b* 680×10^{-6} F
- ❑ *c* 680×10^{-9} F
- ❑ *d* 680×10^{-12} F

D.39 In the diagram the resistance between A and B is
- ❑ *a* 10 Ω
- ❑ *b* 11 Ω
- ❑ *c* 12 Ω
- ❑ *d* 13 Ω

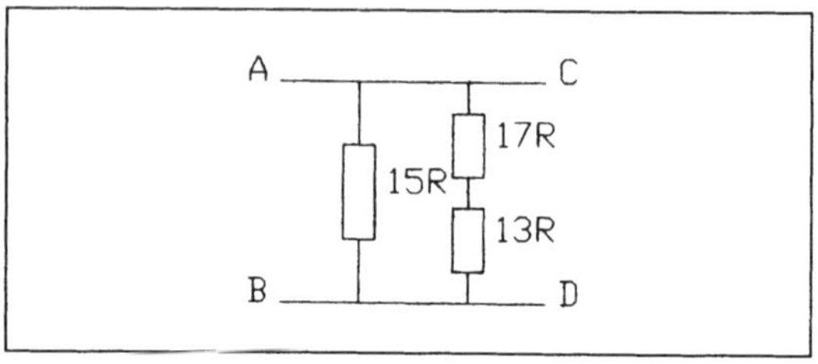

D.40 If in the circuit shown in Question D.39 a 40-Ω resistor is connected between C and D, the resistance between A and B is then
- ❑ *a* 2 Ω
- ❑ *b* 4 Ω
- ❑ *c* 6 Ω
- ❑ *d* 8 Ω

D.41 If in the circuit shown in Question D.39 a short circuit is placed between points C and D, the resistance between A and B is then
- ❑ *a* 0 Ω
- ❑ *b* 15 Ω
- ❑ *c* 30 Ω
- ❑ *d* 45 Ω

D.42 $256^0 =$
- ❑ *a* 0
- ❑ *b* 1
- ❑ *c* 4
- ❑ *d* 256

D.43 $256^0 \times 256^0 =$
- ❑ *a* 0
- ❑ *b* 1
- ❑ *c* 4
- ❑ *d* 256

D.44 The colour grey in the resistor code denotes
- ❑ *a* 5
- ❑ *b* 6
- ❑ *c* 7
- ❑ *d* 8

D.45 Wavelength is denoted by the Greek letter

- ❑ *a* Alpha
- ❑ *b* Lambda
- ❑ *c* Delta
- ❑ *d* Omega

D.46 A frequency of 300 kHz has a wavelength of

- ❑ *a* 10 m
- ❑ *b* 10^2 m
- ❑ *c* 10^3 m
- ❑ *d* 10^4 m

D.47 The work done when a force of 10 N moves through a distance of 0.8 m is

- ❑ *a* 1.8 J
- ❑ *b* 8 J
- ❑ *c* 16 J
- ❑ *d* 80 J

D.48 Electromagnetic radiation has a velocity in vacuum of

- ❑ *a* 3×10^4 m s^{-1}
- ❑ *b* 3×10^5 m s^{-1}
- ❑ *c* 3×10^6 m s^{-1}
- ❑ *d* 3×10^8 m s^{-1}

D.49 Sound has a velocity in air of

- ❑ *a* 330 m s^{-1}
- ❑ *b* 400 m s^{-1}
- ❑ *c* 500 m s^{-1}
- ❑ *d* 660 m s^{-1}

D.50 In the diagram the potential difference between A and B is

- ❑ *a* 0 V
- ❑ *b* 3 V
- ❑ *c* 6 V
- ❑ *d* 9 V

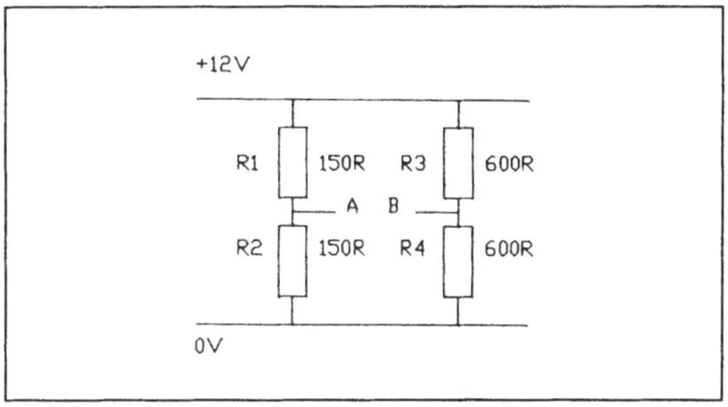

Appendix E Sample test paper

E.1 22% of a certain resistance is 55R. The total value of the resistance is

- ❑ *a* 120R
- ❑ *b* 155R
- ❑ *c* 200R
- ❑ *d* 250R

E.2 In the diagram the peak value of the waveform is

- ❑ *a* 0 V
- ❑ *b* –24 V
- ❑ *c* 12 V
- ❑ *d* 24 V

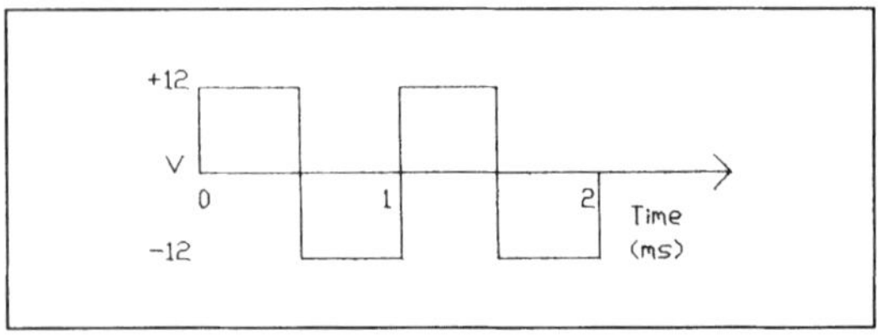

E.3 In Question E.2 the peak to peak value of the waveform is

- ❑ *a* 0 V
- ❑ *b* 12 V
- ❑ *c* –12 V
- ❑ *d* 24 V

E.4 In Question E.2 the frequency of the waveform is

- ❑ *a* 1 Hz
- ❑ *b* 10 Hz
- ❑ *c* 100 Hz
- ❑ *d* 1000 Hz

E.5 In Question E.2 the wavelength is

- ❑ *a* 3×10^4 Hz
- ❑ *b* 3×10^5 Hz
- ❑ *c* 3×10^6 Hz
- ❑ *d* 3×10^8 Hz

E.6 In Question E.2 the mark/space ratio is

- ❑ *a* 0:1
- ❑ *b* 1:0
- ❑ *c* 1:1
- ❑ *d* 1:2

E.7 In Question E.2 the average value is

- ❑ *a* 0 V
- ❑ *b* 6 V
- ❑ *c* 12 V
- ❑ *d* 24 V

E.8 If $\sqrt{(y)} - 4 = 2$, y is

- ❑ *a* 3
- ❑ *b* 4
- ❑ *c* 6
- ❑ *d* 36

E.9 The impedance of a 20-μF capacitor at 100 Hz is

- ❑ *a* 28.68 Ω
- ❑ *b* 35.42 Ω
- ❑ *c* 66.62 Ω
- ❑ *d* 79.57 Ω

E.10 The impedance of a 5-H inductor at 100 Hz is

- ❑ *a* 1608 Ω
- ❑ *b* 3142 Ω
- ❑ *c* 6284 Ω
- ❑ *d* 8080 Ω

E.11 The time constant of a 1-MΩ resistor and a 2-μF capacitor is

- ❑ *a* 1.5 s
- ❑ *b* 2.0 s
- ❑ *c* 4.0 s
- ❑ *d* 4.5 s

E.12 The time constant of a 10-H inductor and a 20-Ω resistor is

❑ *a* 0.5 s
❑ *b* 2.0 s
❑ *c* 20 s
❑ *d* 50 s

E.13 If $3y + z = 6 + 4z$, y is

❑ *a* $z + 6$
❑ *b* $6 - z$
❑ *c* $z + 2$
❑ *d* $3z + 6$

E.14 In Question E.13, z is

❑ *a* $3y + 2$
❑ *b* $3y - 2$
❑ *c* $2 + y$
❑ *d* $y - 2$

E.15 The number of possible inputs to a three-input logic gate is

❑ *a* 2
❑ *b* 4
❑ *c* 6
❑ *d* 8

E.16 The average value over a complete cycle of a voltage sine wave symmetrical about earth is

❑ *a* 0 V
❑ *b* 1 V
❑ *c* π V
❑ *d* 2π V

E.17 The average value over half a cycle of a sine wave is

❑ *a* 0.5 × peak value
❑ *b* 0.637 × peak value
❑ *c* 0.707 × peak value
❑ *d* π × peak value

E.18 In the diagram the resistance between A and B is

❑ *a* 20 Ω
❑ *b* 40 Ω
❑ *c* 60 Ω
❑ *d* 80 Ω

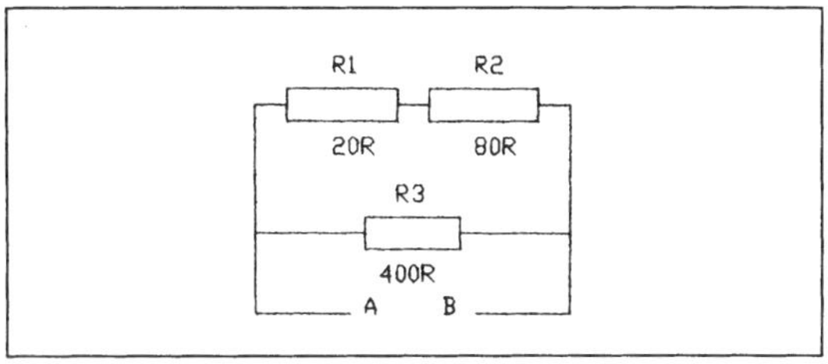

E.19 If 240 V is applied to the circuit in Question E.18, the current in R_1 is

❑ *a* 1.2 A
❑ *b* 2.4 A
❑ *c* 3.6 A
❑ *d* 4.8 A

E.20 If 300 V is applied to the circuit in Question E.18, the current in R_2 is

❑ *a* 1.6 A
❑ *b* 2.2 A
❑ *c* 2.8 A
❑ *d* 3.0 A

E.21 If the resistance between A and B in Question E.18 is 100 Ω, and R_1 is 150 Ω and R_2 is 50 Ω, the value of R_3 is

❑ *a* 50 Ω
❑ *b* 100 Ω
❑ *c* 150 Ω
❑ *d* 200 Ω

E.22 If V_p is the peak value of a sine wave, the r.m.s. value is

❑ *a* π/V_p
❑ *b* V_p/π
❑ *c* $\pi/2V_p$
❑ *d* $2V_p/\pi$

E.23 The gain or the loss of an electronic network can be measured in the

❑ *a* Joule
❑ *b* Farad
❑ *c* Ohm
❑ *d* Decibel

E.24 In the diagram the reactance of the capacitor is approximately
- ❑ *a* 4.83 Ω
- ❑ *b* 6.37 Ω
- ❑ *c* 8.42 Ω
- ❑ *d* 9.33 Ω

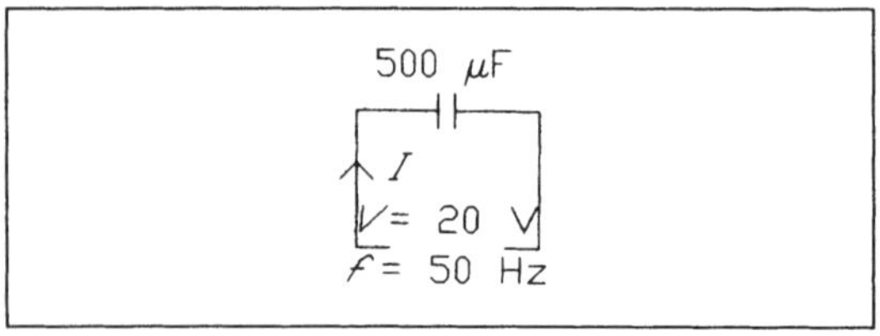

E.25 In Question E.24 the current *I* is approximately
- ❑ *a* 1.42 A
- ❑ *b* 2.28 A
- ❑ *c* 3.14 A
- ❑ *d* 6.88 A

E.26 In Question E.24 the phase angle between the voltage and current is
- ❑ *a* 45°
- ❑ *b* 90°
- ❑ *c* 135°
- ❑ *d* 180°

E.27 If $y = a^4 + a^4 + a^4$, y is
- ❑ *a* a^{12}
- ❑ *b* a^{16}
- ❑ *c* $3a^4$
- ❑ *d* a^4

E.28 In the diagram the reactance of the inductor is
- ❑ *a* 3016 Ω
- ❑ *b* 4244 Ω
- ❑ *c* 5208 Ω
- ❑ *d* 6331 Ω

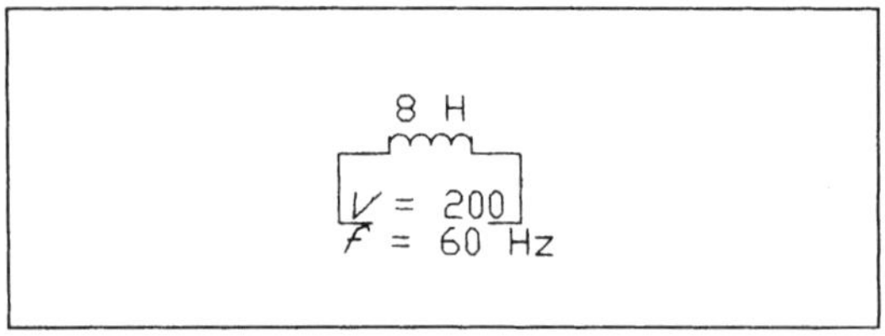

E.29 In Question E.28 the current *I* is
- ❑ *a* 20.4 mA
- ❑ *b* 50.8 mA
- ❑ *c* 66.3 mA
- ❑ *d* 70.4 mA

E.30 In Question E.28 the phase angle between voltage and current is
- ❑ *a* 0°
- ❑ *b* 45°
- ❑ *c* 90°
- ❑ *d* 180°

E.31 The unit for magnetic flux is the
- ❑ *a* Henry
- ❑ *b* Weber
- ❑ *c* Tesla
- ❑ *d* Ampere

E.32 The unit for magnetic flux density is the
- ❑ *a* Henry
- ❑ *b* Ohm
- ❑ *c* Tesla
- ❑ *d* Ampere

E.33 The Greek letter symbol for resistivity is
- ❑ *a* Delta
- ❑ *b* Alpha
- ❑ *c* Lambda
- ❑ *d* Rho

E.34 In the block diagram of a two-stage amplifier, if the gain of stage 1 is 2 and the gain of stage 2 is 8, the overall gain is
- ❑ *a* 2
- ❑ *b* 8
- ❑ *c* 10
- ❑ *d* 16

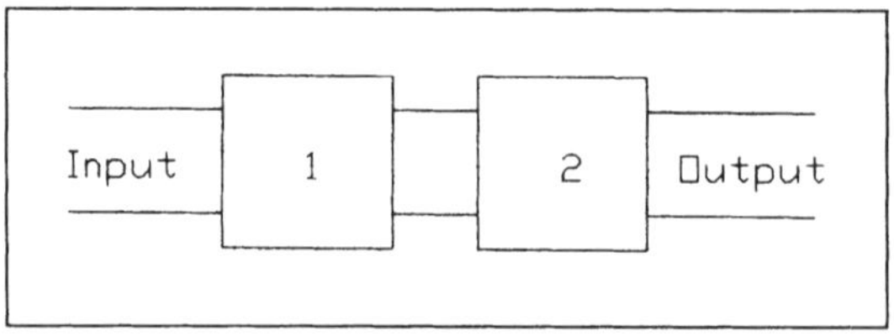

E.35 In Question E.34, if the input to stage 1 is 0.5 V, the output of stage 2 is

- ❑ *a* 1.0 V
- ❑ *b* 2.0 V
- ❑ *c* 8.0 V
- ❑ *d* 16.0 V

E.36 In Question E.35 the overall gain in decibels is

- ❑ *a* 12
- ❑ *b* 24
- ❑ *c* 36
- ❑ *d* 40

E.37 The block diagram shown is for a power supply unit. Block 1 is the

- ❑ *a* Transformer
- ❑ *b* Rectifier
- ❑ *c* Filter
- ❑ *d* Stabilizer

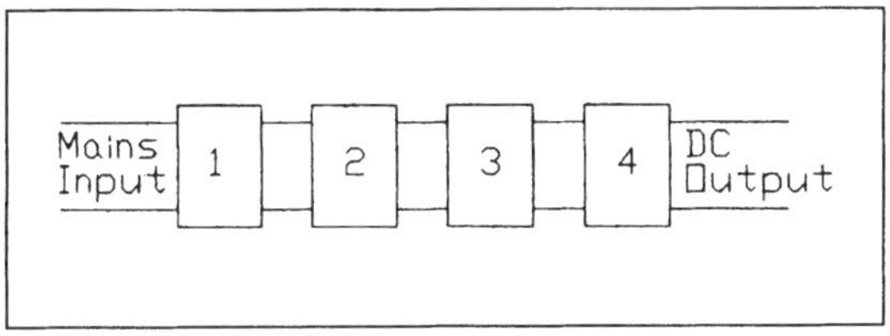

E.38 In Question E.37 block 2 is the

- ❑ *a* Transformer
- ❑ *b* Rectifier
- ❑ *c* Filter
- ❑ *d* Stabilizer

E.39 In Question E.37 block 3 is the

- ❑ *a* Transformer
- ❑ *b* Rectifier
- ❑ *c* Filter
- ❑ *d* Stabilizer

E.40 In Question E.37 block 4 is the

- ❑ *a* Transformer
- ❑ *b* Rectifier
- ❑ *c* Filter
- ❑ *d* Stabilizer

E.41 The ripple frequency in a mains half-wave rectifier system is

- ❑ *a* 25 Hz
- ❑ *b* 50 Hz
- ❑ *c* 75 Hz
- ❑ *d* 100 Hz

E.42 The ripple frequency in a mains full-wave rectifier system is

- ❑ *a* 25 Hz
- ❑ *b* 50 Hz
- ❑ *c* 75 Hz
- ❑ *d* 100 Hz

E.43 The addition of the binary numbers 111 and 110 is

- ❑ *a* 1101
- ❑ *b* 1110
- ❑ *c* 1100
- ❑ *d* 1000

E.44 The base 3 number 121 converted to base 10 is

- ❑ *a* 14
- ❑ *b* 15
- ❑ *c* 16
- ❑ *d* 22

E.45 In the diagram, gate 1 is

- ❑ *a* An EXCLUSIVE–OR
- ❑ *b* A NOR
- ❑ *c* A NOT
- ❑ *d* A NAND

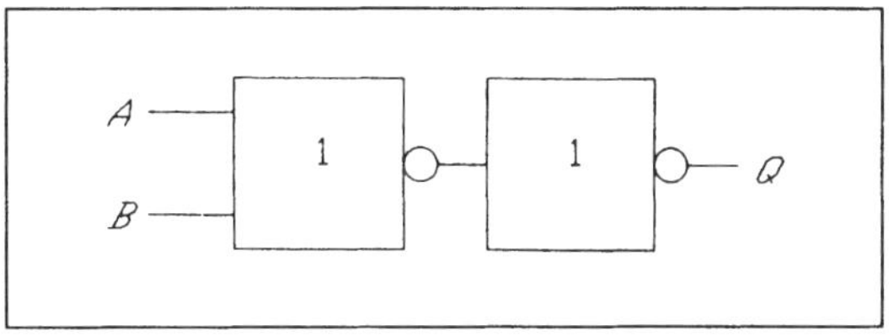

E.46 In Question E.45, gate 2 is

- ❑ *a* An AND
- ❑ *b* A NAND
- ❑ *c* A NOT
- ❑ *d* An EXCLUSIVE–OR

E.47 Which one of the following gates performs the same function as the gate combination shown in Question E.45?

❑ *a* OR
❑ *b* NOT
❑ *c* NAND
❑ *d* AND

E.48 In the diagram the resistance between A and B is

❑ *a* 21 Ω
❑ *b* 24 Ω
❑ *c* 30 Ω
❑ *d* 42 Ω

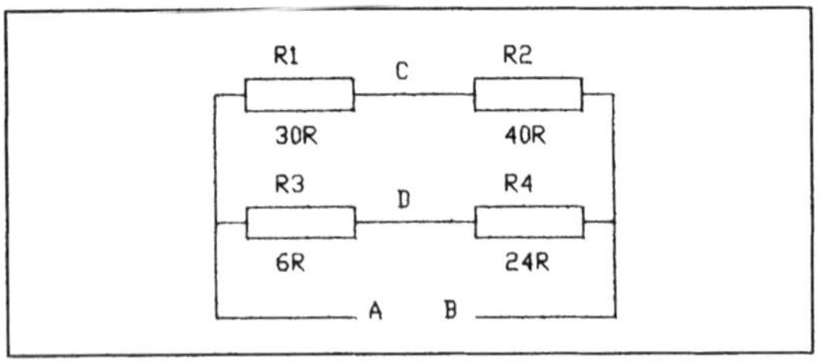

E.49 In Question E.48, if a short circuit is placed between C and D, the resistance between A and B is then

❑ *a* 16 Ω
❑ *b* 20 Ω
❑ *c* 28 Ω
❑ *d* 40 Ω

E.50 If in Question E.48 R_2 is replaced by a 140-Ω resistor, the resistance between A and B becomes

❑ *a* 18.2 Ω
❑ *b* 21.3 Ω
❑ *c* 25.5 Ω
❑ *d* 30.4 Ω

Appendix F Sample test paper

F.1 A voltage amplifier has a gain of 2. The gain expressed in decibels is
- ❑ *a* 2
- ❑ *b* 4
- ❑ *c* 6
- ❑ *d* 12

F.2 If an amplifier has a power gain of 16, the gain in decibels is
- ❑ *a* 2
- ❑ *b* 6
- ❑ *c* 12
- ❑ *d* 24

F.3 Three networks connected in series have individual gains of 18 dB, –6 dB and 9 dB. The overall gain is
- ❑ *a* 15 dB
- ❑ *b* 21 dB
- ❑ *c* 27 dB
- ❑ *d* 33 dB

F.4 The input voltage to a filter is 8 mV, and the output voltage is 1.0 mV. The overall gain in decibels is
- ❑ *a* +6
- ❑ *b* –6
- ❑ *c* +18
- ❑ *d* –18

F.5 A network has a voltage loss of 6 dB. If the input voltage is 10 V, the output voltage is
- ❑ *a* 5 V
- ❑ *b* 6 V
- ❑ *c* 7 V
- ❑ *d* 8 V

F.6 $4^0 \times 4^2 \times 4^{-2} =$
- ❑ *a* 1
- ❑ *b* 4
- ❑ *c* 16
- ❑ *d* 64

F.7 The base for natural logarithms is equal to
- ❑ *a* 1.321
- ❑ *b* 2.718
- ❑ *c* 3.142
- ❑ *d* 6.284

F.8 The time constant of a 2-H inductor and a 100-Ω resistor is
- ❑ *a* 20 ms
- ❑ *b* 0.05 s
- ❑ *c* 20 s
- ❑ *d* 50 s

F.9 In the diagram the overall capacitance of the circuit is
- ❑ *a* 1 μF
- ❑ *b* 2 μF
- ❑ *c* 3 μF
- ❑ *d* 4 μF

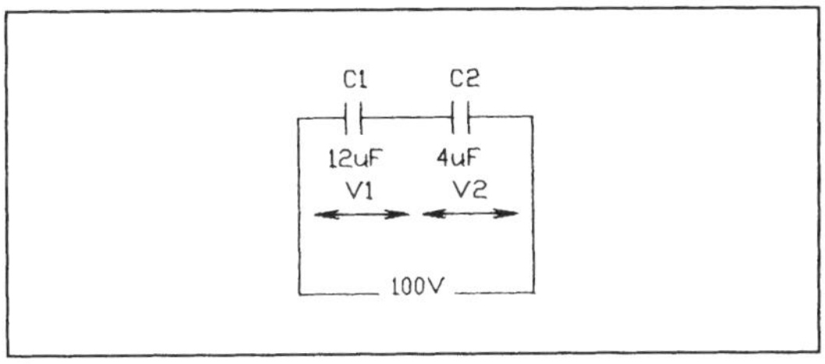

F.10 In Question F.9 the voltage V_1 is
- ❑ *a* 10 V
- ❑ *b* 25 V
- ❑ *c* 75 V
- ❑ *d* 80 V

F.11 In Question F.9 the voltage V_2 is
- ❑ *a* 12 V
- ❑ *b* 25 V
- ❑ *c* 75 V
- ❑ *d* 90 V

F.12 In Question F.9 the charge stored in C_1 is
- ❑ *a* 100 μC
- ❑ *b* 150 μC
- ❑ *c* 250 μC
- ❑ *d* 300 μC

F.13 In Question F.9 the charge stored in C_2 is
- ❑ *a* 100 μC
- ❑ *b* 200 μC
- ❑ *c* 300 μC
- ❑ *d* 400 μC

F.14 In Question F.9 the energy stored in C_1 is
- ❑ *a* 3.55 mJ
- ❑ *b* 3.75 mJ
- ❑ *c* 3.55 μJ
- ❑ *d* 3.75 μJ

F.15 In Question F.9 the energy stored in C_2 is
- ❑ *a* 10.04 mJ
- ❑ *b* 10.18 mJ
- ❑ *c* 10.88 mJ
- ❑ *d* 11.25 mJ

F.16 A capacitor plate has an area of 12 mm^2. The diameter is approximately
- ❑ *a* 1.95 mm
- ❑ *b* 3.9 mm
- ❑ *c* 4.7 mm
- ❑ *d* 6.2 mm

F.17 The product of the binary numbers 100 and 101 is
- ❑ *a* 10010
- ❑ *b* 10011
- ❑ *c* 10100
- ❑ *d* 11101

F.18 If binary 10010 is divided by binary 11, the answer is
- ❑ *a* 101
- ❑ *b* 110
- ❑ *c* 100
- ❑ *d* 111

F.19 Octal 34 in binary is
- ❑ *a* 010011
- ❑ *b* 001001
- ❑ *c* 001111
- ❑ *d* 011100

F.20 The diagram shows a T attenuator. The resistance looking into AB is
- ❑ *a* 20 Ω
- ❑ *b* 30 Ω
- ❑ *c* 50 Ω
- ❑ *d* 70 Ω

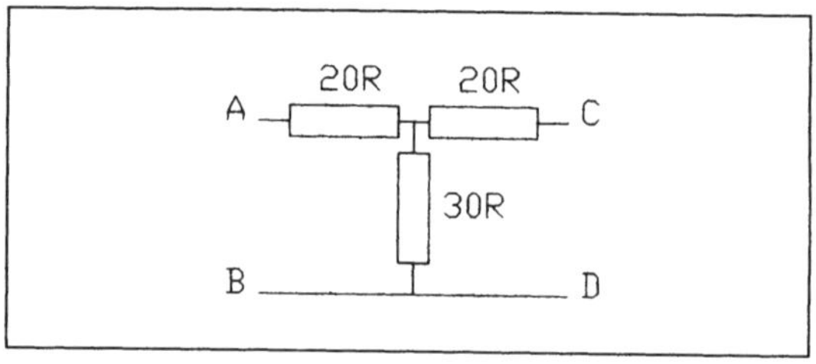

F.21 In Question F.20, if a short circuit is placed between C and D, the resistance looking into AB is
- ❑ *a* 16 Ω
- ❑ *b* 32 Ω
- ❑ *c* 48 Ω
- ❑ *d* 50 Ω

F.22 A sine wave has a peak value of 40 V. The r.m.s. value is approximately
- ❑ *a* 28.29 V
- ❑ *b* 30.33 V
- ❑ *c* 31.46 V
- ❑ *d* 32.52 V

F.23 The average value over half a cycle in Question F.22 is
- ❑ *a* 20 V
- ❑ *b* 25.48 V
- ❑ *c* 28.28 V
- ❑ *d* 56.56 V

F.24 The form factor for the waveform in Question F.22 is
- ❑ *a* 1.01
- ❑ *b* 1.11
- ❑ *c* 2.22
- ❑ *d* 3.11

F.25 If the waveform in Question F.22 has a frequency of 10^{13} Hz, the wavelength is

- ❑ *a* 3×10^7 m
- ❑ *b* 3×10^{-8} m
- ❑ *c* 3×10^5 m
- ❑ *d* 3×10^{-5} m

F.26 If $R = (R_1/R_2) + K$, R_2 is

- ❑ *a* $(R - R_1)/K$
- ❑ *b* $(R - K)/R_1$
- ❑ *c* $R_1/(R - K)$
- ❑ *d* $R_1/(K - R)$

F.27 If in Question F.26 $R = 11$, $R_1 = 60$ and $K = 7$, the value of R_2 is

- ❑ *a* 10
- ❑ *b* 15
- ❑ *c* 20
- ❑ *d* 25

F.28 In Question F.26, K is

- ❑ *a* $(R + R_1)/R_2$
- ❑ *b* $R_2/(R - R_1)$
- ❑ *c* $(R - R_1)/R_2$
- ❑ *d* $(RR_2 - R_1)/R_2$

F.29 In Question F.26, R_1 is

- ❑ *a* $R_2(R - K)$
- ❑ *b* $R_2/(R + K)$
- ❑ *c* $R(R_2 - K)$
- ❑ *d* $R(R_2 + K)$

F.30 The diagram shows a speed control system. Block A is

- ❑ *a* An oscillator
- ❑ *b* A DC generator
- ❑ *c* A DC power amplifier
- ❑ *d* A differential amplifier

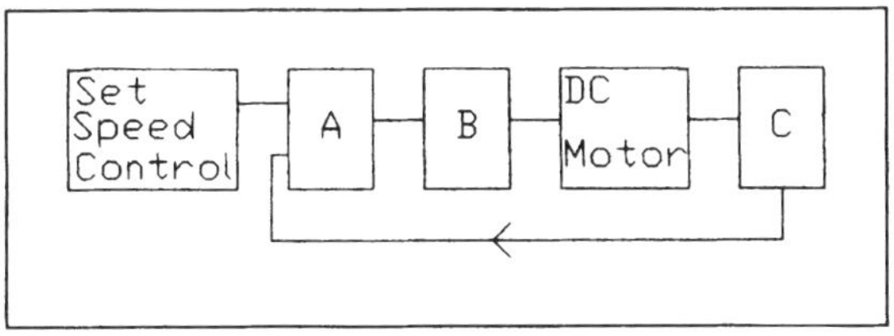

F.31 In Question F.30 block B is

- ❑ *a* An oscillator
- ❑ *b* A DC generator
- ❑ *c* A DC power amplifier
- ❑ *d* A differential amplifier

F.32 In Question F.30 block C is

- ❑ *a* A DC generator
- ❑ *b* An oscillator
- ❑ *c* A differential amplifier
- ❑ *d* A DC power amplifier

F.33 An RF amplifier in a TV is

- ❑ *a* Selective
- ❑ *b* Fixed tune
- ❑ *c* Used to amplify the IF carrier
- ❑ *d* Used for demodulation

F.34 In the diagram the voltage drop across R_4 is

- ❑ *a* 10
- ❑ *b* 20
- ❑ *c* 30
- ❑ *d* 40

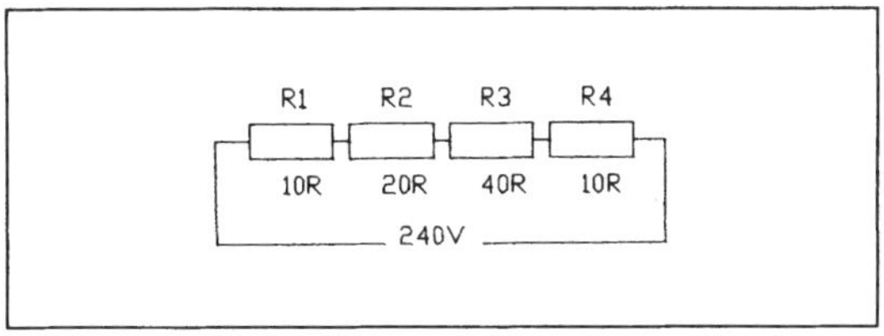

F.35 A 10 : 1 step-down transformer would give a peak to peak mains voltage on the secondary of

- ❑ *a* 24
- ❑ *b* 34
- ❑ *c* 48
- ❑ *d* 68

F.36 In the diagram the logic function satisfied by the lamp is
❑ *a* AND
❑ *b* OR
❑ *c* NOT
❑ *d* NOR

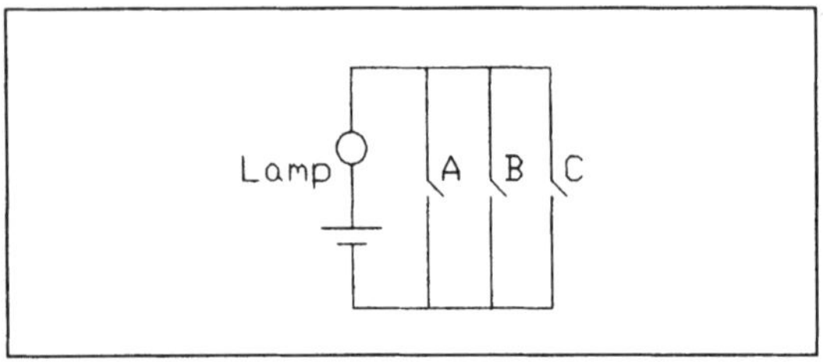

F.37 Information from the output signal of an amplifier fed back to modify the input signal is called
❑ *a* Attenuation
❑ *b* Feedback
❑ *c* Oscillation
❑ *d* Gain

F.38 A 120-mV r.m.s. sine wave will have a peak value of
❑ *a* 169 mV
❑ *b* 189 mV
❑ *c* 200 mV
❑ *d* 240 mV

F.39 In the diagram the value of *R* is
❑ *a* 20 Ω
❑ *b* 40 Ω
❑ *c* 80 Ω
❑ *d* 120 Ω

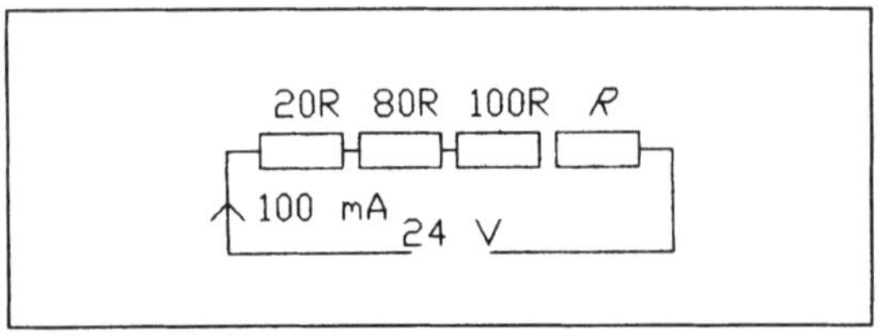

F.40 Decimal 87 in binary is
❑ *a* 1010111
❑ *b* 1111010
❑ *c* 1001101
❑ *d* 1010001

F.41 If an aerial's length is to be 1/4 wavelength, the length for a frequency of 100 MHz will be
❑ *a* 0.5 m
❑ *b* 1.0 m
❑ *c* 0.75 m
❑ *d* 3.0 m

F.42 Given that the power gain of an amplifier is 10^5, and that the input power is 0.1 μW, the output power is
❑ *a* 1 mW
❑ *b* 1 W
❑ *c* 100 mW
❑ *d* 10 mW

F.43 In the diagram the circuit shows the operation of
❑ *a* A NAND
❑ *b* A NOR
❑ *c* A NOT
❑ *d* An AND

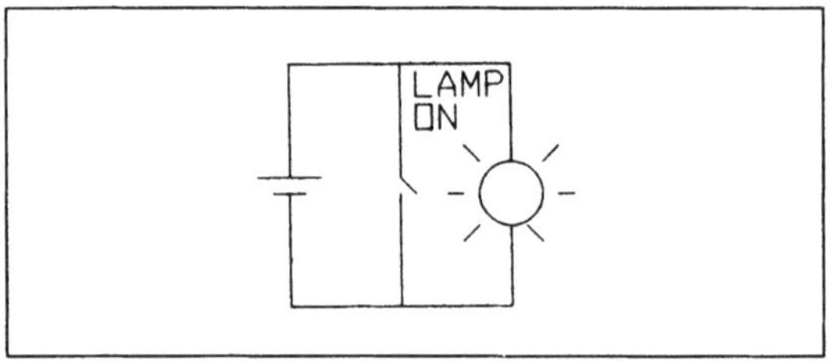

F.44 Binary 110100 in decimal is
❑ *a* 42
❑ *b* 46
❑ *c* 48
❑ *d* 52

F.45 In the diagram, *I* is equal to
❑ *a* 6 A
❑ *b* 60 mA
❑ *c* 600 μA
❑ *d* 1 A

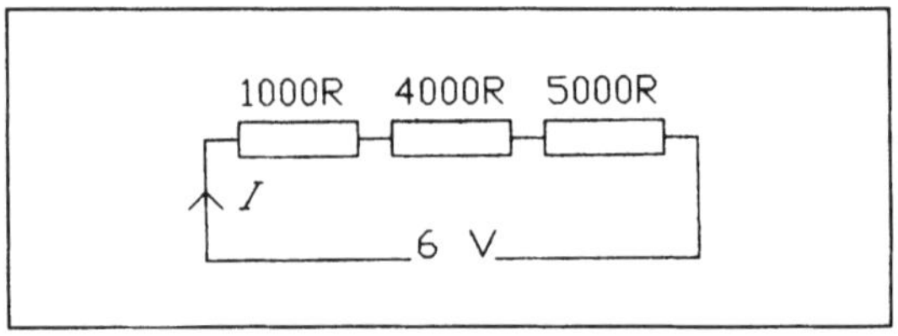

F.46 Materials formed by one type of atom are called

- ❑ *a* Electrons
- ❑ *b* Elements
- ❑ *c* Protons
- ❑ *d* Neutrons

F.47 Four resistors of 68 Ω each connected in parallel will have a combined resistance of

- ❑ *a* 4 Ω
- ❑ *b* 6.8 Ω
- ❑ *c* 17 Ω
- ❑ *d* 68 Ω

F.48 Manganese dioxide is used in a dry cell to act as

- ❑ *a* An electrode
- ❑ *b* A source of electrons
- ❑ *c* A conductor
- ❑ *d* A depolarizer

F.49 In colour television mixing, red + green + blue in suitable proportions gives

- ❑ *a* Yellow
- ❑ *b* White
- ❑ *c* Cyan
- ❑ *d* Magenta

F.50 The unit quantity of electrical charge is

- ❑ *a* The ampere
- ❑ *b* The joule
- ❑ *c* The watt
- ❑ *d* The coulomb

Appendix G Sample test paper

G.1 A 5-kW mains-operated electric fire will draw a current of

- ❑ *a* 5 A
- ❑ *b* 10.8 A
- ❑ *c* 15.8 A
- ❑ *d* 20.8 A

G.2 In a TV receiver the field timebase runs at

- ❑ *a* 50 Hz
- ❑ *b* 25 Hz
- ❑ *c* 625 Hz
- ❑ *d* 15 625 Hz

G.3 Given that a resistor connected to a 240-V supply dissipates 480 W, the value is

- ❑ *a* 60 Ω
- ❑ *b* 120 Ω
- ❑ *c* 240 Ω
- ❑ *d* 480 Ω

G.4 For the circuit shown to oscillate, block X would need to be

- ❑ *a* Negative feedback
- ❑ *b* Positive feedback
- ❑ *c* A filter network
- ❑ *d* An attenuator

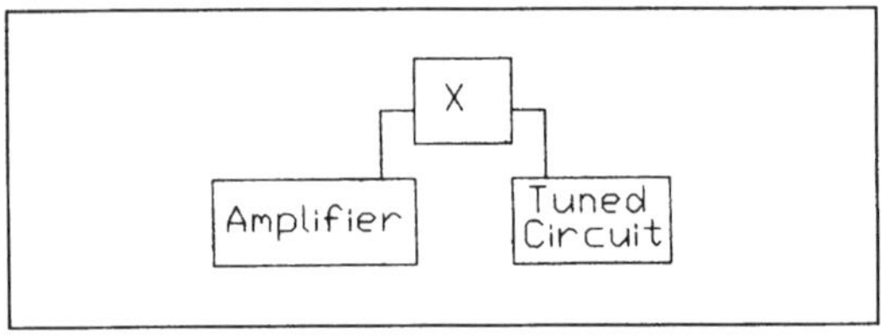

G.5 The body of a portable dry powder fire extinguisher is normally painted

- ❑ *a* Blue
- ❑ *b* Black
- ❑ *c* Red
- ❑ *d* Green

G.6 In the diagram, X would be

- ❑ *a* A frequency changer
- ❑ *b* A buffer amplifier
- ❑ *c* A modulator
- ❑ *d* A demodulator

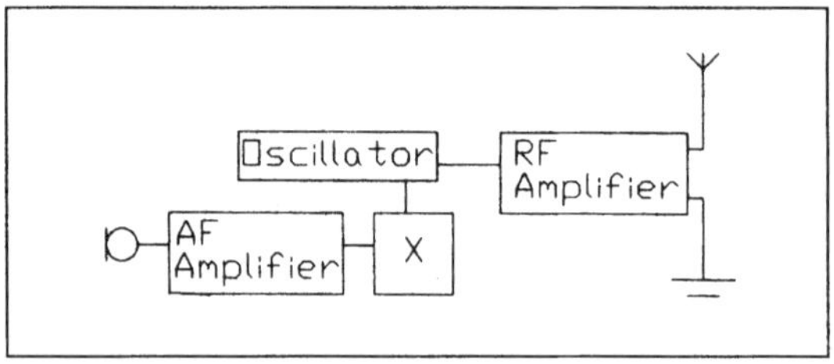

G.7 If $a = b\sqrt{(36)}$, b is

- ❑ *a* $36a$
- ❑ *b* $6a$
- ❑ *c* $a/6$
- ❑ *d* $a/36$

G.8 Denary 65 expressed in binary is

- ❑ *a* 1000001
- ❑ *b* 1100011
- ❑ *c* 1011010
- ❑ *d* 1110011

G.9 The diagram shows a battery with an internal resistance *r*, connected to a load *R*. The value of *r* is

❑ *a* 0.05 Ω
❑ *b* 0.08 Ω
❑ *c* 0.1 Ω
❑ *d* 1.0 Ω

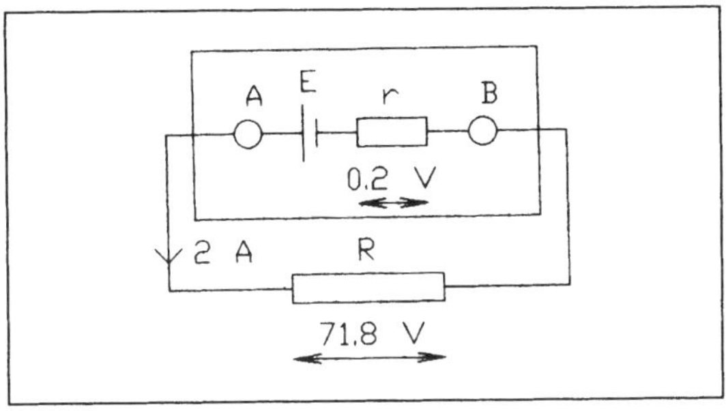

G.10 In Question G.9 the value of resistor *R* is

❑ *a* 33.8 Ω
❑ *b* 35.9 Ω
❑ *c* 40.6 Ω
❑ *d* 47.9 Ω

G.11 The diagram shows a half-wave rectifier circuit and a 24 : 1 step down transformer. If the input voltage is 120 V r.m.s., and assuming no losses, the peak voltage across the load is

❑ *a* 4.62 V
❑ *b* 5.32 V
❑ *c* 7.07 V
❑ *d* 9.61 V

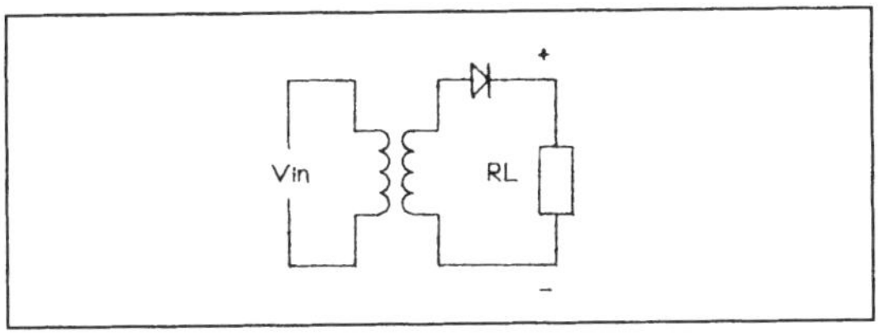

G.12 If in Question G.11 the input voltage V_{in} is increased to 480 V, the peak voltage across R_L is

❑ *a* 14.14
❑ *b* 28.28
❑ *c* 56.56
❑ *d* 60.67

G.13 If $a = b\sqrt{(25)}$, *b* is

❑ *a* 25/*a*
❑ *b* 0.2*a*
❑ *c* *a*/25
❑ *d* *a* + 5

G.14 Denary 85 expressed in binary is

❑ *a* 1000100
❑ *b* 1000101
❑ *c* 1010101
❑ *d* 1110111

G.15 20 °C converted to degrees Fahrenheit is

❑ *a* 36 °F
❑ *b* 40 °F
❑ *c* 59 °F
❑ *d* 68 °F

G.16 The diagram shows a three-stage network. The gain of stage 2 is

❑ *a* 0.2
❑ *b* 0.5
❑ *c* 2.0
❑ *d* 5.0

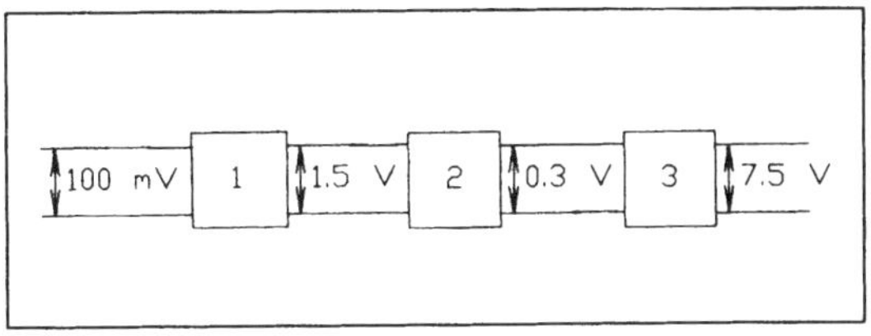

G.17 The overall gain in Question G.16 is

❑ *a* 15
❑ *b* 30
❑ *c* 75
❑ *d* 80

G.18 In Question G.16 the gain of stage 2 in decibels is approximately

❑ *a* −7
❑ *b* 7
❑ *c* 14
❑ *d* −14

G.19 In Question G.16 the overall gain in decibels is
- ❑ *a* 18.75
- ❑ *b* 37.5
- ❑ *c* 75.0
- ❑ *d* 80.2

G.20 If $x = b/(atc)$, t is
- ❑ *a* ax/bc
- ❑ *b* bc/ax
- ❑ *c* ab/cx
- ❑ *d* $b/(acx)$

G.21 Denary 96 expressed in binary is
- ❑ *a* 1100000
- ❑ *b* 1110000
- ❑ *c* 1111000
- ❑ *d* 1111100

G.22 The diagram shows a half-wave rectifier circuit connected to a 240-V mains supply. If the peak output voltage from the rectifier is 120 V, the peak current is
- ❑ *a* 1.5 A
- ❑ *b* 2.0 A
- ❑ *c* 2.5 A
- ❑ *d* 3.0 A

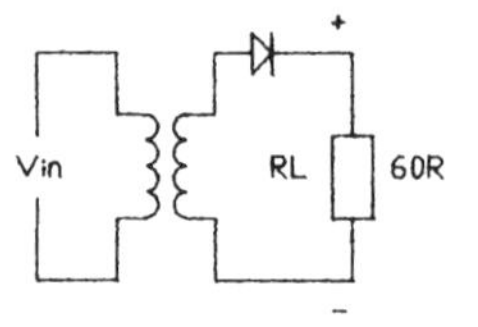

G.23 In Question G.22 the DC current is approximately
- ❑ *a* 482 mA
- ❑ *b* 580 mA
- ❑ *c* 637 mA
- ❑ *d* 707 mA

G.24 In Question G.22 the r.m.s. current is
- ❑ *a* 0.6 A
- ❑ *b* 0.7 A
- ❑ *c* 0.9 A
- ❑ *d* 1.0 A

G.25 In Question G.22 the DC voltage across R_L is approximately
- ❑ *a* 19.81 V
- ❑ *b* 20.31 V
- ❑ *c* 38.19 V
- ❑ *d* 48.82 V

G.26 In Question G.22 the r.m.s. voltage across R_L is
- ❑ *a* 60 V
- ❑ *b* 70 V
- ❑ *c* 75 V
- ❑ *d* 80 V

G.27 If $C/LN = 1/B$, B is
- ❑ *a* L/NC
- ❑ *b* NC/L
- ❑ *c* LN/C
- ❑ *d* C/LN

G.28 Denary 17 expressed in binary is
- ❑ *a* 10001
- ❑ *b* 10011
- ❑ *c* 10100
- ❑ *d* 11101

G.29 In the Fahrenheit temperature scale the temperature of melting ice is taken as
- ❑ *a* 0°
- ❑ *b* 32°
- ❑ *c* 40°
- ❑ *d* 42°

G.30 The diagram shows a single stage. The output power W_2 is approximately
- ❑ *a* 3.62 W
- ❑ *b* 4.88 W
- ❑ *c* 6.32 W
- ❑ *d* 7.54 W

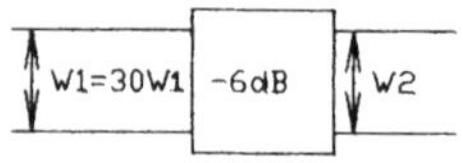

G.31 If in Question G.30 the gain is 6 dB, the output power is approximately

- ❑ *a* 160 W
- ❑ *b* 240 W
- ❑ *c* 120 W
- ❑ *d* 150 W

G.32 If in Question G.30 the gain is –30 dB and the output power is 2.0 W, the input power is

- ❑ *a* 500 W
- ❑ *b* 1 kW
- ❑ *c* 2 kW
- ❑ *d* 4 kW

G.33 The denary number 14 converted to base 4 is

- ❑ *a* 12
- ❑ *b* 22
- ❑ *c* 30
- ❑ *d* 32

G.34 If $W = I^2R$ and $W = 30$ and $R = 120$, I is

- ❑ *a* 0.25
- ❑ *b* 0.5
- ❑ *c* 2.5
- ❑ *d* 5.5

G.35 Denary 102 expressed in binary is

- ❑ *a* 1000010
- ❑ *b* 1000011
- ❑ *c* 1000110
- ❑ *d* 1100110

G.36 A four-band resistor has colour bands orange, white, red and red. The ohmic value is

- ❑ *a* 1.8 kΩ ± 5%
- ❑ *b* 2.7 kΩ ± 2%
- ❑ *c* 3.9 kΩ ± 2%
- ❑ *d* 4.7 kΩ ± 10%

G.37 Given a 0.47-Ω ± 2% resistor, band 1 is coloured

- ❑ *a* Orange
- ❑ *b* Yellow
- ❑ *c* Blue
- ❑ *d* Green

G.38 In Question G.37 band 2 is coloured

- ❑ *a* Blue
- ❑ *b* Red
- ❑ *c* Green
- ❑ *d* Violet

G.39 In Question G.37 band 3 is coloured

- ❑ *a* Gold
- ❑ *b* Silver
- ❑ *c* Red
- ❑ *d* Yellow

G.40 In Question G.37 band 4 is coloured

- ❑ *a* Brown
- ❑ *b* Black
- ❑ *c* Red
- ❑ *d* Gold

G.41 If $a^2 + b^2 = c^2$, where $a = 12$ and $b = 16$, c is

- ❑ *a* 12
- ❑ *b* 14
- ❑ *c* 16
- ❑ *d* 20

G.42 Denary 111 expressed in binary is

- ❑ *a* 1101111
- ❑ *b* 1110000
- ❑ *c* 1110011
- ❑ *d* 1111001

G.43 An electronic unit dissipating 1.2 kW from a 240-V mains supply takes a current of

- ❑ *a* 2 A
- ❑ *b* 3 A
- ❑ *c* 5 A
- ❑ *d* 8 A

G.44 In amplitude modulation, if the peak signal is 0.6 V and the carrier peak is 2.4 V, the depth of modulation is

- ❑ *a* 20%
- ❑ *b* 25%
- ❑ *c* 33%
- ❑ *d* 50%

G.45 If in Question G.44 the carrier peak is reduced to 1.5 V, the depth of modulation is then

- ❑ *a* 20%
- ❑ *b* 35%
- ❑ *c* 40%
- ❑ *d* 50%

G.46 A carrier has a peak value of 1.2 V. If the depth of modulation is 60%, the peak signal is

- ❑ *a* 600 mV
- ❑ *b* 680 mV
- ❑ *c* 720 mV
- ❑ *d* 800 mV

G.47 If ${}^4\surd(x) = 0.5$, x is

- ❑ *a* 0.5
- ❑ *b* 0.06
- ❑ *c* 0.055
- ❑ *d* 0.0625

G.48 If $x = 4/\surd(y)$, y is

- ❑ *a* $4/x$
- ❑ *b* $x/4$
- ❑ *c* $x^2/16$
- ❑ *d* $16/x^2$

G.49 Denary 13 expressed in binary is

- ❑ *a* 1001
- ❑ *b* 1101
- ❑ *c* 1110
- ❑ *d* 1111

G.50 $\surd(-1)$ is

- ❑ *a* 0
- ❑ *b* 1
- ❑ *c* j
- ❑ *d* -1

Appendix H Sample test paper

H.1 In the diagram the value of the overall resistance R is

- ❑ *a* 50R
- ❑ *b* 60R
- ❑ *c* 70R
- ❑ *d* 80R

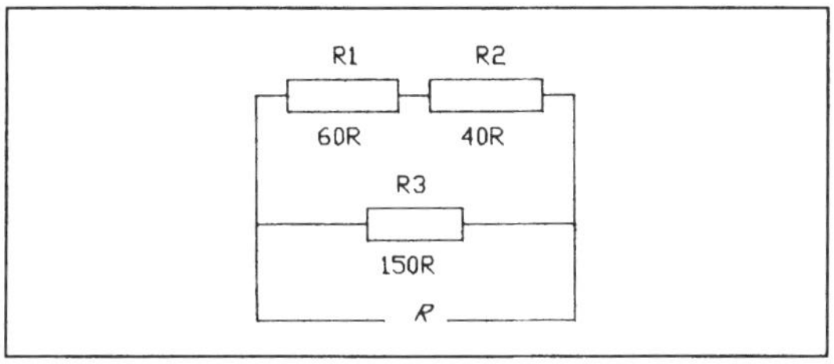

H.2 In Question H.1, R_2 is changed in order to make the value of R 75R. The new value of R_2 is

- ❑ *a* 90R
- ❑ *b* 95R
- ❑ *c* 100R
- ❑ *d* 120R

H.3 In Question H.1, R_1 is changed in order to make the value of R 37R5. The new value of R_1 is

- ❑ *a* 10R
- ❑ *b* 20R
- ❑ *c* 40R
- ❑ *d* 50R

H.4 A sinusoidal oscillator produces 20 cycles in 400 ms. The frequency is

- ❑ *a* 25 Hz
- ❑ *b* 50 Hz
- ❑ *c* 75 Hz
- ❑ *d* 80 Hz

H.5 Denary 19 converted to base 8 is

- ❑ *a* 15
- ❑ *b* 16
- ❑ *c* 20
- ❑ *d* 23

H.6 If $y = a + (x/b)$, x is

- ❑ *a* $b(y + a)$
- ❑ *b* $b(y - a)$
- ❑ *c* $y(b - a)$
- ❑ *d* $y(b + a)$

H.7 Denary 10 expressed in binary is

- ❑ *a* 1000
- ❑ *b* 1001
- ❑ *c* 1010
- ❑ *d* 1100

H.8 In the diagram, if each battery has an internal resistance of 0.25 Ω, the current I is

- ❑ *a* 1.5 A
- ❑ *b* 1.6 A
- ❑ *c* 1.8 A
- ❑ *d* 2.0 A

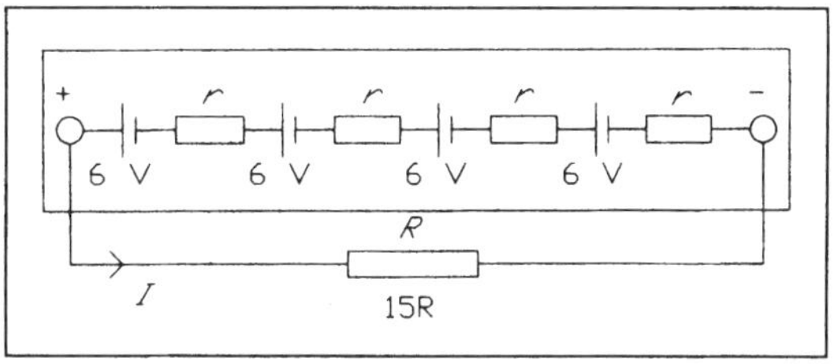

H.9 In Question H.8 the voltage across R is

- ❑ *a* 21.0 V
- ❑ *b* 21.5 V
- ❑ *c* 22.0 V
- ❑ *d* 22.5 V

H.10 In Question H.8, if one battery is reversed, the current I is then

❑ *a* 0.5 A
❑ *b* 0.75 A
❑ *c* 1.0 A
❑ *d* 1.5 A

H.11 In Question H.10 the voltage across R is

❑ *a* 5.25 V
❑ *b* 5.75 V
❑ *c* 11.25 V
❑ *d* 17.25 V

H.12 If $c = \sqrt{(abd)}$, d is

❑ *a* c/ab
❑ *b* bc/a
❑ *c* c^2/ab
❑ *d* abc

H.13 Denary 117 expressed in binary is

❑ *a* 1000100
❑ *b* 1001010
❑ *c* 1010101
❑ *d* 1110101

H.14 A power ratio of 10 mW input and 12.6 mW output expressed in decibels is

❑ *a* 1
❑ *b* 2
❑ *c* −1
❑ *d* −6

H.15 A network has an input of 1 mW and an output of 2 W. The gain in decibels is

❑ *a* 26
❑ *b* 33
❑ *c* 40
❑ *d* 66

H.16 If the output of the network in Question H.15 is 800 μW, the gain in decibels is

❑ *a* 0
❑ *b* −1
❑ *c* 1
❑ *d* 10

H.17 Four resistors connected in parallel have a combined resistance of 7.5 Ω. Three of the resistors have values of 15 Ω, 30 Ω and 40 Ω. The value of the fourth is

❑ *a* 100 Ω
❑ *b* 105 Ω
❑ *c* 110 Ω
❑ *d* 120 Ω

H.18 If $^3\sqrt{(x)} = 9$, x is

❑ *a* 3
❑ *b* 81
❑ *c* 243
❑ *d* 729

H.19 If $3x + 4y = 11$, and $x + 7y = 15$, y is

❑ *a* 2
❑ *b* 4
❑ *c* 6
❑ *d* 8

H.20 In Question H.19, x is

❑ *a* 1
❑ *b* 4
❑ *c* 6
❑ *d* 7

H.21 Denary 127 expressed in base 2 is

❑ *a* 1000000
❑ *b* 1100000
❑ *c* 1111000
❑ *d* 1111111

H.22 The diagram shows a 12 : 1 step-up transformer. If V_{in} is 110 V r.m.s. and assuming no losses, the peak to peak value of V_{out} is approximately

❑ *a* 1320 V
❑ *b* 1866 V
❑ *c* 3733 V
❑ *d* 7466 V

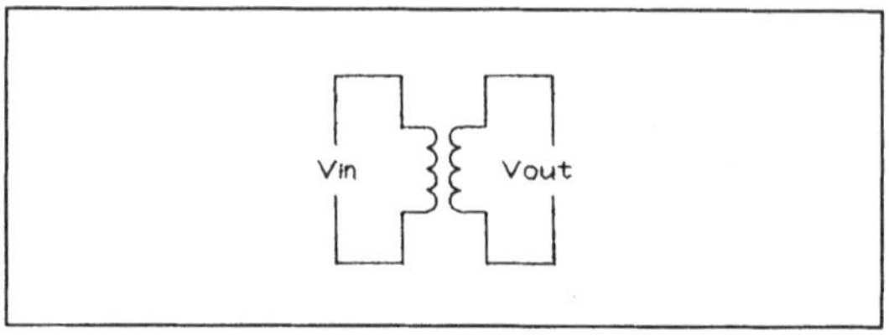

H.23 A temperature of 250 K expressed in degrees Celsius is

- ❑ *a* –23 °C
- ❑ *b* 23 °C
- ❑ *c* –50 °C
- ❑ *d* 50 °C

H.24 90 °C expressed in the Kelvin scale is

- ❑ *a* 300 K
- ❑ *b* 330 K
- ❑ *c* 363 K
- ❑ *d* 400 K

H.25 The temperature in Question H.24 expressed in degrees Fahrenheit is

- ❑ *a* 110 °F
- ❑ *b* 122 °F
- ❑ *c* 160 °F
- ❑ *d* 194 °F

H.26 The diagram shows a four-resistor network. The value of the current flowing in the 8R resistor is

- ❑ *a* 1.5 A
- ❑ *b* 2.0 A
- ❑ *c* 4.0 A
- ❑ *d* 8.0 A

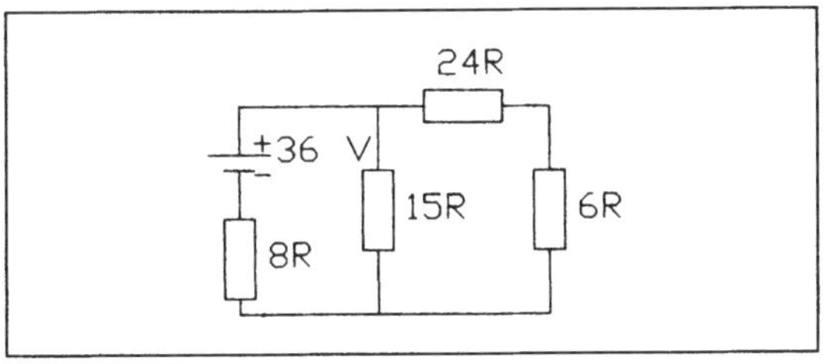

H.27 In Question H.26 the current flowing in the 15R resistor is

- ❑ *a* 1.12 A
- ❑ *b* 1.33 A
- ❑ *c* 1.67 A
- ❑ *d* 2.33 A

H.28 In Question H.26 the current flowing in the 24R resistor is

- ❑ *a* 0.67 A
- ❑ *b* 0.96 A
- ❑ *c* 1.67 A
- ❑ *d* 2.67 A

H.29 If the 15R resistor in Question H.26 is replaced by a 60R resistor, the total current supplied by the battery is approximately

- ❑ *a* 1.29 A
- ❑ *b* 3.61 A
- ❑ *c* 3.82 A
- ❑ *d* 4.93 A

H.30 $20^{1.8}$ is approximately

- ❑ *a* 198.3
- ❑ *b* 202.2
- ❑ *c* 219.7
- ❑ *d* 300.4

H.31 If $5x^4 = 1280$, x is

- ❑ *a* 2
- ❑ *b* 4
- ❑ *c* 5
- ❑ *d* 8

H.32 2.5% of 800 is

- ❑ *a* 20
- ❑ *b* 100
- ❑ *c* 120
- ❑ *d* 160

H.33 A 6-V 12-W bulb has a filament resistance of

- ❑ *a* 0.3 Ω
- ❑ *b* 1.5 Ω
- ❑ *c* 3.0 Ω
- ❑ *d* 6.0 Ω

H.34 A 12-V 3-W bulb has a filament resistance of

- ❑ *a* 12 Ω
- ❑ *b* 20 Ω
- ❑ *c* 24 Ω
- ❑ *d* 48 Ω

H.35 A 125-Ω resistor is passing a current of 200 mA. The applied voltage is

- ❑ *a* 25 V
- ❑ *b* 50 V
- ❑ *c* 75 V
- ❑ *d* 80 V

H.36 The power dissipated in the resistor in Question H.35 is

- ❑ *a* 2 W
- ❑ *b* 3 W
- ❑ *c* 5 W
- ❑ *d* 7 W

H.37 The product of the binary numbers 100 and 101 is

- ❑ *a* 10001
- ❑ *b* 10010
- ❑ *c* 10011
- ❑ *d* 10100

H.38 The product of the binary numbers 1000 and 1010 is

- ❑ *a* 1001110
- ❑ *b* 1010000
- ❑ *c* 1010111
- ❑ *d* 1011000

H.39 The denary number 127 expressed in binary form is

- ❑ *a* 1000000
- ❑ *b* 1100000
- ❑ *c* 1110000
- ❑ *d* 1111111

H.40 In the diagram the current in the 2R resistor is

- ❑ *a* 600 mA
- ❑ *b* 720 mA
- ❑ *c* 824 mA
- ❑ *d* 910 mA

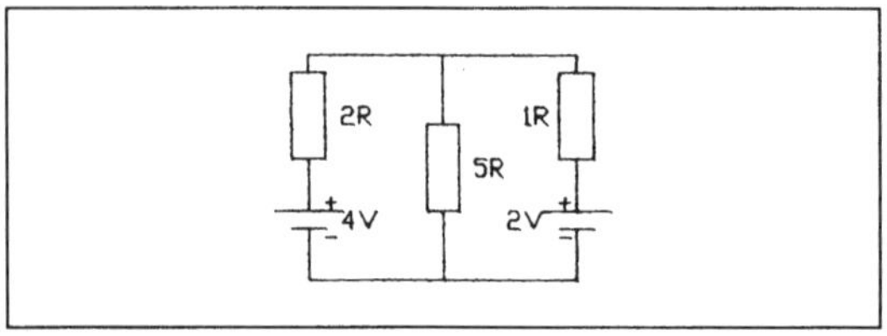

H.41 In Question H.40 the current in the 5R resistor is

- ❑ *a* 332 mA
- ❑ *b* 471 mA
- ❑ *c* 603 mA
- ❑ *d* 821 mA

H.42 In Question H.40 the current in the 1R resistor is

- ❑ *a* 353 mA
- ❑ *b* 406 mA
- ❑ *c* 608 mA
- ❑ *d* 735 mA

H.43 In Question H.40 the voltage between the top and bottom rails is

- ❑ *a* 1.65 V
- ❑ *b* 2.35 V
- ❑ *c* 5.65 V
- ❑ *d* 6.35 V

H.44 The binary number 1100 converted to hexadecimal is

- ❑ *a* A
- ❑ *b* B
- ❑ *c* C
- ❑ *d* F

H.45 The denary number 29.75 converted to hexadecimal is

- ❑ *a* 8A.A
- ❑ *b* 3D.4
- ❑ *c* 4A.C
- ❑ *d* 1D.C

H.46 The denary number 335 converted to hexadecimal is

- ❑ *a* 13C
- ❑ *b* 14F
- ❑ *c* 15D
- ❑ *d* 17B

H.47 The diagram shows three capacitors connected in series. The value of C is

- ❑ *a* 20 μF
- ❑ *b* 30 μF
- ❑ *c* 60 μF
- ❑ *d* 80 μF

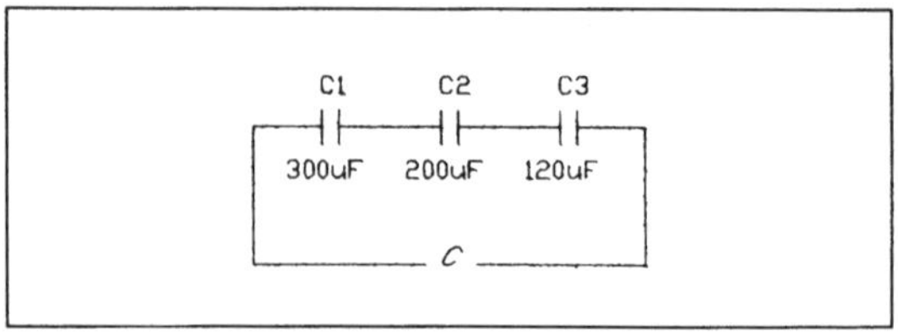

H.48 If a 120-μF capacitor is connected in parallel with C_3 in Question H.47, the value of C is then

- ❑ *a* 20 μF
- ❑ *b* 25 μF
- ❑ *c* 30 μF
- ❑ *d* 80 μF

H.49 If a 100-μF capacitor is connected in parallel with C_2 in Question H.47, the value of C is then

- ❑ *a* 33.33 μF
- ❑ *b* 40 μF
- ❑ *c* 66.67 μF
- ❑ *d* 80 μF

H.50 If a 300-μF capacitor is connected in parallel with C_1 in Question H.47, the value of C is then

- ❑ *a* 33.33 μF
- ❑ *b* 66.67 μF
- ❑ *c* 80.33 μF
- ❑ *d* 90.67 μF

Appendix J Sample test paper

J.1 The binary number 1000110 divided by the binary number 1010 gives

❑ *a* 100
❑ *b* 101
❑ *c* 111
❑ *d* 1011

J.2 The answer to Question J.1 in denary is

❑ *a* 7
❑ *b* 10
❑ *c* 14
❑ *d* 28

J.3 $3 + j4$ expressed in polar form is

❑ *a* $3\,\underline{/30^\circ}$
❑ *b* $5\,\underline{/53.12^\circ}$
❑ *c* $6\,\underline{/68.2^\circ}$
❑ *d* $7\,\underline{/45^\circ}$

J.4 $(2 + j1)^2 =$

❑ *a* $4 + j5$
❑ *b* $4 + j3$
❑ *c* $3 + j4$
❑ *d* $3 + j5$

J.5 $(3 + j4) \times (1 - j1) =$

❑ *a* $7 + j1$
❑ *b* $4 + j4$
❑ *c* $3 - j5$
❑ *d* $2 - j2$

J.6 Which single gate could be used to replace the combination shown?

❑ *a* OR
❑ *b* NOR
❑ *c* AND
❑ *d* NAND

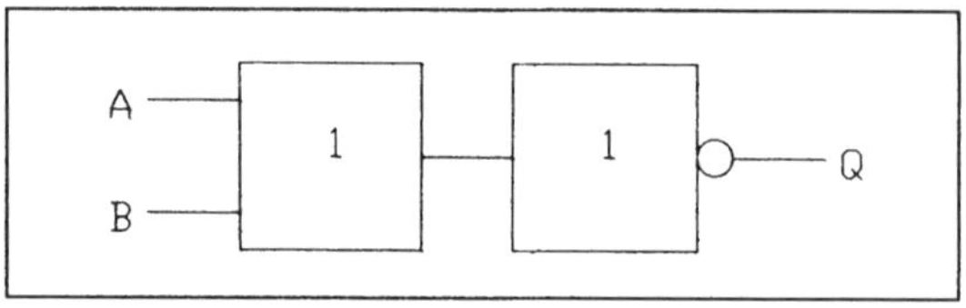

J.7 Octal 24 converted to binary is

❑ *a* 010100
❑ *b* 010101
❑ *c* 100101
❑ *d* 100111

J.8 The binary number 1101.10111 converted to octal is

❑ *a* 5.5
❑ *b* 15.56
❑ *c* 16.56
❑ *d* 16.66

J.9 In the diagram the resistance between A and B is

❑ *a* 2.25 Ω
❑ *b* 3.25 Ω
❑ *c* 3.5 Ω
❑ *d* 3.75 Ω

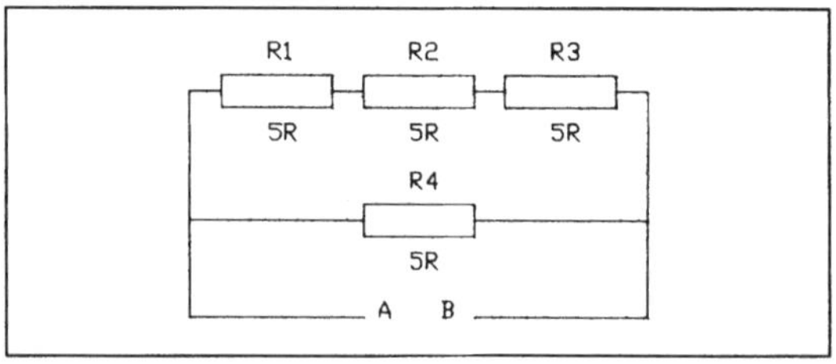

J.10 If 60 V is applied to the circuit in Question J.9, the input power is

- ❑ *a* 100 W
- ❑ *b* 192 W
- ❑ *c* 840 W
- ❑ *d* 960 W

J.11 In Question J.10 the power dissipated in R_4 is

- ❑ *a* 600 W
- ❑ *b* 720 W
- ❑ *c* 820 W
- ❑ *d* 880 W

J.12 In Question J.10 the power dissipated in R_3 is

- ❑ *a* 20 W
- ❑ *b* 60 W
- ❑ *c* 80 W
- ❑ *d* 90 W

J.13 In the diagram, how many combinations give a logic 1 output?

- ❑ *a* 3
- ❑ *b* 4
- ❑ *c* 5
- ❑ *d* 6

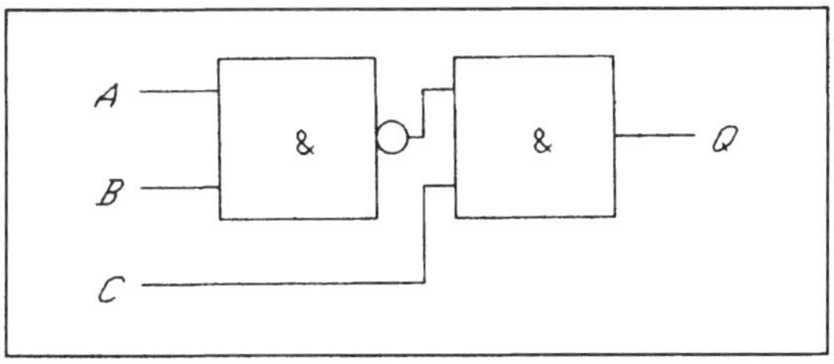

J.14 Denary 41 expressed in base 5 is

- ❑ *a* 120
- ❑ *b* 124
- ❑ *c* 131
- ❑ *d* 201

J.15 Denary 41 expressed in base 7 is

- ❑ *a* 54
- ❑ *b* 56
- ❑ *c* 57
- ❑ *d* 60

J.16 In the diagram the combined inductance is

- ❑ *a* 1.0 H
- ❑ *b* 8.0 H
- ❑ *c* 10.0 H
- ❑ *d* 12.0 H

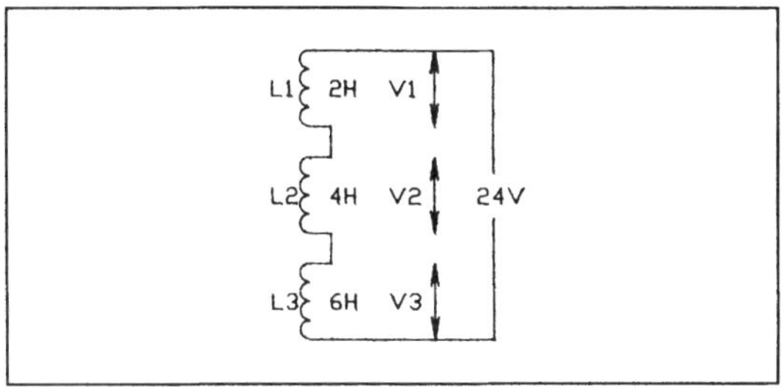

J.17 In Question J.16 the current increases at a rate of

- ❑ *a* 2 A s^{-1}
- ❑ *b* 4 A s^{-1}
- ❑ *c* 5 A s^{-1}
- ❑ *d* 6 A s^{-1}

J.18 In Question J.16 the voltage V_1 is

- ❑ *a* 2 V
- ❑ *b* 4 V
- ❑ *c* 8 V
- ❑ *d* 9 V

J.19 In Question J.16 the voltage V_2 is

- ❑ *a* 2 V
- ❑ *b* 4 V
- ❑ *c* 8 V
- ❑ *d* 9 V

J.20 In Question J.16 the circuit current after 4 s is

- ❑ *a* 2 A
- ❑ *b* 4 A
- ❑ *c* 6 A
- ❑ *d* 8 A

J.21 The energy stored in L_1 after 4 s in Question J.16 is

- ❑ *a* 32 J
- ❑ *b* 64 J
- ❑ *c* 80 J
- ❑ *d* 95 J

J.22 In Question J.16 the total energy stored after 4 s is

- ❑ *a* 128 J
- ❑ *b* 200 J
- ❑ *c* 384 J
- ❑ *d* 400 J

J.23 Given that the formula for a straight line graph is $y = mx + c$, x is

- ❑ *a* $(y + c)/m$
- ❑ *b* $(y - c)/m$
- ❑ *c* $yc + m$
- ❑ *d* $cm + y$

J.24 In Question J.23, m is

- ❑ *a* $(y + c)/x$
- ❑ *b* $(c + x)/y$
- ❑ *c* $(x - c)/y$
- ❑ *d* $(y - c)/x$

J.25 In Question J.23, c is

- ❑ *a* $y + mx$
- ❑ *b* $y - mx$
- ❑ *c* $my + x$
- ❑ *d* $my - x$

J.26 In Question J.23, if $y = 20$, $c = 4$ and $x = 8$, m is

- ❑ *a* 1
- ❑ *b* 2
- ❑ *c* 3
- ❑ *d* 4

J.27 The reactance of a 16-μF capacitor at 50 Hz is approximately

- ❑ *a* 163.42 Ω
- ❑ *b* 184.61 Ω
- ❑ *c* 198.94 Ω
- ❑ *d* 240.62 Ω

J.28 A capacitor has a reactance of 20 Ω at a frequency of 750 Hz. The capacitor value is

- ❑ *a* 8.33 μF
- ❑ *b* 9.61 μF
- ❑ *c* 10.61 μF
- ❑ *d* 12.43 μF

J.29 At what frequency is the reactance of a 68-pF capacitor equal to 470 kΩ?

- ❑ *a* 4980 Hz
- ❑ *b* 6324 Hz
- ❑ *c* 7428 Hz
- ❑ *d* 8000 Hz

J.30 In the diagram the current in R_1 is

- ❑ *a* 0.1 A
- ❑ *b* 0.2 A
- ❑ *c* 0.6 A
- ❑ *d* 0.8 A

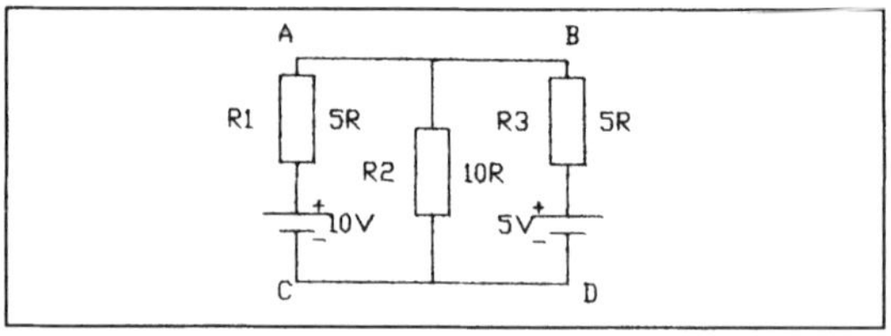

J.31 In Question J.30 the current in R_2 is

- ❑ *a* 0.5 A
- ❑ *b* 0.6 A
- ❑ *c* 0.7 A
- ❑ *d* 0.9 A

J.32 In Question J.30 the current in R_3 is

- ❑ *a* 0.2 A
- ❑ *b* 0.3 A
- ❑ *c* 0.6 A
- ❑ *d* 0.8 A

J.33 In Question J.30 the voltage between point A and the positive plate of the 10-V battery is

- ❑ *a* +4 V
- ❑ *b* −4 V
- ❑ *c* +6 V
- ❑ *d* −6 V

J.34 The voltage drop across R_3 in Question J.30 is

- ❑ *a* 1.0 V
- ❑ *b* 1.5 V
- ❑ *c* 2.5 V
- ❑ *d* 4.5 V

J.35 In Question J.30 the voltage between points A and C is

- ❑ *a* 2 V
- ❑ *b* 4 V
- ❑ *c* 6 V
- ❑ *d* 9 V

J.36 If $2x + 3y = 40$, and $5x + y = 35$, x is

- ❑ *a* 2
- ❑ *b* 3
- ❑ *c* 4
- ❑ *d* 5

J.37 In Question J.36, y is

- ❑ *a* 10
- ❑ *b* 12
- ❑ *c* 16
- ❑ *d* 20

J.38 A 100-mH inductor has a reactance at 50 Hz of

- ❑ *a* 15.71 Ω
- ❑ *b* 20.33 Ω
- ❑ *c* 31.42 Ω
- ❑ *d* 62.84 Ω

J.39 An inductor has a reactance of 200 Ω at 200 Hz. The value of the inductance is

- ❑ *a* 106 mH
- ❑ *b* 124 mH
- ❑ *c* 141 mH
- ❑ *d* 159 mH

J.40 In the diagram the reactance of the capacitor is approximately

- ❑ *a* 240.6 Ω
- ❑ *b* 318.3 Ω
- ❑ *c* 483.6 Ω
- ❑ *d* 561.8 Ω

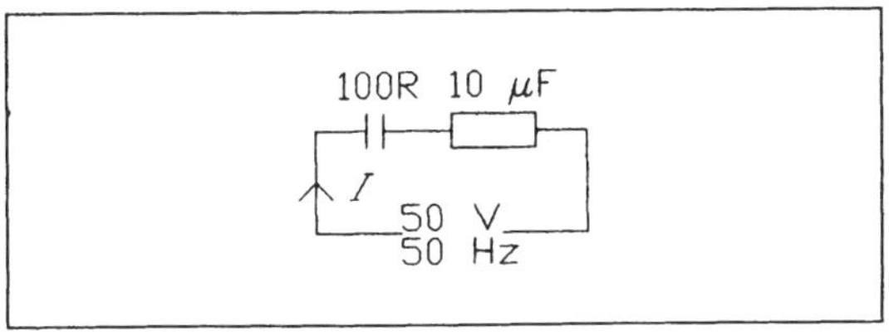

J.41 In Question J.40 the circuit impedance Z is

- ❑ *a* 124.54 Ω
- ❑ *b* 166.81 Ω
- ❑ *c* 333.64 Ω
- ❑ *d* 667.27 Ω

J.42 In Question J.40 the current I is approximately

- ❑ *a* 150 mA
- ❑ *b* 165 mA
- ❑ *c* 170 mA
- ❑ *d* 190 mA

J.43 The time constant of a 4-μF capacitor and a 100-kΩ resistor is

- ❑ *a* 0.4 μs
- ❑ *b* 4 μs
- ❑ *c* 4 s
- ❑ *d* 0.4 s

J.44 In the diagram the time constant is

- ❑ *a* 100 μs
- ❑ *b* 200 μs
- ❑ *c* 150 ms
- ❑ *d* 200 ms

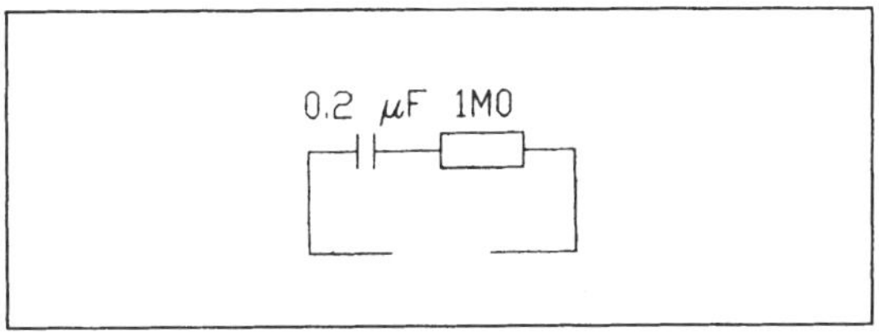

J.45 If 25 V DC is applied to the circuit in Question J.44, the initial current is

- ❑ *a* 25 μA
- ❑ *b* 250 μA
- ❑ *c* 25 mA
- ❑ *d* 250 mA

J.46 In Question J.45 the voltage across the capacitor after one time constant is approximately

- ❑ *a* 12.4 V
- ❑ *b* 14.1 V
- ❑ *c* 15.8 V
- ❑ *d* 19.3 V

J.47 If $10 \log x/200 = -20$, x is

- ❑ *a* 2
- ❑ *b* 4
- ❑ *c* 8
- ❑ *d* 10

J.48 If $20 \log x/200 = -40$, x is

- ❑ *a* 1.5
- ❑ *b* 2.0
- ❑ *c* 4.0
- ❑ *d* 8.5

J.49 If $10 \log 1000/x = 20$, x is

- ❑ *a* 8
- ❑ *b* 10
- ❑ *c* 12
- ❑ *d* 14

J.50 If $x^4 - 9x^2 = 0$, x is

- ❑ *a* 0
- ❑ *b* 1
- ❑ *c* 2
- ❑ *d* 3

Appendix K Sample test paper

K.1 The denary number 109 converted to binary is

- ❑ *a* 1001110
- ❑ *b* 1100101
- ❑ *c* 1101101
- ❑ *d* 1110000

K.2 2^7 expressed in binary is

- ❑ *a* 10001
- ❑ *b* 110001
- ❑ *c* 11110110
- ❑ *d* 10000000

K.3 The multiple giga means

- ❑ *a* 10^3
- ❑ *b* 10^6
- ❑ *c* 10^9
- ❑ *d* 10^{12}

K.4 The octal number 2 expressed in binary is

- ❑ *a* 001
- ❑ *b* 011
- ❑ *c* 010
- ❑ *d* 101

K.5 Which single gate could be used to replace the combination shown?

- ❑ *a* OR
- ❑ *b* NOR
- ❑ *c* AND
- ❑ *d* NAND

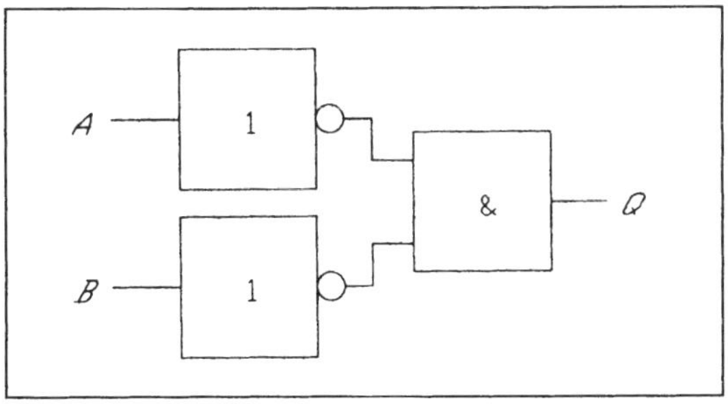

K.6 In the diagram, the inductive reactance XL is approximately

- ❑ *a* 31.42 Ω
- ❑ *b* 44.46 Ω
- ❑ *c* 62.84 Ω
- ❑ *d* 75.67 Ω

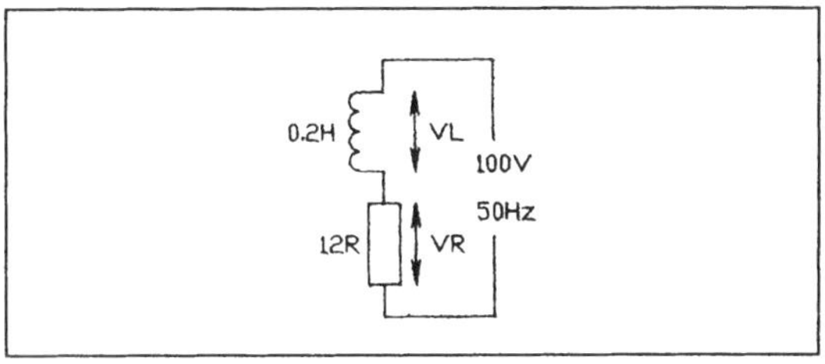

K.7 In Question K.6 the circuit impedance Z is

- ❑ *a* 42.81 Ω
- ❑ *b* 48.33 Ω
- ❑ *c* 50.62 Ω
- ❑ *d* 63.98 Ω

K.8 In Question K.6 the current I is

- ❑ *a* 1.12 A
- ❑ *b* 1.56 A
- ❑ *c* 2.22 A
- ❑ *d* 3.46 A

K.9 In Question K.6 the voltage V_R is

- ❑ *a* 11.31 V
- ❑ *b* 18.72 V
- ❑ *c* 28.44 V
- ❑ *d* 30.67 V

K.10 In Question K.6 the voltage V_L is

- ❑ *a* 44 V
- ❑ *b* 62 V
- ❑ *c* 80 V
- ❑ *d* 98 V

K.11 π rad expressed in degrees is

❑ *a* 110°
❑ *b* 150°
❑ *c* 180°
❑ *d* 220°

K.12 120° expressed in radians is

❑ *a* $2\pi/3$ rad
❑ *b* $2\pi/6$ rad
❑ *c* $5\pi/6$ rad
❑ *d* $4\pi/3$ rad

K.13 Ten cycles of a voltage waveform occur in 1.0 ms. The frequency is

❑ *a* 2 kHz
❑ *b* 10 kHz
❑ *c* 100 kHz
❑ *d* 150 kHz

K.14 The diagram shows a parallel *R–C* circuit. The current I_R is

❑ *a* 2.1 A
❑ *b* 4.2 A
❑ *c* 6.8 A
❑ *d* 7.3 A

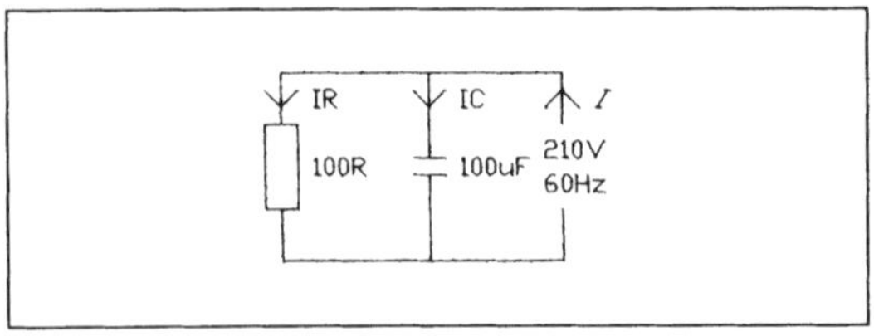

K.15 In Question K.14 the current I_C is

❑ *a* 6.32 A
❑ *b* 6.88 A
❑ *c* 7.92 A
❑ *d* 8.74 A

K.16 In Question K.14 the current I is

❑ *a* 4.22 A
❑ *b* 4.33 A
❑ *c* 6.12 A
❑ *d* 8.19 A

K.17 If $-4(y + x) = -68$, x is

❑ *a* $(68 + 4y)/4$
❑ *b* $(68 - 4y)/4$
❑ *c* $(68y + 4)/4$
❑ *d* $(68y - 4)/4$

K.18 In Question K.17, if $x = 5$, y is

❑ *a* 10
❑ *b* 11
❑ *c* 12
❑ *d* 14

K.19 If $3x^3 = 2187$, x is

❑ *a* 2
❑ *b* 3
❑ *c* 6
❑ *d* 9

K.20 The diagram shows three inductors connected in parallel. The combined inductance is

❑ *a* 1.0 H
❑ *b* 1.5 H
❑ *c* 2.0 H
❑ *d* 11.0 H

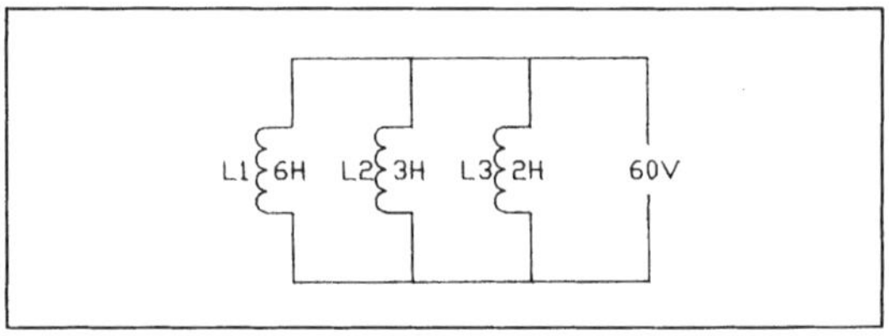

K.21 The rate of current growth in L1 in Question K.20 is

❑ *a* $10\ \text{A s}^{-1}$
❑ *b* $15\ \text{A s}^{-1}$
❑ *c* $20\ \text{A s}^{-1}$
❑ *d* $30\ \text{A s}^{-1}$

K.22 In Question K.20 the rate of current growth in L2 is

❑ *a* $10\ \text{A s}^{-1}$
❑ *b* $20\ \text{A s}^{-1}$
❑ *c* $30\ \text{A s}^{-1}$
❑ *d* $50\ \text{A s}^{-1}$

K.23 In Question K.20 the rate of current growth in L3 is

❑ *a* $20\ \text{A s}^{-1}$
❑ *b* $30\ \text{A s}^{-1}$
❑ *c* $40\ \text{A s}^{-1}$
❑ *d* $50\ \text{A s}^{-1}$

K.24 In Question K.20 the total rate of current growth is

- ❑ *a* 10 A s^{-1}
- ❑ *b* 20 A s^{-1}
- ❑ *c* 50 A s^{-1}
- ❑ *d* 60 A s^{-1}

K.25 An ammeter reads 5.61 A when the true reading is 5.5 A. The percentage error is

- ❑ *a* 0.5%
- ❑ *b* 1.0%
- ❑ *c* 2.0%
- ❑ *d* 2.2%

K.26 $(3 + j5) \times (2 - j)$ is

- ❑ *a* $5 - j6$
- ❑ *b* $11 + j7$
- ❑ *c* $4 - j3$
- ❑ *d* $12 + j2$

K.27 The diagram shows three lamps connected in series. The current *I* is

- ❑ *a* 0.15 A
- ❑ *b* 0.25 A
- ❑ *c* 0.55 A
- ❑ *d* 0.75 A

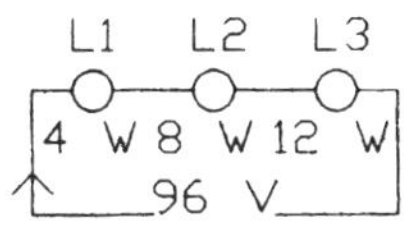

K.28 The resistance of lamp L1 in Question K.27 is

- ❑ *a* 10 Ω
- ❑ *b* 16 Ω
- ❑ *c* 32 Ω
- ❑ *d* 64 Ω

K.29 In Question K.27 the resistance of L2 is

- ❑ *a* 100 Ω
- ❑ *b* 120 Ω
- ❑ *c* 128 Ω
- ❑ *d* 150 Ω

K.30 In Question K.27 the total circuit resistance is

- ❑ *a* 240 Ω
- ❑ *b* 340 Ω
- ❑ *c* 354 Ω
- ❑ *d* 384 Ω

K.31 In Question K.27 the voltage across L1 is

- ❑ *a* 12 V
- ❑ *b* 16 V
- ❑ *c* 32 V
- ❑ *d* 48 V

K.32 In Question K.27 the voltage across L3 is

- ❑ *a* 16 V
- ❑ *b* 32 V
- ❑ *c* 48 V
- ❑ *d* 60 V

K.33 $(1/4)^0$ is

- ❑ *a* 0
- ❑ *b* 0.25
- ❑ *c* 1
- ❑ *d* 4

K.34 In the network shown, the resistance between A and B is

- ❑ *a* 100 Ω
- ❑ *b* 125 Ω
- ❑ *c* 130 Ω
- ❑ *d* 170 Ω

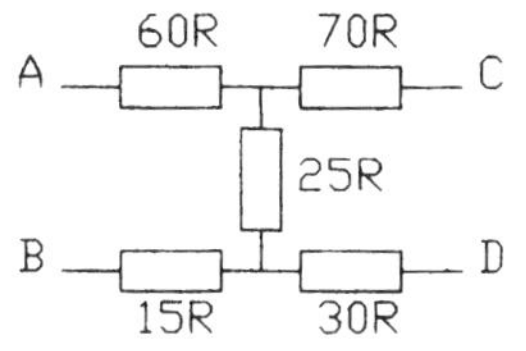

K.35 In Question K.34, if a short circuit is placed between C and D, the resistance between A and B is

- ❑ *a* 60 Ω
- ❑ *b* 75 Ω
- ❑ *c* 80 Ω
- ❑ *d* 95 Ω

K.36 In Question K.34 the resistance between C and D is

❑ *a* 100 Ω
❑ *b* 105 Ω
❑ *c* 125 Ω
❑ *d* 185 Ω

K.37 In Question K.34, if a short circuit is placed between A and B, the resistance between C and D is

❑ *a* 105.65 Ω
❑ *b* 118.75 Ω
❑ *c* 140.25 Ω
❑ *d* 150.55 Ω

K.38 A power output/input ratio of 1 : 16 expressed in decibels is

❑ *a* +12
❑ *b* +3
❑ *c* −9
❑ *d* −12

K.39 A voltage output/input ratio of 1 : 16 expressed in decibels is

❑ *a* +18
❑ *b* +6
❑ *c* −24
❑ *d* −30

K.40 $\sqrt[3]{27} \times \sqrt[3]{27} \times \sqrt[3]{27}$ is

❑ *a* 3
❑ *b* 6
❑ *c* 9
❑ *d* 27

K.41 For the gate combination shown, which input gives a logic 1 output?

❑ *a* 00
❑ *b* 01
❑ *c* 10
❑ *d* 11

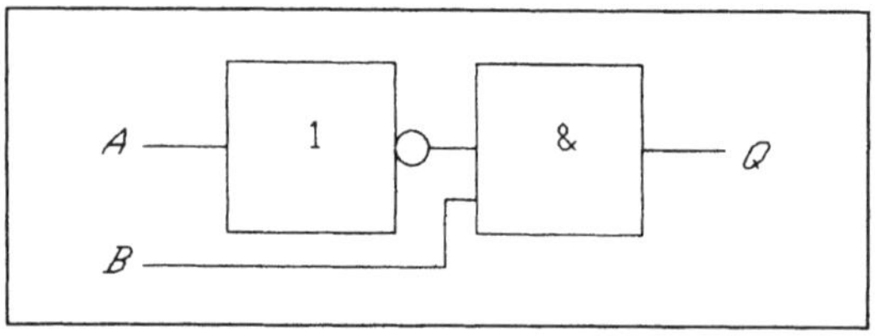

K.42 Hexadecimal 3BC converted to denary is

❑ *a* 462
❑ *b* 580
❑ *c* 678
❑ *d* 956

K.43 Binary 11001111 converted to hexadecimal is

❑ *a* AB
❑ *b* CF
❑ *c* AA
❑ *d* CB

K.44 An 82-Ω ± 10% resistor has a lower limit value of

❑ *a* 72.2 Ω
❑ *b* 72.4 Ω
❑ *c* 73.8 Ω
❑ *d* 78.8 Ω

K.45 The resistor in Question K.44 has an upper value of

❑ *a* 88.6 Ω
❑ *b* 90.2 Ω
❑ *c* 92.2 Ω
❑ *d* 93.4 Ω

K.46 In the network shown, the resistance between A and B is

❑ *a* 21 Ω
❑ *b* 28 Ω
❑ *c* 30 Ω
❑ *d* 60 Ω

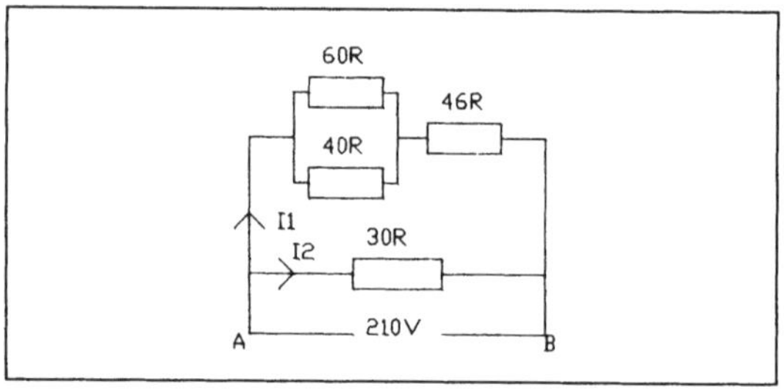

K.47 In Question K.46 the current I_1 is

❑ *a* 3 A
❑ *b* 8 A
❑ *c* 12 A
❑ *d* 15 A

K.48 In Question K.46 the current I_2 is

- ❑ *a* 3 A
- ❑ *b* 7 A
- ❑ *c* 14 A
- ❑ *d* 21 A

K.49 In the network in Question K.46 the input power is

- ❑ *a* 630 W
- ❑ *b* 1470 W
- ❑ *c* 2000 W
- ❑ *d* 2100 W

K.50 In Question K.46 the power consumed by the 40R resistor is

- ❑ *a* 129.6 W
- ❑ *b* 138.4 W
- ❑ *c* 182.2 W
- ❑ *d* 206.2 W

Appendix L Symbols, abbreviations and definitions

Greek letters used as symbols

Letter	Upper case		Lower case	
Alpha			α	angle, temperature coefficient of resistance
Delta	Δ	large increment	δ	small increment
Epsilon			ϵ	permittivity
Theta			θ	angle
Lambda			λ	wavelength
Mu			μ	micro, permeability
Pi			π	circumference/diameter
Rho			ρ	resistivity
Phi	Φ	magnetic flux	ϕ	angle
Omega	Ω	ohm	ω	angular velocity

Multiples and submultiples

T	tera	10^{12}
G	giga	10^{9}
M	mega	10^{6}
k	kilo	10^{3}
d	deci	10^{-1}
c	centi	10^{-2}
m	milli	10^{-3}
μ	micro	10^{-6}
n	nano	10^{-9}
p	pico	10^{-12}

Miscellaneous

Quantity	Unit
Length	Metre (m)
Area	Square metre (m^2)
Velocity	Metres per second ($m\,s^{-1}$)
Acceleration	Metres per second per second ($m\,s^{-2}$)
Mass	Kilogram (kg)
Force	Newton (N)
Pressure	Pascal (Pa)

Table of quantities and units

Quantity	Unit
Capacitance (*C*)	Farad (F)
Charge or quantity of electricity (*Q*)	Coulomb (C)
Current (*I*)	Ampere (A)
	Milliampere (mA)
	Microampere (μa)
Electric field strength (*E*)	Volts per metre ($\mathrm{V\,m^{-1}}$)
Electromotive force (*E*)	Volt (V)
Energy (*W*)	Joule (J)
Flux density, electric (*D*)	Coulombs per square metre ($\mathrm{C\,m^{-2}}$)
Force (*F*)	Newton (N)
Frequency (*f*)	Hertz (Hz)
Impedance (*Z*)	Ohm (Ω)
Inductance, self (*L*)	Henry (H)
Inductance, mutual (*M*)	Henry (H)
Magnetic field strength (*H*)	Ampere per metre ($\mathrm{A\,m^{-1}}$)
Magnetic flux (Φ)	Weber (Wb)
Magnetic flux density (*B*)	Tesla (T)
Magnetomotive force (*F*)	Ampere (A)
Permeability of free space (μ_o)	Henry per metre ($\mathrm{H\,m^{-1}}$)
Permeability, relative (μr)	
Permeability, absolute (μ)	Henry per metre ($\mathrm{H\,m^{-2}}$)
Permittivity of free space (ϵ_o)	Farad per metre ($\mathrm{F\,m^{-2}}$)
Permittivity, relative (ϵr)	
Permittivity, absolute (ϵ)	Farad per metre ($\mathrm{F\,m^{-1}}$)
Power (*P*)	Watt (W)
Reactance (*X*)	Ohm (Ω)
Reluctance (R_m)	Ampere per weber ($\mathrm{A\,Wb^{-1}}$)
Resistance (*R*)	Ohm (Ω)
Resistivity (ρ)	Ohm per metre ($\Omega\,\mathrm{m^{-2}}$)
Time (*t*)	Second (s)
Wavelength (λ)	Metre (m)

Appendix M BS 1852 resistance code

The table gives examples showing how the code is used to indicate resistance values on circuit diagrams

Value	BS 1852 code
100 Ω	100R
1.2 kΩ	1k2
0.47 Ω	R47
1 kΩ	1k0
1.2 MΩ	1M2

Tolerances are indicated by letters:

F = ± 1 %; G = ± 2%; J = ± 5%;
K = ± 10%; M = ± 20%.

Examples:
3.9 kΩ ± 10% = 3k9K
0.47 Ω ± 2% = R47G
1.2 MΩ ± 1% = 1M2F
68 kΩ ± 10% = 68kK

Answers to questions

Chapter 1

1.1 c	1.2 a	1.3 c	1.4 c	1.5 b	1.6 b
1.7 b	1.8 b	1.9 b	1.10 d	1.11 a	1.12 b
1.13 d	1.14 c	1.15 b	1.16 c	1.17 c	1.18 c
1.19 b	1.20 c	1.21 c	1.22 b	1.23 d	1.24 a
1.25 c	1.26 c	1.27 b	1.28 a	1.29 c	1.30 b
1.31 b	1.32 c	1.33 b	1.34 d	1.35 a	1.36 d
1.37 a	1.38 d	1.39 b	1.40 d	1.41 b	1.42 b
1.43 c	1.44 b	1.45 a	1.46 d	1.47 d	1.48 b
1.49 a	1.50 b	1.51 c	1.52 d	1.53 c	1.54 d
1.55 b	1.56 d	1.57 c	1.58 b	1.59 d	1.60 a
1.61 d	1.62 a	1.63 d	1.64 b	1.65 a	1.66 b
1.67 c	1.68 b	1.69 d	1.70 b	1.71 c	1.72 d
1.73 a	1.74 c	1.75 b			

Chapter 2

2.1 d	2.2 c	2.3 a	2.4 b	2.5 c	2.6 b
2.7 d	2.8 a	2.9 b	2.10 a	2.11 d	2.12 b
2.13 d	2.14 c	2.15 a	2.16 a	2.17 a	2.18 d
2.19 d	2.20 d	2.21 c	2.22 b	2.23 d	2.24 a
2.25 b	2.26 c	2.27 a	2.28 d	2.29 b	2.30 a
2.31 d	2.32 c	2.33 a	2.34 b	2.35 c	2.36 c
2.37 d	2.38 a	2.39 d	2.40 b	2.41 a	2.42 c
2.43 d	2.44 d	2.45 a	2.46 d	2.47 b	2.48 c
2.49 b	2.50 a	2 51 c	2.52 d	2.53 b	2.54 b
2.55 c	2.56 d	2.57 a	2.58 d	2.59 d	2.60 b
2.61 a	2.62 d	2.63 d	2.64 b	2.65 d	2.66 d
2.67 c	2.68 a	2.69 b	2.70 d	2.71 d	2.72 a
2.73 d	2.74 c	2.75 d			

Chapter 3

3.1 b	3.2 c	3.3 a	3.4 b	3.5 a	3.6 b
3.7 d	3.8 d	3.9 b	3.10 c	3.11 b	3.12 d
3.13 a	3.14 c	3.15 d	3.16 b	3.17 a	3.18 c
3.19 c	3.20 d	3.21 c	3.22 d	3.23 a	3.24 d
3.25 d	3.26 b	3.27 c	3.28 b	3.29 a	3.30 c
3.31 a	3.32 c	3.33 a	3.34 d	3.35 c	3.36 a
3.37 c	3.38 d	3.39 d	3.40 c	3.41 b	3.42 a
3.43 d	3.44 c	3.45 b	3.46 d	3.47 a	3.48 b
3.49 d	3.50 c	3.51 c	3.52 a	3.53 d	3.54 c
3.55 b	3.56 c	3.57 a	3 58 b	3.59 d	3.60 c
3.61 d	3.62 a	3.63 a	3.64 b	3.65 d	3.66 c
3.67 a	3.68 d	3.69 b	3.70 a	3.71 c	3.72 a
3.73 d	3.74 b	3.75 d			

Chapter 4

4.1 d	4.2 a	4.3 c	4.4 d	4.5 b	4.6 d
4.7 b	4.8 b	4.9 c	4.10 c	4.11 b	4.12 d
4.13 c	4.14 a	4.15 c	4.16 b	4.17 d	4.18 b
4.19 b	4.20 c	4.21 a	4.22 d	4.23 c	4.24 c
4.25 b	4.26 c	4.27 a	4.28 c	4.29 b	4.30 c
4.31 a	4.32 d	4.33 a	4.34 b	4.35 d	4.36 c
4.37 a	4.38 d	4.39 b	4.40 c	4.41 a	4.42 c
4.43 c	4.44 d	4.45 b	4.46 d	4.47 a	4.48 b
4.49 d	4.50 c	4.51 b	4.52 d	4.53 b	4.54 a
4.55 d	4.56 d	4.57 b	4.58 c	4.59 a	4.60 b
4.61 c	4.62 b	4.63 b	4.64 a	4.65 d	4.66 a
4.67 b	4.68 a	4.69 d	4.70 b	4.71 c	4.72 b
4.73 a	4.74 c	4.75 a			

Chapter 5

5.1 c	5.2 d	5.3 a	5.4 c	5.5 c	5.6 b
5.7 d	5.8 a	5.9 c	5.10 b	5.11 a	5.12 c
5.13 d	5.14 b	5.15 c	5.16 a	5.17 b	5.18 c
5.19 a	5.20 b	5.21 d	5.22 c	5.23 a	5.24 c
5.25 a	5.26 d	5.27 c	5.28 d	5.29 b	5.30 b
5.31 a	5.32 b	5.33 c	5.34 a	5.35 c	5.36 d
5.37 b	5.38 d	5.39 d	5.40 a	5.41 b	5.42 a
5.43 d	5.44 b	5.45 b	5.46 d	5.47 a	5.48 c
5.49 c	5.50 d	5.51 c	5.52 b	5.53 a	5.54 d
5.55 b	5.56 a	5.57 b	5.58 b	5.59 d	5.60 c
5.61 d	5.62 a	5.63 b	5.64 d	5.65 c	5.66 d
5.67 a	5.68 c	5.69 a	5.70 c	5.71 d	5.72 a
5.73 d	5.74 c	5.75 b			

Chapter 6

6.1 b	6.2 d	6.3 b	6.4 c	6.5 a	6.6 c
6.7 d	6.8 a	6.9 d	6.10 c	6.11 c	6.12 d
6.13 a	6.14 c	6.15 d	6.16 b	6.17 c	6.18 a
6.19 d	6.20 a	6.21 c	6.22 a	6.23 c	6.24 b
6.25 b	6.26 d	6.27 a	6.28 c	6.29 b	6.30 d
6.31 b	6.32 c	6.33 d	6.34 a	6.35 b	6.36 a
6.37 d	6.38 b	6.39 a	6.40 c	6.41 d	6.42 a
6.43 d	6.44 d	6.45 b	6.46 b	6.47 c	6.48 d
6.49 a	6.50 c	6.51 c	6.52 d	6.53 b	6.54 a
6.55 a	6.56 d	6.57 b	6.58 c	6.59 a	6.60 a
6.61 c	6.62 d	6.63 b	6.64 b	6.65 d	6.66 a
6.67 c	6.68 b	6.69 d	6.70 b	6.71 d	6.72 a
6.73 c	6.74 d	6.75 b			

Chapter 7

7.1 a	7.2 d	7.3 b	7.4 c	7.5 d	7.6 c
7.7 d	7.8 a	7.9 b	7.10 c	7.11 b	7.12 d
7.13 a	7.14 c	7.15 d	7.16 d	7.17 b	7.18 c
7.19 d	7.20 d	7.21 a	7.22 b	7.23 c	7.24 d
7.25 b	7.26 c	7.27 a	7.28 d	7.29 b	7.30 c
7.31 b	7.32 a	7.33 c	7.34 b	7.35 a	7.36 c
7.37 d	7.38 b	7.39 c	7.40 a	7.41 b	7.42 c
7.43 b	7.44 d	7.45 a	7.46 c	7.47 d	7.48 a
7.49 b	7.50 c	7.51 a	7.52 d	7.53 b	7.54 d
7.55 b	7.56 c	7.57 c	7.58 d	7.59 b	7.60 a
7.61 d	7.62 b	7.63 c	7.64 a	7.65 a	7.66 a
7.67 c	7.68 b	7.69 b	7.70 c	7.71 d	7.72 a
7.73 c	7.74 d	7.75 c			

Chapter 8

8.1 c	8.2 a	8.3 d	8.4 b	8.5 b	8.6 a
8.7 d	8.8 b	8.9 c	8.10 d	8.11 a	8.12 c
8.13 a	8.14 b	8.15 d	8.16 c	8.17 c	8.18 d
8.19 a	8.20 b	8.21 b	8.22 d	8.23 d	8.24 a
8.25 b	8.26 b	8.27 c	8.28 d	8.29 a	8.30 d
8.31 a	8.32 b	8.33 c	8.34 b	8.35 d	8.36 b
8.37 d	8.38 c	8.39 d	8.40 c	8.41 b	8.42 d
8.43 b	8.44 a	8.45 a	8.46 d	8.47 c	8.48 b
8.49 a	8.50 c	8.51 b	8.52 c	8.53 d	8.54 d
8.55 a	8.56 c	8.57 b	8.58 c	8.59 a	8.60 c
8.61 d	8.62 c	8.63 b	8.64 a	8.65 d	8.66 a
8.67 d	8.68 b	8.69 b	8.70 a	8.71 c	8.72 b
8.73 a	8.74 d	8.75 b			

Chapter 9

9.1 d	9.2 b	9.3 b	9.4 a	9.5 b	9.6 a
9.7 d	9.8 b	9.9 d	9.10 c	9.11 b	9.12 a
9.13 c	9.14 c	9.15 a	9.16 d	9.17 b	9.18 a
9.19 d	9.20 b	9.21 c	9.22 d	9.23 c	9.24 a
9.25 b	9.26 a	9.27 d	9.28 c	9.29 b	9.30 a
9.31 c	9.32 a	9.33 b	9.34 d	9.35 b	9.36 c
9.37 a	9.38 d	9.39 b	9.40 d	9.41 a	9.42 c
9.43 a	9.44 d	9.45 b	9.46 b	9.47 c	9.48 a
9.49 c	9.50 c	9.51 a	9.52 b	9.53 c	9.54 d
9.55 b	9.56 a	9.57 b	9.58 c	9.59 a	9.60 d
9.61 b	9.62 a	9.63 d	9.64 c	9.65 c	9.66 b
9.67 d	9.68 d	9.69 a	9.70 b	9.71 c	9.72 c
9.73 d	9.74 b	9.75 a			

Chapter 10

10.1 d	10.2 b	10.3 c	10.4 a	10.5 c	10.6 a
10.7 b	10.8 d	10.9 b	10.10 c	10.11 d	10.12 a
10.13 b	10.14 c	10.15 d	10.16 b	10.17 c	10.18 b
10.19 d	10.20 a	10.21 c	10.22 d	10.23 b	10.24 d
10.25 b	10.26 a	10.27 c	10.28 b	10.29 d	10.30 b
10.31 d	10.32 b	10.33 c	10.34 a	10.35 c	10.36 d
10.37 a	10.38 d	10.39 b	10.40 c	10.41 d	10.42 b
10.43 a	10.44 c	10.45 a	10.46 d	10.47 b	10.48 c
10.49 a	10.50 b	10.51 d	10.52 a	10.53 c	10.54 b
10.55 a	10.56 b	10.57 c	10.58 d	10.59 b	10.60 c
10.61 a	10.62 c	10.63 d	10.64 b	10.65 a	10.66 b
10.67 d	10.68 a	10.69 b	10.70 c	10.71 d	10.72 a
10.73 c	10.74 c	10.75 b			

Chapter 11

11.1 a	11.2 b	11.3 b	11.4 d	11.5 c	11.6 b
11.7 a	11.8 d	11.9 b	11.10 c	11.11 a	11.12 d
11.13 b	11.14 c	11.15 d	11.16 a	11.17 a	11.18 b
11.19 c	11.20 a	11.21 d	11.22 a	11.23 c	11.24 a
11.25 b	11.26 d	11.27 a	11.28 b	11.29 a	11.30 c
11.31 c	11.32 d	11 33 a	11.34 b	11.35 d	11.36 b
11.37 a	11.38 b	11.39 d	11.40 c	11.41 b	11 42 c
11.43 a	11.44 d	11.45 a	11.46 b	11.47 c	11.48 d
11.49 d	11.50 b	11.51 b	11.52 c	11.53 a	11.54 c
11.55 d	11.56 b	11.57 b	11.58 d	11.59 b	11.60 c
11.61 a	11.62 c	11.63 d	11.64 b	11.65 d	11.66 a
11.67 c	11.68 c	11.69 d	11.70 a	11.71 b	11.72 a
11.73 d	11.74 b	11.75 a			

Chapter 12

12.1 a	12.2 c	12.3 d	12.4 d	11.5 b	12.6 d
12.7 a	12.8 c	12.9 a	12.10 c	12.11 b	12.12 d
12.13 c	12 14 d	12.15 b	12.16 a	12.17 c	12.18 a
12.19 d	12.20 b	12.21 a	12.22 c	12.23 d	12.24 b
12 25 b	12.26 c	12.27 d	12.28 b	12.29 a	12.30 c
12.31 d	12.32 b	12.33 c	12.34 a	12.35 b	12.36 b
12.37 b	12.38 d	12.39 b	12.40 a	12.41 c	12.42 c
12.43 d	12.44 b	12.45 c	12.46 d	12.47 a	12.48 a
12.49 a	12.50 c	12.51 a	12.52 d	12.53 b	12.54 c
12.55 d	12.56 d	12.57 a	12.58 c	12.59 b	12.60 d
12.61 c	12.62 a	12.63 b	12.64 c	12.65 d	12.66 b
12.67 c	12.68 a	11.69 a	12.70 d	12.71 b	12.72 c
12.73 a	12.74 b	12.75 c			

Chapter 13

13.1 d	13.2 c	13.3 a	13.4 c	13.5 b	13.6 d
13.7 a	13,8 c	13.9 c	13.10 a	13.11 c	13.12 a
13.13 c	13 14 d	13 15 b	13 16 a	13.17 b	13.18 b
13.19 c	13.20 a	13.21 d	13.22 b	13.23 d	13.24 d
13.25 d	13.26 b	13.27 b	13.28 c	13.29 d	13.30 c
13.31 d	13.32 b	13.33 d	13.34 b	13.35 a	13.36 b
13.37 d	13.38 b	13.39 a	13.40 a	13.41 c	13.42 d
13.43 b	13.44 c	13.45 d	13.46 c	13.47 b	13.48 d
13.49 c	13.50 d	13.51 a	13.52 c	13.53 b	13.54 b
13.55 c	13.56 a	13.57 b	13.58 b	13.59 d	13.60 a
13.61 b	13.62 c	13.63 b	13.64 a	13.65 c	13.66 b
13.67 c	13.68 d	13.69 d	13.70 a	13.71 c	13.72 d
13.73 b	13.74 c	13.75 a			

Chapter 14

14.1 b	14.2 d	14 3 c	14.4 b	14.5 a	14.6 c
14.7 c	14.8 a	14.9 d	14.10 a	14.11 c	14.12 b
14.13 c	14.14 a	14.15 d	14.16 c	14.17 c	14.18 d
14.19 c	14.20 a	14.21 c	14.22 b	14.23 c	14.24 b
14.25 d	14.26 a	14.27 d	14.28 d	14.29 c	14.30 b
14.31 a	14.32 b	14.33 c	14.34 a	14.35 c	14.36 d
14.37 b	14.38 b	14.39 c	14.40 c	14.41 d	14.42 a
14.43 b	14.44 c	14.45 a	14.46 b	14.47 d	14.48 c
14.49 b	14.50 b	14.51 a	14.52 d	14.53 b	14.54 c
14.55 a	14.56 a	14.57 d	14.58 b	14.59 b	14.60 d
14.61 b	14.62 a	14.63 d	14.64 c	14.65 c	14.66 b
14.67 d	14.68 c	14.69 a	14.70 c	14.71 b	14.72 d
14.73 a	14.74 c	14.75 b			

Chapter 15

15.1 c	15.2 b	15.3 a	15.4 d	15.5 c	15.6 c
15.7 a	15.8 b	15.9 b	15.10 d	15.11 c	15 12 c
15.13 a	15 14 d	15.15 b	14.16 b	15.17 d	15.18 b
15.19 c	15.20 d	15.21 a	15.22 b	15.23 d	15.24 a
15.25 b	15.26 a	15.27 c	15,28 a	15.29 d	15.30 b
15.31 d	15 32 c	15.33 b	15.34 a	15.35 c	15.36 b
15.37 b	15.38 d	15.39 a	15.40 c	15.41 d	15.42 b
15.43 a	15.44 b	15.45 c	14.46 d	15.47 b	15.48 c
15.49 a	15.50 c	15.51 b	15.52 b	15.53 d	15.54 c
15.55 a	15.56 c	15.57 d	15.58 a	15.59 c	15.60 b
15.61 a	15.62 b	15.63 a	15.64 c	15.65 d	15.66 a
15.67 c	15.68 a	15.69 b	15.70 c	15.71 d	15.72 d
15.73 c	15.74 a	15.75 d			

Chapter 16

16.1 c	16.2 a	16.3 d	16.4 b	16.5 c	16.6 d
16.7 c	16.8 a	16.9 b	16.10 d	16.11 a	16.12 c
16.13 c	16.14 b	16.15 a	16.16 d	16.17 b	16.18 a
16.19 c	16.20 b	16.21 a	16.22 b	16.23 d	16.24 a
16.25 c	16.26 d	16.27 b	16.28 c	16.29 d	16.30 a
16.31 c	16.32 d	16.33 b	16.34 b	16.35 d	16.36 a
16.37 d	16.38 b	16.39 c	16.40 b	16.41 d	16.42 c
16.43 b	16.44 d	16.45 a	16.46 c	16.47 b	16.48 a
16.49 c	16.50 d	16.51 c	16.52 b	16.53 a	16.54 c
16.55 c	16.56 d	16.57 c	16.58 b	16.59 c	16.60 a
16.61 b	16.62 d	16.63 b	16.64 d	16.65 b	16.66 c
16.67 d	16.68 a	16.69 c	16.70 b	16.71 d	16.72 b
16.73 a	16.74 d	16.75 a			

Chapter 17

17.1 d	17.2 c	17.3 a	17.4 a	17.5 b	17.6 d
17.7 c	17 8 b	17.9 b	17.10 a	17.11 d	17.12 c
17.13 c	17.14 a	17.15 c	17.16 a	17.17 d	17.18 c
17.19 a	17.20 d	17.21 a	17.22 b	17.23 a	17.24 c
17.25 b	17.26 d	17.27 a	17.28 c	17.29 b	17.30 b
17.31 a	17.32 b	17.33 a	17.34 d	17.35 c	17.36 b
17.37 b	17.38 a	17.39 d	17.40 a	17.41 a	17.42 c
17.43 c	17.44 d	17.45 b	17.46 b	17.47 a	17.48 d
17.49 c	17.50 a	17.51 b	17.52 c	17.53 d	17.54 a
17.55 a	17.56 c	17.57 b	17.58 b	17.59 c	17.60 d
17.61 c	17.62 a	17.63 a	17.64 c	17.65 c	17.66 b
17.67 d	17.68 a	17.69 b	17.70 c	17.71 d	17.72 c
17.73 d	17.74 b	17.75a			

Chapter 18

18.1 c	18.2 d	18.3 a	18.4 d	18.5 d	18.6 b
18.7 b	18.8 c	18.9 a	18.10 c	19.11 b	18.12 d
18.13 b	18.14 c	18.15 a	18.16 c	18.17 b	18.18 a
18.19 d	18.20 c	18.21 b	18.22 a	18.23 c	18.24 d
18.25 b	18.26 a	18.27 b	18.28 b	18.29 d	18.30 a
18.31 c	18.32 c	18.33 a	18.34 d	18.35 c	18.36 c
18.37 b	18.38 c	18.39 d	18.40 b	18.41 c	18.42 d
18.43 b	18.44 a	18.45 b	18.46 d	18.47 b	18.48 c
18.49 a	18.50 c	18.51 a	18.52 b	18.53 b	18.54 c
18.55 a	18.56 d	18.57 c	18.58 a	18.59 c	18.60 b
18.61 d	18.62 b	18.63 c	18.64 c	18.65 a	18.66 d
18.67 d	18.68 b	18.69 c	18.70 c	18.71 b	18.72 d
18.73 a	18.74 b	18.75 d			

Chapter 19

19.1 d	19.2 b	19.3 b	19.4 d	19.5 b	19.6 c
19.7 b	19.8 d	19.9 a	19.10 c	19.11 c	19.12 d
19.13 d	19.14 c	19.15 d	19.16 a	19.17 c	19.18 d
19.19 c	19.20 a	19.21 b	19.22 d	19.23 c	19.24 b
19.25 a	19.26 b	19.27 a	19.28 d	19.29 b	19.30 b
19.31 c	19.32 d	19.33 b	19.34 b	19.35 a	19.36 b
19.37 c	19.38 a	19.39 d	19.40 b	19.41 d	19.42 a
19.43 c	19.44 c	19.45 a	19.46 d	19.47 b	19.48 d
19.49 c	19.50 d	19.51 b	19.52 a	19.53 c	19.54 a
19.55 b	19.56 d	19.57 c	19.58 b	19.59 d	19.60 b
19.61 a	19.62 d	19.63 c	19.64 a	19.65 d	19.66 c
19.67 c	19.68 a	19.69 b	19.70 c	19.71 b	19.72 d
19.73 a	19.74 b	19.75 d			

Chapter 20

20.1 a	20.2 d	20.3 b	20.4 c	20.5 a	20.6 c
20.7 d	20.8 b	20.9 b	20.10 a	20.11 c	20.12 c
20.13 d	20.14 a	20.15 b	20.16 d	20.17 a	20.18 c
20.19 c	20.20 a	20.21 c	20.22 d	20.23 a	20.24 c
20.25 b	20.26 d	20.27 a	20.28 b	20.29 d	20.30 c
20.31 b	20.32 c	20.33 d	20.34 a	20.35 a	20.36 b
20.37 d	20.38 c	20.39 b	20.40 a	20.41 c	20.42 b
20.43 d	20.44 c	20.45 a	20.46 d	20.47 b	20.48 d
20.49 c	20.50 a	20.51 b	20.52 b	20.53 c	20.54 a
20.55 d	20.56 d	20.57 a	20.58 c	20.59 b	20.60 d
20.61 b	20.62 b	20.63 a	20.64 b	20.65 d	20.66 a
20.67 c	20.68 d	20.69 c	20.70 a	20.71 b	20.72 a
20.73 d	20.74 d	20.75 b			

Appendix A

A.1 c	A.2 d	A.3 b	A.4 d	A.5 b	A.6 d
A.7 c	A.8 c	A.9 a	A.10 b	A.11 c	A.12 b
A.13 b	A.14 c	A.15 c	A.16 c	A.17 c	A.18 a
A.19 b	A.20 d	A.21 c	A.22 a	A.23 d	A.24 d
A.25 c	A.26 d	A.27 d	A.28 c	A.29 c	A.30 c
A.31 b	A.32 a	A.33 c	A.34 b	A.35 b	A.36 c
A.37 b	A.38 c	A.39 c	A.40 a	A.41 b	A.42 c
A.43 c	A.44 d	A.45 a	A.46 b	A.47 b	A.48 d
A.49 c	A.50 d				

Appendix B

B.1 a	B.2 a	B.3 d	B.4 b	B.5 a	B.6 d
B.7 c	B.8 c	B.9 c	B.10 d	B.11 a	B.12 b
B.13 c	B.14 d	B.15 c	B.16 b	B.17 a	B.18 d
B.19 c	B.20 b	B.21 a	B.22 c	B.23 b	B.24 d
B.25 a	B.26 a	B.27 c	B.28 d	B.29 c	B.30 b
B.31 d	B.32 c	B.33 a	B.34 d	B.35 c	B.36 b
B.37 c	B.38 b	B.39 a	B.40 b	B.41 d	B.42 c
B.43 a	B.44 d	B.45 b	B.46 c	B.47 c	B.48 b
B.49 b	B.50 a				

Appendix C

C.1 d	C.2 d	C.3 c	C.4 d	C.5 b	C.6 c
C.7 c	C.8 b	C.9 c	C.10 a	C.11 c	C.12 b
C.13 d	C.14 c	C.15 d	C.16 b	C.17 a	C.18 d
C.19 b	C.20 d	C.21 b	C.22 c	C.23 b	C.24 a
C.25 d	C.26 b	C.27 c	C.28 b	C.29 c	C.30 d
C.31 b	C.32 c	C.33 d	C.34 b	C.35 c	C.36 a
C.37 b	C.38 a	C.39 d	C.40 c	C.41 d	C.42 c
C.43 a	C.44 d	C.45 c	C.46 c	C.47 b	C.48 c
C.49 a	C.50 b				

Appendix D

D.1 b	D.2 b	D.3 b	D.4 c	D.5 d	D.6 b
D.7 d	D.8 a	D.9 a	D.10 c	D.11 d	D.12 a
D.13 c	D.14 a	D.15 d	D.16 a	D.17 b	D.18 b
D.19 c	D.20 a	D.21 d	D.22 b	D.23 c	D.24 c
D.25 d	D.26 b	D.27 b	D.28 c	D.29 b	D.30 b
D.31 a	D.32 c	D.33 a	D.34 d	D.35 b	D.36 c
D.37 c	D.38 d	D.39 a	D.40 d	D.41 a	D.42 b
D.43 b	D.44 d	D.45 b	D.46 c	D.47 b	D.48 d
D.49 a	D.50 a				

Appendix E

E.1 d	E.2 c	E.3 d	E.4 d	E.5 b	E.6 c
E.7 a	E.8 d	E.9 d	E.10 b	E.11 b	E.12 a
E.13 c	E.14 d	E.15 d	E.16 a	E.17 b	E.18 d
E.19 b	E.20 d	E.21 d	E.22 b	E.23 d	E.24 b
E.25 c	E.26 b	E.27 c	E.28 a	E.29 c	E.30 c
E.31 b	E.32 c	E.33 d	E.34 d	E.35 c	E.36 b
E.37 a	E.38 b	E.39 c	E.40 d	E.41 b	E.42 d
E.43 a	E.44 c	E.45 b	E.46 c	E.47 a	E.48 a
E.49 b	E.50 c				

Appendix F

F.1 c	F.2 c	F.3 b	F.4 d	F.5 a	F.6 a
F.7 b	F.8 a	F.9 c	F.10 b	F.11 c	F.12 d
F.13 c	F.14 b	F.15 d	F.16 b	F.17 c	F.18 b
F.19 d	F.20 c	F.21 b	F.22 a	F.23 b	F.24 b
F.25 d	F.26 c	F.27 b	F.28 d	F.29 a	F.30 d
F.31 c	F.32 a	F.33 a	F.34 c	F.35 d	F.36 b
F.37 b	F.38 a	F.39 b	F.40 a	F.41 c	F.42 d
F.43 c	F.44 d	F.45 c	F.46 b	F.47 c	F.48 d
F.49 b	F.50 d				

Appendix G

G.1 d	G.2 a	G.3 b	G.4 b	G.5 a	G.6 c
G.7 c	G.8 a	G.9 c	G.10 b	G.11 c	G.12 b
G.13 b	G.14 c	G.15 d	G.16 a	G.17 c	G.18 d
G.19 b	G.20 d	G.21 a	G.22 b	G.23 c	G.24 d
G.25 c	G.26 a	G.27 c	G.28 a	G.29 b	G.30 d
G.31 c	G.32 c	G.33 d	G.34 b	G.35 d	G.36 c
G.37 b	G.38 d	G.39 b	G.40 c	G.41 d	G.42 a
G.43 c	G.44 b	G.45 c	G.46 c	G.47 d	G.48 d
G.49 b	G.50 c				

Appendix H

H.1 b	H.2 a	H.3 a	H.4 b	H.5 d	H.6 b
H.7 c	H.8 a	H.9 d	H.10 b	H.11 c	H.12 c
H.13 d	H.14 a	H.15 b	H.16 b	H.17 d	H.18 d
H.19 a	H.20 a	H.21 d	H.22 c	H.23 a	H.24 c
H.25 d	H.26 b	H.27 b	H.28 a	H.29 a	H.30 c
H.31 b	H.32 a	H.33 c	H.34 d	H.35 a	H.36 c
H.37 d	H.38 b	H.39 d	H.40 c	H.41 b	H.42 a
H.43 b	H.44 c	H.45 d	H.46 b	H.47 c	H.48 d
H.49 c	H.50 b				

Appendix J

J.1 c	J.2 a	J.3 b	J.4 c	J.5 a	J.6 b
J.7 a	J.8 b	J.9 d	J.10 d	J.11 b	J.12 c
J.13 a	J.14 c	J.15 b	J.16 d	J.17 a	J.18 b
J.19 c	J.20 d	J.21 b	J.22 c	J.23 b	J.24 d
J.25 b	J.26 b	J.27 c	J.28 c	J.29 a	J.30 d
J.31 b	J.32 a	J.33 b	J.34 a	J.35 c	J.36 d
J.37 a	J.38 c	J.39 d	J.40 b	J.41 c	J.42 a
J.43 d	J.44 d	J.45 a	J.46 c	J.47 a	J.48 b
J.49 b	J.50 d				

Appendix K

K.1 c	K.2 d	K.3 c	K.4 c	K.5 b	K.6 c
K.7 d	K.8 b	K.9 b	K.10 d	K.11 c	K.12 a
K.13 b	K.14 a	K.15 c	K.16 d	K.17 b	K.18 c
K.19 d	K.20 a	K.21 a	K.22 b	K.23 b	K.24 d
K.25 c	K.26 b	K.27 b	K.28 d	K.29 c	K.30 d
K.31 b	K.32 c	K.33 c	K.34 a	K.35 d	K.36 c
K.37 b	K.38 d	K.39 c	K.40 d	K.41 b	K.42 d
K.43 b	K.44 c	K.45 b	K.46 a	K.47 a	K.48 b
K.49 d	K.50 a				

Made in United States
Orlando, FL
04 November 2026

90201574R00116